Tadpole Hunter

Tadpole Hunter

A Personal History of Amphibian Conservation and Research

Arnold Cooke

PELAGIC PUBLISHING

First published in 2023 by
Pelagic Publishing
20–22 Wenlock Road
London N1 7GU

www.pelagicpublishing.com

Tadpole Hunter: A Personal History of Amphibian Conservation and Research

https://doi.org/10.53061/BGOE4941

A CIP record for this book is available from the British Library

ISBN 978-1-78427-448-1 Pbk
ISBN 978-1-78427-449-8 ePub
ISBN 978-1-78427-450-4 PDF

Cover photograph: Common frogs in amplexus, Peak District,
Derbyshire © Jack Perks

Frontispiece: Tadpole hunting – netting smooth newt tadpoles in 2021.
Photo by Rosemarie Cooke

Printed and bound in Great Britain by
TJ Books Limited, Padstow, Cornwall

Contents

Foreword

The United Kingdom has a mere seven native species of amphibians, all of which have received substantial study and conservation efforts over the past half-century. Arnold Cooke has been at the forefront of these activities and one of Britain's leading contributors throughout an exciting period of enlightenment. This book provides a very personal account of Arnold's work, mostly in the field but also in the laboratory. As a member of Britain's statutory nature conservation organisation (from the Nature Conservancy up to English Nature) he was well placed to liaise with non-government conservation organisations and provide practical advice, activities which he also carried out with aplomb.

It is remarkable how much time and effort went into a wide range of groundbreaking studies, all of which are detailed in this book. Early work included national and regional surveys of common frog and toad distributions and abundance, in response to widespread concern about declines in the countryside. Attempts to understand the causes of population trends involved studies of pesticide effects, artificial fertilisers and predation. Arnold recognised early on the increasing importance of garden ponds, including his own, especially for frogs. He pioneered the use of road-casualty counts to demonstrate the devastating effects of road traffic on local toad populations, and successfully introduced frogs and toads to garden ponds and to a large new nature reserve. Later he engaged with natterjack conservation, helping to monitor populations at several sites around the country. Nor did newts escape Arnold's attention. His investigations of great crested newt ecology covered population dynamics and basic ecology at several locations, making substantial and valuable contributions to our knowledge of this enchanting animal. Some of these studies generated data extending over more than 30 years, among the longest time series for amphibians anywhere and invaluable for detecting long-term population trends. Throughout all of this, Arnold found time to publish in scientific journals, speak at national and international meetings, and liaise extensively with like-minded friends and colleagues.

This book is unusual in that rather than documenting general information about amphibians, as most similar texts do, it provides a wonderful insight into the motives and accomplishments of one person. Arnold's passion for wildlife and its conservation is evident throughout. His contributions have not been exclusively for amphibians, and in more recent years he has also made fascinating studies of local deer populations, resulting in a highly informative book on that subject in 2019.

In the late eighteenth century, Gilbert White combined shrewd observations of wildlife with clever investigations in and around his parish of Selborne in Hampshire. I find a comparison with Arnold's approach, some 200 years later, quite compelling. Technology has changed beyond recognition, but the fascination for nature and an enquiring attitude towards it by both men seem strikingly similar. I hope that there will always be such people in our countryside.

Trevor Beebee

Acknowledgements

The following people have helped, in some cases considerably, by commenting on chapters or on the concept of the book: John Baker, Marc Baldwin, Trevor Beebee, Roy Bradley, John Buckley, Alastair Burn, John Fellowes, Jim Foster, Tony Gent, Richard Griffiths, Matt Hamilton, Tom Langton, Terry and Helen Moore, Steve Parnwell, Silviu Petrovan and Kathy Wormald. Any mistakes, omissions or misunderstandings are down to me. I am particularly grateful to Trevor Beebee for writing the foreword.

John Baker, Marc Baldwin, John Buckley, Rosemarie Cooke, Lynne Farrell, Jim Foster, Richard Griffiths, Mihai Leu and Jonathan Webster kindly contributed photographs and are acknowledged in the relevant places. All other photographs are mine.

Those whom I worked or rubbed shoulders with on amphibian research and conservation included many of those just listed plus Henry Arnold, Brian Banks, Maggie Beebee, Tony Bell, Robert Bray, Robin Buxton, Pete Carty, John Comont, Keith Corbett, Clive Cummins, Gary Dean, Jonty Denton, Robert Duff, Peter Ferguson, Deryk Frazer, Sue Freestone, William Fulford, Howard Ginn, Paul Gittins, Tim Halliday, Paul Harding, Geoff and Beth Haslewood, John Heath, Howard Hillier, Derek Hilton-Brown, Jon Hutton, Rob Oldham, Roger and Sarah Orbell, Dan Osborn, Chris and Pam Newbold, Mark Nicholson, Madeleine Parnwell, Frank Perring, Ian Prestt, Chris Reading, Maxwell Savage, Al Scorgie, Fred Slater, Phil Smith, Tim Sparks, Harry Spoelstra, Henry Stanier, Bob Stebbings, Mary Swan, Peter Tinning, Richard Tinsley, Paul Verrell and Peter Walker.

I am also grateful to the many people who helped in my various surveys; some of them are mentioned personally in the text. While I was with the Nature Conservancy Council, I met many individuals during site visits, but I regret their names have largely slipped my mind over the decades. I wish to acknowledge the commitment and good humour of those reserve staff who have encouraged or put up with me doing strange things in and around their ponds, and also the dedication of other regional staff of the Nature Conservancy, the Nature Conservancy Council, English Nature and various conservation bodies. And I have frequently received assistance from research, advisory and support staff in these organisations and in the Institute of Terrestrial Ecology and the Centre for Ecology and Hydrology.

Special thanks go to my wife Rosemarie for her help in many ways and our son Steven for his company and input on numerous field trips. Finally, I am conscious that I have not mentioned by name many people who have contributed since the beginning of this century because I have become progressively less active and we have probably never met; I thank you now for what you have done already and will do in the future.

Finally, I wish to thank Nigel Massen and David Hawkins at Pelagic Publishing, Hugh Brazier the copy-editor, and Arun Rajakumar and the team at Deanta for their help, encouragement and patience in the production of this book.

Abbreviations

ARC	Amphibian and Reptile Conservation
ARG	Amphibian and Reptile Group
ARG UK	Amphibian and Reptile Groups of the United Kingdom
ASA	Amphibian Survival Alliance
BAP	Biodiversity Action Plan
BCI	Body Condition Index
BHS	British Herpetological Society
BRC	Biological Records Centre
BTO	British Trust for Ornithology
DAPTF	Declining Amphibian Populations Task Force
Defra	Department for Environment, Food and Rural Affairs
eDNA	Environmental DNA
FFPS	Fauna and Flora Preservation Society
FHT	Freshwater Habitats Trust
HCEAC	House of Commons Environmental Audit Committee
HCI	Herpetofauna Consultants International
HCIL	Herpetofauna Conservation International Limited
HCT	Herpetological Conservation Trust
HGBI	Herpetofauna Groups of Britain and Ireland
HSI	Habitat Suitability Index
ITE	Institute of Terrestrial Ecology
JNCC	Joint Nature Conservation Committee
LPA	Local Planning Authority
MAFF	Ministry of Agriculture, Fisheries and Food
NARRS	National Amphibian and Reptile Recording Scheme
NBN	National Biodiversity Network
NCC	Nature Conservancy Council
NCP	Newt Conservation Partnership
NNR	National Nature Reserve
OEP	Office for Environmental Protection
PDC	Peterborough Development Corporation
RSPB	Royal Society for the Protection of Birds
SAC	Special Area of Conservation
SAP	Species Action Plan
SSSI	Site of Special Scientific Interest
VCH	Victoria County History
WIIS	Wildlife Incident Investigation Scheme

CHAPTER 1

Introduction

I have seen and caught countless amphibians in my life. And for some of the time I was fortunate enough to be paid to do so. I started working on them more than 50 years ago and may well have been the first professional amphibian conservationist in Britain. This book is not a comprehensive account of amphibian natural history and conservation. Instead, it focuses on how I fitted in to the first few decades of the overall story – and what has happened since. It tells how and why decisions were made, what was done and what the outcomes were. It will probably contain a few surprises for readers who are only familiar with what has happened during the early years of this century. For instance, are you aware of the massive declines that occurred in amphibian populations between 1940 and 1970, and how little we knew about the status and ecology of most species at that time?

This introductory chapter is in several parts. First, how knowledge slowly accumulated up until the 1960s is briefly sketched. Second, my life and work are outlined to illustrate the influences, including luck, which allowed me to start a career in wildlife conservation and then develop my interests along several different paths. Third, the aims, contents and layout of the book are explained in more detail.

1.1 Historical development of knowledge about British amphibians

By the time that Shakespeare wrote *Macbeth* in the early years of the seventeenth century, the witches and other country folk appreciated that there were creatures they called frogs and newts, but they seemed to know little about some of the common amphibians. Even a century and a half later, when Gilbert White was alive, the newts and the common toad (*Bufo bufo*) were poorly known. *The Natural History of Selborne* appeared in 1789 and reproduced a large number of letters written by White to Thomas Pennant and Daines Barrington. In one letter to Pennant in 1768, White stated that how 'toads procreate' seemed to be unknown (White 1978). He was evidently familiar with the mating of common frogs (*Rana temporaria*), but had never seen or heard about toads coupling. And he was surprised to learn that newts lived on land as well as in water. Gilbert White referred to amphibians as vile reptiles. At that time, most people would probably have agreed that amphibians were disgusting animals. Also the words 'amphibian' and 'reptile' were interchangeable and would be used for more than another century to describe either or both of these classes of vertebrate.

Herpetology is the study of amphibians and reptiles. In order to bring the narrative forward to when my herpetological career began in the late 1960s, I am summarising the lives of six men who, among other attributes, were active herpetologists from the first part of the nineteenth century up to the second half of the twentieth century: Thomas Bell (1792–1880), Leonard Jenyns (1800–1893), Mordecai Cubitt Cooke (1825–1914), George Albert Boulenger (1858–1937), Malcolm Smith (1875–1958) and Ronald Maxwell Savage (1900–1985). All enjoyed long, productive lives. It is interesting to record what these authors knew (and what they did not) and to pick out some of their more striking achievements and observations. Only from the early 1800s could any of the observations be regarded as field research, and it was not until the middle of the twentieth century that much thought was given to whether populations had decreased, or much consideration to the need for conservation.

Professor Bell was already a distinguished zoologist when, in 1839 at the age of 47, he published his most important herpetological work, *A History of British Reptiles*. Bell knew the common frog well, describing it as 'one of our most common vertebrates'. Around that time there had evidently been considerable discussion about whether a second species of frog occurred in Scotland and whether that might be the edible frog (now *Pelophylax* kl. *esculentus*). Bell compared a large skeleton of a Scottish frog with those of a common frog and an edible frog and declared it more closely resembled the former but was distinct from it. He provisionally named it *Rana scotica*, but Boulenger (1893) pointed out that common frogs in Scotland could reach an unusual size and in addition could be atypically marked. The situation was in fact even more complex, as we now consider that Bell's edible frogs were pool frogs (now *Pelophylax lessonae*; Beebee and Griffiths 2000, Beebee *et al.* 2005; Section 12.6.1).

The common toad was clearly pitied by Bell; he described how people still hated toads and subjected them to 'barbarous acts of cruelty'. The natterjack (now *Epidalea calamita*) was not common, but could be found in considerable numbers in some places; for instance, he was aware of it occurring at Southerness on the North Solway coast and at Blackheath and Deptford in London. Bell took to task early classifiers of wildlife who tended to group the newts with the lizards rather than the frogs and toads. The 'common warty-newt or great water-newt' (now great crested newt, *Triturus cristatus*) was widespread in ponds and large ditches, as was the 'common smooth-newt' (now smooth newt, *Lissotriton vulgaris*). Bell argued that the 'palmated smooth-newt' (now palmate newt, *Lissotriton helveticus*) was 'very common in this country' but had been previously overlooked because of its similarity to the common smooth-newt. On the basis of a specimen in the collection of the Zoological Society, Bell also proposed the existence in this country of another species closely related to the common warty-newt: the 'strait-lipped warty-newt' (*Triton bibronii*). I cannot recall another reference to such a species. Despite the forgivable speculation, Bell's account included extensive descriptions of life histories of the better-known species and, overall, considerably advanced knowledge of our amphibians. Nearly 20 years after the publication of *A History of British Reptiles*, he chaired the meeting of the Linnean Society at which were read the famous initial papers on natural selection from Charles Darwin and Alfred Russel Wallace. Apparently, he either failed to recognise their significance or did not wish to do so in view of his religious beliefs (Smith 1969).

Bell knew the Reverend Leonard Jenyns and referred to his accuracy and acumen. The latter's *Manual of British Vertebrate Animals* was published in 1835, four years before Bell's more famous book, but I have chosen to draw attention to his observations of amphibian species in Cambridgeshire between 1820 and 1870 because of my long association with the county. These and accounts of vertebrate and molluscan fauna were published and updated by Richard Preece and Tim Sparks (2012). The Cambridgeshire of Jenyns was just vice-county 29 and so was much more restricted in size than today's county. Common frogs, great crested newts and smooth newts were generally widespread and abundant, but it was what he had to say about the edible frog and natterjack toad that is more interesting. 'Edible frogs' had been discovered at Fowlmere Fen in 1843, but these are now considered to have been pool frogs. The original discovery generated arguments about whether they had been introduced. The fen was drained in 1847, but that was not the end of the debate, as museum specimens existed (see Section 12.6.1). Jenyns and his brother-in-law, John Stevens Henslow, were the first to observe and describe natterjacks on Gamlingay Heath in 1824. Later, specimens were found in central Cambridge, perhaps having been translocated from Gamlingay. Despite destruction of their habitat at Gamlingay by 1862, some natterjacks apparently survived into the twentieth century. Charles Darwin was one of Henslow's pupils at Cambridge University, and Jenyns and Henslow were both invited to take part in the *Beagle* voyage of 1831–1836 prior to the post of naturalist being offered to Darwin. One can only guess at what might have happened had Jenyns or Henslow sailed instead.

Mordecai Cubitt Cooke was a well-known botanist and mycologist, who published *Our Reptiles* in 1865 because of a passion for amphibians and reptiles that dated back to childhood. By then it was clear that the common frog, great crested newt and smooth newt were generally familiar and people had a grasp of their distribution and abundance. The palmate newt and natterjack toad tended to be more local and consequently less well known and described. The common toad, despite being widespread and abundant, still seemed to retain an air of mystery – and writers appeared reluctant to comment on its ubiquity and the nature of its breeding haunts. Cooke's book contained many extensive quotes from the letters and publications of other people, perhaps because of his position as editor of the magazine *Science-Gossip*. On the toad, he covered its poisonous skin, habituation and feeding behaviour, stories of toads embedded in stone and the etymology of the word 'toadstool', but said nothing about toads breeding in the countryside.

Some of the best features of Cooke's book were hand-coloured prints of the various species of amphibians and reptiles. Most were signed 'E. Cooke'; the author thanked staff at the British Museum for 'procuring figures to illustrate the work' but said no more about the identity of the artist. The coloured prints included those of the edible frog and 'Gray's banded newt', the latter being a striking black and white beast which in shape and size (relative to the palmate newt in the same illustration) most closely resembled a great crested newt. The story of Gray's banded newt is strange and complicated. At this point I should say that Cooke, Bell and Jenyns not surprisingly knew one another. The banded newt was described to the scientific community by Dr J. E. Gray and Jenyns in 1835 and given the Latin name *Triton vittatus* (now *Ommatotriton* spp.). Specimens had been found preserved with other newts in a jar in the British Museum marked simply 'England'. Bell (1839) examined these unusual specimens and declared them to be a 'variety' of palmate newt, despite Gray believing they were related to the great crested newt. Thus, banded newts which occur naturally in the Middle East (Griffiths 1996) were first described from specimens thought to have been collected in England. The saga was summarised by Steward (1969): to the disappointment of British herpetologists, no further specimens were found here and it was generally accepted that the newts in the jar had come from abroad and been mislabelled.

George Albert Boulenger was born in Brussels, but came to work in the British Museum in 1881 at the age of 23 and stayed until he retired in 1925. He had a prodigious output in the field of herpetology, but here I concentrate on the two volumes he wrote for the Ray Society in 1897 and 1898, *The Tailless Batrachians of Europe.* Essentially, this was a single account with the Introduction and the Discoglossidae and Pelobatidae in Part I and the Bufonidae, Hylidae and Ranidae in Part II. Boulenger referred to Bell (1839) and Cooke (1865) but made no mention of Jenyns. It is an amazingly impressive and learned work and contains nuggets of useful information, such as how to distinguish between tadpoles of different species by examining their mouthparts (see Section 7.2.2). And the coloured illustrations of the species, the work of P. J. Smit, have surely never been bettered. Nevertheless, it was not of course written with conservation in mind. For instance, all that might be gleaned from the paragraph about the 'habitat' of the common toad is that it 'inhabits nearly the whole of Europe' and also ranges as far east as Japan and as far south as North Africa. Malcolm Smith worked with Boulenger before the latter retired, and complimented his considerable knowledge of life histories of the species, much of which was acquired during holidays taken on the European mainland (Smith 1969).

Malcolm Smith had a lifelong interest in reptiles and amphibians. He decided to qualify as a medical doctor because it would provide the opportunity to collect and study animals in countries where they were more numerous and exotic (Bellairs 1959). This plan worked extremely well, as he settled in Bangkok and eventually became Physician to the Siamese Court, which enabled him to retire comfortably back to England in the 1920s when he was about 50 years old. In his tribute to Smith after his death in 1958, Angus Bellairs recounted how his interest then turned increasingly to reptiles and amphibians in Britain, and in particular to understanding and documenting their life histories and distributions. His New Naturalist volume, *The British Amphibians and Reptiles*, was first published in 1951. He revised a second edition, which appeared in 1954 and included a short new chapter on the influence of climate on the distribution and habits of the British species. After his death, a third edition was published in 1964, in which Bellairs and Deryk Frazer inserted some new information and a separate list of references dating from 1956 to 1962. A fourth edition was published in 1969 but nowhere was it stated how it might differ from the third edition; this is the version I have owned for more than 50 years. Smith was, as my tattered dust jacket states, 'a happy combination' of a field naturalist and a museum man. It is still possible to find facts in this book that cannot be easily located elsewhere. The book also contains hints of concern that amphibian populations might be declining, as there is reference to the loss of several natterjack colonies.

Smith was primarily responsible for founding the British Herpetological Society (BHS) in 1947 and became its first president. He will have known Maxwell Savage, as his book lists five of Savage's papers published between 1934 and 1950, and Savage was editor of the BHS's *British Journal of Herpetology* during the mid-1950s. Trevor Beebee (2010) investigated the latter's life story and wrote a tribute because he rightly felt that Maxwell Savage was 'a remarkable pioneer of amphibian research in Britain' but was in danger of being forgotten. Savage worked as a chemist in the pharmaceutical industry. His 'spare-time studies' concentrated primarily on the common frog and, despite never receiving any funding, he investigated issues such as the distribution of breeding ponds in relation to geology, how frogs might navigate to their breeding ponds, the influence of climate and other factors on spawning date, the thermal function of spawn jelly and the nutrition of tadpoles. In addition to his papers, he drew together his information on the frog in a monograph published in 1961: *The Ecology and Life History of the Common Frog*. He selected the frog for study in the late 1920s precisely because it was so common, but by the late 1950s the species had become rarer. In 1974, I visited him at home at Welwyn in Hertfordshire because my enquiry into changes in status of the frog and toad had indicated that population declines were widespread and severe in some regions (Section 2.2), and I wanted to ask him for more details about what had happened in his study area. He was self-effacing about his achievements and genuinely delighted that I was interested in his work. I met him again in 1975 when he helped during a study of breeding success in various sites (Figure 1.1; Section 3.4.2).

The *British Journal of Herpetology* is referred to above, and that periodical carried in its early years two important papers on the distribution of amphibians and reptiles that at last illustrated where they occurred across the British Isles (Taylor 1948, 1963). Maps in Taylor (1948) were on a vice-county basis and these appeared in all editions of Malcolm Smith's book. The later maps (Taylor 1963) had dots indicating locations from which each species was reported; the brief text drew attention to population declines for both the natterjack toad and common frog. The national Biological Records Centre (BRC) was set up at Monks Wood in Cambridgeshire in 1964, and future recording of amphibian distribution would be by 10 km squares on the national grid. This takes the story more or less to the point in the late 1960s when I first started working on amphibians.

1.2 My interest in amphibians

1.2.1 *The early years up to 1968*

I have read that herpetologists are born and not made. For instance, Angus Bellairs (1959) said it of Malcolm Smith – and many dedicated individuals began their long-term interest in amphibians or reptiles during childhood. The same

Figure 1.1 Maxwell Savage at one of his frog study sites at Welwyn, Hertfordshire, May 1975.

was not true for me, however, mainly I suspect because of lack of much opportunity. I was a boy in the wrong part of the country at a time when amphibian populations were declining in the face of development of various kinds as the nation recovered after the Second World War. If there had been ready access to ponds with breeding amphibians, I do not doubt that I would have emerged dripping from them as many times as I stumbled out of the embryonic River Lea, which had its source close to my home.

Growing up in the late 1940s and the 1950s in south Bedfordshire and north Hertfordshire, I was hardly aware of ponds and pond life. Roaming the hills of the north end of the Chilterns around Luton with friends, I was in a landscape that was largely chalky, scrubby grassland, with arable farming dominating the flatter parts. Many relatives lived in the Hertfordshire village of Redbourn, and there the countryside seemed more traditionally rural with mixed farming. Some of my relatives farmed and one uncle was a milkman, bottling the milk and delivering it from carts drawn by horses. Redbourn was the only place where I could find frogs.

The best newt pond I knew was in the Luton garden of one of my friends. It was a small sheer-sided concrete pool less than 2 m across – and it teemed with newts. They were so numerous that one pastime was seeing how many could be caught by hand within a short space of time. The newts were of two distinct types: one had a wavy crest along its back, the other did not. So what species were they? Most children today with a keen interest in wildlife and the countryside would probably know the answer, but in those far-off days information on such subjects was much harder to acquire. By then, Malcolm Smith had published his New Naturalist volume on the British amphibians and reptiles. Even if I had been aware of the book's existence, I would have considered it too technical and detailed to buy or ask for as a present. Unbeknown to me there were books written for the young naturalist, such as Ford (1954).

My only source of help was an album of cigarette cards issued by John Player & Sons in 1939 and passed on to me by an uncle. This was *Animals of the Countryside*, and he also gave me albums on *Birds and their Young* and *British Freshwater Fishes*. These three were so well illustrated that I still treasure them today. *Animals of the Countryside* covered mammals, reptiles and amphibians. All three newt species had a picture of an adult male in breeding attire, so it was obvious that one type in the pond were male 'common or smooth' newts. The brief description under the card stated that the male of this species lost his nuptial dress in mid-summer, the implication being that the female looked like the male without the wavy crest. The trouble was that the second type in my friend's pond was much plainer than a crestless male would be. Nevertheless, I tended to assume because the two types occurred in roughly equal numbers that the second type must be female 'common or smooth' newts, but a nagging doubt remained that some of them at least might be something else: perhaps palmate newts? Many years later, I realised they were definitely all smooth newts.

Up to the age of about 15, I had a very general interest in the countryside. This was perhaps driven more by a desire to be outside than by a special attraction to plants, animal groups or geology. I collected and pressed wild flowers, and gradually acquired very modest collections of fossils, skulls and birds' eggs. In his autobiography, Richard Fortey (2021) describes pangs of guilt when collecting eggs from nests. My interest in eggs was probably fired by the cigarette-card album showing the nest of each species, and I do not recall any such qualms. I remember being determined to find the nest of a skylark (*Alauda arvensis*), and I moved on once that was accomplished. At the time, I was unaware of the important role eggshells would have in shaping my future career. I do recollect feeling sad or disappointed when attempting to keep various unfortunate invertebrates in jars; these activities invariably failed, owing to my lack of knowledge about the creatures' daily requirements, normal life expectancy or ability to escape. My mother had a dread of frogs and toads so I never kept amphibians of any life stage or species. She never lost the fear and prejudice that was evidently so ingrained in previous generations. Then, in my mid-teens, I made a conscious decision to specialise. I asked myself what group of organisms would be most enjoyable to study and what had a good diversity of readily visible species? The answer might have been plants or insects, but instead I chose birds for, I admit, not entirely logical reasons. Neither amphibians nor reptiles were even in the frame.

Over the next few years, I absorbed as much information as possible about birds and began doing my own simple research on their behaviour, which was eventually published much later (Cooke 1980). Because I was unable to study biology at school beyond the age of 14, I began to teach myself the subject, and this is still something that occupies me some 60 years later. I left school in 1961 and worked in the chemical industry prior to going to Leeds University in the autumn of 1962. Chemistry was my main subject at university, but I was able to take agricultural chemistry as a subsidiary subject, and this turned out to be the first in a series of lucky breaks that saw me claw my way towards a biological career.

Venturing into agricultural science opened the door for me to go to Reading University three years later to undertake a biochemical PhD on the composition and role of the organic matrix in chickens' eggshells. In 1967, while I was at Reading, Derek Ratcliffe of the Nature Conservancy's Monks Wood Experimental Station published his seminal paper on eggshell thinning by DDT in British birds of prey. Soon after, I happened to meet Dr Ratcliffe's colleague Ian Prestt. As a result of that meeting, I was invited to talk to staff at Monks Wood about eggshells. Suddenly my prospects of finding an interesting biology-based job seemed very much brighter.

Nevertheless, I put the Nature Conservancy to the back of my mind when trying to secure employment in 1968 as my course was reaching its conclusion. The situation then was very different to how it is now. In the late 1960s, there were very few jobs in conservation research, but if you could find one, there was relatively little competition for it. My wife, Rosemarie, and I assumed we would probably move to a Commonwealth country, but a new post of biochemist at Monks Wood was advertised. I applied, was interviewed and offered the job. I duly started in October 1968 as a member of the Toxic Chemicals and Wildlife Section to work on pesticide issues under the leadership of Dr Norman Moore.

1.2.2 *The years after 1968*

In the following chapters, I explain how various decisions were made and acted upon, and now will quickly sketch out my career. Monks Wood was a vibrant and stimulating place during the 10 years I was there. We were really interested in what we were doing, and so were other people. As I recall, most years there were open days or open weekends when the public would descend on us in huge numbers to see exhibits and displays and talk to staff. We also had visits from dignitaries such as Prince Charles: I was one of those selected to have lunch with him, as I was closer to him in age than most of the scientific staff. Harold Wilson, who was then Prime Minister, came to visit with his Chief Scientific Adviser, Sir Solly Zuckerman, and his Minister, Anthony Crosland. I have taken flak over the years from friends who have seen a well-scrubbed, much younger version of me in a photograph in one of Crosland's books.

There was a great deal of freedom afforded to scientists at Monks Wood in the heady days of the 1960s and early 1970s. Once, a colleague commented over lunch that we did not know how lucky we were. The staff generally responded positively and with enormous enthusiasm and commitment. In the sphere of toxic chemical research, there were real issues to be tackled, particularly over the use of organochlorine insecticides. Norman Moore was a respected and benign presence, encouraging, complimenting and suggesting rather than being critical. Much of my work was on birds, including looking at the phenomenon of eggshell thinning by asking the question: what can the altered structure of the shell tell us about what happened to processes in the bird to cause thinning? This involved laboratory studies on eggs collected by colleagues. But when I could manage it, I would make field trips to help collect eggs, for instance visiting Lincolnshire heronries with Ian Prestt's assistant, Tony Bell.

When I had been at Monks Wood for a few months, I realised, mainly from conversations in the canteen, that there was concern about the common frog: it was believed to be declining, but no one seemed to be trying to find out why. Here was a project that I could approach from the pesticide angle, but at the same time I would be able to learn much more about frogs and their needs. It soon expanded to include other amphibians.

In 1973, the old Nature Conservancy was split into the new Nature Conservancy Council (NCC), which looked after practical conservation in the field, and the Institute of Terrestrial Ecology (ITE), which continued with research. We were now two separate organisations and ITE had to attract 'customers' for some of its funding. Suddenly, those of us who opted for ITE were thrust into the modern world of having to concentrate more on why and how we were doing research. By the mid-1970s, I found myself gradually feeling less enthusiastic about some of the work I was doing. This particularly applied to the eggshell studies, where I felt I was learning more and more about a progressively narrower field in which a diminishing number of people appeared to be interested. In 1978, I applied for a new post in the NCC; this was as Toxic Chemical Specialist in the Chief Scientist Team, led by Derek Ratcliffe, the man who, among many other achievements, had first alerted the world to eggshell thinning. I got the job, which was based in Huntingdon, and it was explained to me that, as I had been working on amphibians, I would also be adviser on amphibians and reptiles. The assumption was that if I knew about amphibians, I must know about reptiles too! That assumption was definitely incorrect, but luckily a friend, Keith Corbett of the BHS's Conservation Committee, very kindly spent a week taking me around the principal reptile sites in southern England, which was an extremely useful crash course in the subject (Figures 1.2 and 12.3).

Initially, the bulk of the herpetological work was on the three endangered species: the sand lizard (*Lacerta agilis*), smooth snake (*Coronella austriaca*) and natterjack toad. Keith had educated me well on the first two species. I was more familiar with natterjacks, having first seen them in 1974 when visiting Saltfleetby–Theddlethorpe on the Lincolnshire coast with Trevor and Maggie Beebee, also members of the BHS's Conservation Committee (Figures 1.3 and 12.1). These three species had protection under the Conservation of Wild Creatures and Wild Plants Act 1975, but the remaining species had no protection. This was partially rectified by the Wildlife and Countryside Act of 1981, which gave the same level of protection to the great crested newt. This move was intended to be helpful as regards conservation, but we then had a protected species which was widespread and little studied. Much more of my herpetological input in the rest of the 1980s concentrated on contracting out work to provide better information on status, location of breeding sites and natural history of this and other widespread species of amphibians and reptiles. Because I was interested and because it helped me with my daily work, I became very involved with local survey and research on crested newts and other species in the evenings and at weekends.

In the early 1990s, NCC was split into the country agencies and the Joint Nature Conservation Committee (JNCC), which oversaw the work of the agencies and represented them. I ended up as Toxic Chemicals Adviser to both English Nature and the JNCC, and no longer had any official responsibility for amphibians or reptiles. Nevertheless, I continued to carry out my long-term surveillance of local amphibian populations in my own time, but my interest in herpetological research and

Figure 1.2 The author on a reptile site on heathland in 1978. Photo by Lynne Farrell.

policy became narrower. In 1984, the office had moved from Huntingdon to Peterborough, and this city remained the home of English Nature's headquarters and the JNCC.

I had been interested in deer since my time at Reading, and by the 1990s a significant chunk of my spare time was taken up with long-term monitoring of deer populations in National Nature Reserves (NNRs) close to home. This monitoring tended to be in the autumn and winter so dovetailed fairly well with my

Figure 1.3 A natterjack at Saltfleetby–Theddlethorpe, Lincolnshire, 1974.

spare-time studies on amphibians. As a result of my work on deer, I was granted a 12-month secondment back to ITE in 1993–1994 to study deer damage in Monks Wood NNR and to advise English Nature on remedial management of deer and their woodland habitat more widely. On my return to the office, I was permitted to have a small amount of official time to continue studying deer. In 1998, I accepted an offer of early retirement in order to carry on with committee work advising government departments on the environmental safety of pesticides, as well as to enjoy virtually complete freedom to work on amphibians, deer and other forms of wildlife.

Those plans were, however, wrecked almost immediately as I was diagnosed with a form of leukaemia in January 1999. Initial treatment was tough and I withdrew from the pesticide committees and gave up some of the more arduous fieldwork projects. My prognosis was not good and I was galvanised into completing and writing up a number of projects. Between 1999 and 2004, I published about 10 articles on my local amphibian fieldwork. By that time, my long-term projects on amphibians were limited to systematic surveillance of a newt colony, recording dead toads on roads around local breeding sites, and trying to understand the population dynamics of frogs in my garden ponds. Long-term fieldwork on amphibians finally stopped in 2015 for various reasons. While I have tended to operate towards the research end of the conservation/research spectrum, my research, surveillance and monitoring has always been undertaken for overt or underlying conservation reasons. Fortunately, since 2002, my leukaemia has been successfully controlled by the first of the 'magic bullet' cancer drugs.

1.3 The book

The rest of this book describes why decisions were made, what was done, what happened as a result and how approaches changed over a period of more than 50 years. The literature quoted is egocentric, but that was the aim of the book – and a more balanced, briefer historical summary is available elsewhere (Beebee 2014). The focus is generally on British issues and British research, but the problems facing our amphibians echo those from many parts of the world. Or perhaps it is more correct to say that theirs echo ours, as many of our problems happened earlier. The book is not restricted to formal publications: unpublished material and articles from the 'grey' literature are included where they help to explain the various topics, as are thoughts, feelings and impressions. Up until the 1990s, I was working at a national level, but since then I have been less aware of broader developments. Nevertheless, some of the more significant recent advances in research and conservation are outlined.

No doubt some readers will wonder why reptiles are not included in the book. So many books include both classes of animals, and it is true that I was responsible for a time for advising NCC on reptile conservation. The simple answer is that I have had much less experience with reptiles, especially of the hands-on type. I have always felt that one reason for the attraction of amphibians for young and old alike is that the species can be caught more easily than any other class of vertebrate (Figure 1.4). I grew up hardly ever seeing a reptile, and my significant fieldwork was restricted to a lengthy study of sand lizards at Merseyside (Cooke 1991, 1993a). As I wanted to be able to write primarily from the point of view of my

Figure 1.4 Amphibians are of interest to young people and present a relatively easy opportunity to get close to wild animals. (a) The author with son Steven during a study of smooth newts in 1984. (b) With grandson Billy, Steven's son, catching frog tadpoles in a garden pond in 2014 (photo by Rosemarie Cooke). Regrettably, because of the biosecurity risk, current advice is to avoid handling amphibians unnecessarily (Section 3.6.2).

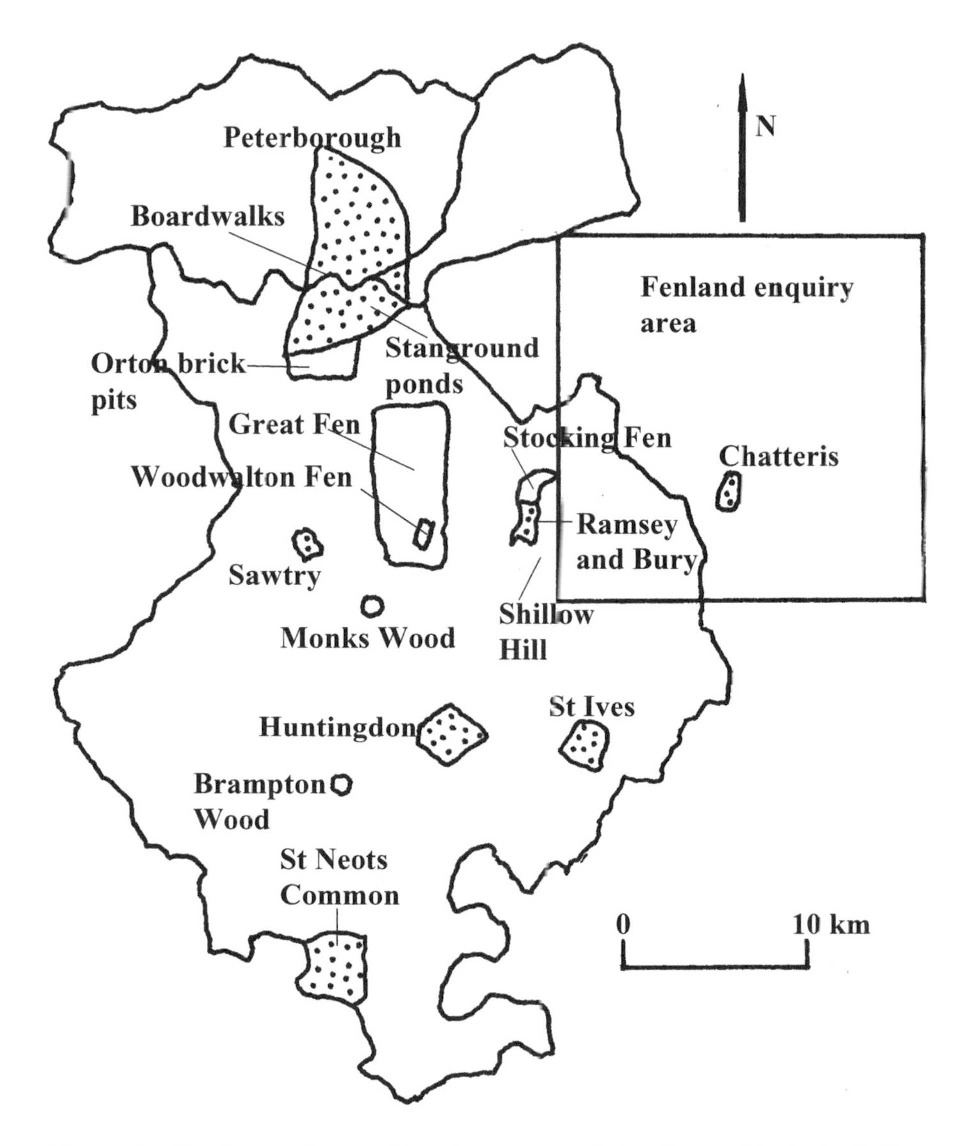

Figure 1.5 Sketch map of some of the places, areas and sites discussed in the book. The main outline is of the former county of Huntingdon and Peterborough, divided into Huntingdonshire in the south, the Soke of Peterborough in the north-west and a fragment of old Cambridgeshire in the north-east (see Section 2.3). All of this land is now part of the new county of Cambridgeshire. The Fenland enquiry area is described in Section 2.4. Relevant developed areas are stippled. Stippling around St Neots Common represents the townships of St Neots, Eaton Ford, Eaton Socon and Eynesbury.

own work, including reptiles would have produced an account that was so biased towards amphibians that it would have drawn queries as to why I had bothered with the reptiles. Had I grown up or worked in an area rich in reptile sites, the outcome might have been very different. However, reptiles are brought into the discussion where appropriate.

This then leads to an explanation of why I have concentrated on my own work in this book. The simple answer is similar to that given by Maxwell Savage in the first paragraph of his monograph on the frog in 1961: initially, at least, I was writing for myself. Rosemarie and I shielded during the worst stages of the Covid-19 pandemic in 2020–2022. When I started writing in March 2020, the pandemic was not forecast to be either so severe or so prolonged, and I did not imagine it might be a book for publication. As I became immersed in it, though, I wondered whether my story might be of some interest both to older readers who lived through the earlier decades with me and to those tadpole hunters who have discovered the joys and tribulations of amphibian conservation more recently.

There are 11 more chapters: two on changes in status and reasons, four on frogs and toads, three on newts, and two concluding chapters. Chapters 2–11 are restricted to our six most widespread species: common frog, common toad, natterjack toad, smooth newt, great crested newt and palmate newt. Regrettably, while I was working for the NCC, the pool frog was not recognised as a native species and attracted relatively modest interest; later, this lack of attention was shown to be misguided (Beebee *et al.* 2005), and the story of the pool frog is one of the topics covered in a historical review (Chapter 12). Very little is included here about introduced species. Each chapter is written with a historical slant describing how populations, methods, attitudes and knowledge changed over the decades. Ideally, I wanted each chapter to be stand-alone, but inevitably have had to include cross-referencing to minimise repetition.

In addition to the text, attempts have been made to explain techniques, results and conclusions with tables, graphs, photographs and maps. A map of a number of local places and sites discussed in the text is shown in Figure 1.5. Tables and graphs have been updated, adapted and reconstructed from the original published versions so as to present material in the most relevant and straightforward manner. References to statistical tests are virtually absent. However, whenever attention is drawn to relationships between variables, then a statistically significant result will have been found. All work with amphibians was performed with appropriate licences from the Home Office, NCC, English Nature or Natural England (which succeeded English Nature in 2006). Weather data were extracted from records of the Royal Meteorological Society's stations at Monks Wood and the National Institute of Agricultural Botany in Cambridge.

CHAPTER 2

Ups and downs in the twentieth century

2.1 General introduction

A thread running through much of my work on amphibians has been 'change'. That might mean change in size of a single population, change in status at a broader level, or change in time of breeding. It could also refer to the ways in which studies or enquiries were initiated, undertaken, published and acted on, because those processes changed over the decades. This chapter concentrates on changes in status at a broad level and looks, as far as possible, at what happened to the five widespread species of amphibians during the twentieth century. More recent developments in the twenty-first century are discussed at the end of the chapter. Surveys and enquiries are dealt with in the order in which they were undertaken. As some involved dodging back in time, one of the last sections (2.6.2) attempts to review events during the twentieth century in chronological order. Changes in size of individual populations are covered in later chapters, with changes in time of breeding in Chapter 11. Natterjacks are discussed in Chapter 7 and Section 11.3.

2.2 Where have all the frogs gone?

2.2.1 *Introduction*

Frank Perring was head of the national BRC at Monks Wood during my early years at the Experimental Station. In 1966, he published an article entitled 'Where have all the frogs gone?' and in it he suggested that pesticides might be implicated in the scarcity of frogs in parts of Britain. This article and my conversations with him helped reinforce my initial interest in frogs at the end of the 1960s; and one of the first things I wanted to do was to obtain information on any changes in status of the species, as virtually nothing was available at the time. Prior to Frank's article, Malcolm Smith (1953) had noted a 'shortage' of frogs and was concerned about the number taken for use in laboratories (Section 3.6.5). After completing a distribution survey of British amphibians and reptiles, R. H. R. Taylor (1963) commented that the frog was 'not as abundant in many districts as it used to be'. He considered that catching frogs for laboratory work and filling in farm ponds were important factors in the decline.

I was a member of a team of research scientists based at Monks Wood who were tasked with determining the effects of toxic chemicals on British wildlife. I had settled upon the frog as a species that might have been affected by pesticides. At the same time, there were a couple of issues concerning the toxicological effects of pesticides on birds that interested me, and in 1969 I began laboratory dosing studies on both Japanese quail (*Coturnix coturnix japonica*) and frog tadpoles. I would not go so far as to say research was especially competitive in those days, but I knew that American institutions were throwing extravagant sums of money at similar avian studies, whereas no one else was doing anything with the common frog. Some people in this country did care about frogs, but they were probably not in a position of being able to undertake detailed studies, especially because of a lack of facilities for experiments and chemical analysis. At Monks Wood we had good analytical support, at least for a range of potential pollutants. This helped to make working on frogs a more worthwhile proposition, but I was also aware that an essential early stage of the work should be to find out whether the common frog really had declined. If it had not become rarer since modern pesticides were introduced from the late 1940s, there would little point in continuing with that line of enquiry.

With no hard information, how should I attempt to discover what had happened in the recent and more distant past? The only course of action seemed to be to ask people what they had observed or what they believed had happened. This approach has of course various inherent weaknesses. In an ideal world, it would be achieved by canvasing information from knowledgeable field naturalists who had worked in an area for many years and had kept detailed records about frogs. In the real world, results had to depend to varying degrees on accepting the perceptions, rather than recorded observations, of people with a range of skills, experience and possibly bias. In 1970, when the first surveys were undertaken, there were few observers who fulfilled the ideal criteria. Nevertheless, the approach and the risks seemed worth taking – at least to see what happened.

Two separate enquiries were undertaken across the British Isles: the Breeding Sites Survey and the Schools Survey (Cooke 1972a). Information on the common toad was also requested, in part to try to ensure that participants were able to distinguish between the two species. In the Breeding Sites Survey, questionnaires were sent to contributors to BRC's Amphibian and Reptile Recording Scheme, to university biological societies, and to selected nature reserve wardens and other naturalists who were or might be interested. They were asked for details about sites that contained breeding frogs or toads at some time during 1950–1970, and not to limit their selection to sites with breeding in 1970, to avoid missing information on colonies that had died out during the period. Information was requested on where the site occurred, the nature of the site and surrounding land (e.g. a garden pond or an agricultural ditch), population trends for the species, and whether the site had been created, modified or destroyed between 1950 and 1970.

The Schools Survey was aimed primarily at biology teachers in schools and colleges. In order to participate, they were required to have been living in the same locality for at least five years, and they were asked to provide information on how long they had been searching for spawn or tadpoles, whether the species were harder or easier to find, and what were the dates of any changes. Locality size was not requested and presumably varied widely. In order to be able to combine locality data from the Schools Survey with site data from the Breeding Sites Survey, each site was assigned to a locality, defined as a 10 km square of the National Grid; several sites may have been recorded in a single square.

2.2.2 *Changes in status*

Replies to the Schools Survey came from 1,115 localities, but 137 were discarded because of inadequate information. Of the remaining 978 completed questionnaires, 691 related to frogs, 14 to toads and 273 to both species. Questionnaires from the Breeding Sites Survey referred to a total of 101 localities (10 km squares), with 49 relating to frogs, 2 to toads and 50 to both species. Information was therefore available for 1,063 localities with frogs and 339 with toads. Geographically, response was patchy. However, time for dealing with correspondence and handling the data was limited, so no further attempts were made to fill in gaps. I considered that sufficient data had been received from a range of geographically and topographically different counties to provide a reasonable picture of past events. In order to present information in a readily digestible form, (1) neighbouring counties were grouped into regions from which replies suggested that population trends were similar and (2) information was assembled into four periods of five years: 1951–1955, 1956–1960, 1961–1965 and 1966–1970.

A simple index of change was calculated for counties, regions and countries by subtracting the number of replies stating decrease from the number stating increase and dividing by the total of a localities where the animals existed during that time period (Table 2.1). An index of zero indicated no overall change, whereas an index of –1.00 meant decreases in all localities. Indices of change for the frog during 1966–1970 are shown in Figure 2.1 for regions and countries in the British Isles. Each index on the map is based on information from more than 30 localities apart from northern England (17), the Republic of Ireland (11) and northern Scotland (6).

In England, indices showed a worsening situation from 1951–1955 up to 1961–1965 for both species, with signs of a deceleration in losses in 1966–1970. In Scotland, rates of loss worsened throughout the 20 years, but were less than in England for the first 15 years. Losses of frogs in Northern Ireland had a similar pattern to those in Scotland, while declines of both species in Wales seemed to accelerate through the 20 years, but the amount of information was less than that from other countries. The declines for the two species in 1966–1970 were very similar to those during 1961–1965 for England, but worse for Scotland and Northern Ireland.

In all regions and countries, the change was negative for the frog in the late 1960s (Figure 2.1). The map does not include data for the Isle of Man, where information from 11 localities produced an index of –0.09. Indices for the toad were similar, apart from in northern England and northern Scotland, where indices were zero but were based on small numbers of replies. The greatest decline for any region was for the frog in what I termed 'south-east Midlands, northern Home Counties and London'. For that region, reports of 'no longer found' occurred in 10 localities, decreases in 84, no change in 18 and increases in 11, giving an index of –0.65.

I have concentrated here on changes in the 1950s and 1960s, but some schoolteachers provided information as far back as the 1940s, and the overall indices

Table 2.1 Combined locality information from the Schools Survey and Breeding Sites Survey: indices of change (with number of localities in brackets) for four time periods in different countries (summarising data from Cooke 1972a). The toad does not occur naturally in Ireland.

Species and Country	Time period			
	1951–1955	1956–1960	1961–1965	1966–1970
Common frog				
England	−0.15 (163)	−0.17 (246)	−0.43 (445)	−0.38 (719)
Scotland	−0.09 (32)	−0.10 (42)	−0.13 (85)	−0.36 (146)
Wales	0.00 (10)	−0.08 (12)	−0.24 (21)	−0.21 (33)
Isle of Man	−0.10 (10)	−0.10 (10)	−0.20 (10)	−0.09 (11)
Northern Ireland	0.00 (30)	−0.02 (47)	−0.13 (94)	−0.26 (143)
Republic of Ireland	0.00 (3)	0.00 (3)	−0.33 (9)	−0.45 (11)
Overall total	−0.12 (248)	−0.13 (360)	−0.30 (664)	−0.35 (1063)
Common toad				
England	−0.14 (68)	−0.15 (101)	−0.40 (190)	−0.36 (279)
Scotland	−0.14 (16)	−0.17 (18)	−0.25 (29)	−0.33 (47)
Wales	0.00 (3)	−0.25 (4)	−0.14 (7)	−0.38 (13)
Overall total	−0.14 (87)	−0.15 (123)	−0.37 (226)	−0.35 (339)

of change for the frog in the British Isles suggested an accelerating decline from the early 1940s: 1941–1945, −0.04 (based on 45 replies); 1946–1950, −0.10 (71 replies). Figures for the next four five periods were −0.12, −0.13, −0.30 and −0.35 respectively (Table 2.1).

Within the 101 localities, contributors to the Breeding Sites Survey provided information on 388 sites, but data on 74 were judged to be insufficient to be of use. Of the remaining 314 sites, 202 had frogs, 22 had toads and 90 had both species. Only 14% of these sites were from towns or cities with populations of more than 10,000 people, suggesting that the survey had a more rural bias than the Schools Survey, as will be revealed later. Indices for various types of site are given in Table 2.2. 'Other' sites included canals, streams, rivers, reservoirs, flooded quarries and ditches, and referred to the following surrounding habitats: moorland, heathland, woodland, parkland, common land and waste land. There were no clear differences in population trends between these various types of site, and they were grouped together.

The frog increased in garden ponds in each of the periods, and the percentage of these ponds in the sample doubled from 9% in 1951–1955 to 18% in 1966–1970 (Figure 2.2). The toad fared less well in garden ponds and badly in agricultural ponds in the 1960s. The frog declined in agricultural ponds throughout the 20 years but did better than the toad, especially in the 1960s. In other types of ponds it declined more than the toad, particularly in the late 1960s. Breeding populations

Figure 2.1 Indices of change for the frog during 1966–1970 calculated for regions or countries of the British Isles (adapted from Cooke 1972a). Brackets indicate an index based on fewer than 30 localities.

had disappeared by 1970 from 22% of frog sites and 24% of toad sites. Nevertheless, the lower negative indices for sites overall in 1966–1970, when compared with those for 1961–1965, were a cause for optimism for both species.

2.2.3 *Reasons given for changes in status*

Participants in both surveys volunteered reasons for change. Total numbers were: Breeding Sites Survey, 88 explanations offered for decreases and 37 for increases for the frog, with 32/11 for the toad; Schools Survey, 264/38 for the frog and 67/5 for the toad. There were relatively minor differences between reasons for losses of the two species, so the list below is restricted to frog losses.

- Habitat destruction: 31% of reasons in the Breeding Sites Survey and 57% in the Schools Survey. This included because of housing (9% and 32% respectively), other forms of urbanisation (5% and 4%), loss of agricultural ponds (7% and 7%) and unspecified destruction (8% and 14%).

Table 2.2 Information on sites from the Breeding Sites Survey: indices of change (with number of sites in brackets) for four time periods (summarising data from Cooke 1972a).

	Time period			
Type of site	1951–1955	1956–1960	1961–1965	1966–1970
Common frog				
Garden	+0.20 (10)	+0.29 (14)	+0.20 (30)	+0.30 (52)
Agricultural	−0.14 (37)	−0.13 (39)	−0.28 (58)	−0.06 (96)
Other	−0.04 (68)	−0.12 (77)	−0.27 (101)	−0.30 (144)
All sites	−0.05 (115)	−0.08 (130)	−0.21 (189)	−0.11 (292)
Toad				
Garden	+0.33 (3)	+0.33 (3)	−0.17 (6)	+0.19 (17)
Agricultural	−0.15 (13)	−0.14 (14)	−0.45 (23)	−0.21 (33)
Other	−0.12 (29)	−0.17 (29)	−0.22 (45)	−0.11(62)
All sites	−0.09 (45)	−0.13 (46)	−0.29 (74)	−0.08 (112)

- Habitat physically modified by natural processes or by human activity: 28% of reasons in the Breeding Sites Survey and 14% in the Schools Survey.
- Habitat affected by pollution: 19% of reasons from both surveys, including reference to pesticides (3% and 5% respectively in the two surveys).
- Increased mortality due to animals, humans and cars (9% and 7% respectively in the two surveys).
- Collection of different life stages: 13% in the Breeding Sites Survey and 3% in the Schools Survey, comprising collection of spawn and tadpoles by children (9% and 3% respectively) and collection of adults for biological experiments (4% and 0.4% respectively).

Because of the small number of reasons given for increases in toads, only reasons for increases in frogs are reported. More than half of the reasons given for increases in both surveys related to breeding site creation: garden ponds comprised 46% of such comments in the Breeding Sites Survey and 37% in the Schools Survey, while creation of other types of site amounted to 22% and 16% respectively. Other, more minor reasons given for increases in frog numbers were: colonisation of new sites after being forced out of former sites, introduction, extension of range, recovery after the bad winter of 1962/63 and reduced predation.

The overriding reasons given for change in both surveys were habitat loss and habitat creation. Loss of breeding ponds added up to more than 50% of the reasons given in the Schools Survey, with house building and other forms of urbanisation being mainly responsible. In this survey, some of the replies that mentioned loss of farm ponds blamed urbanisation. The Schools Survey provided a more suburban view of frog losses than did the Breeding Sites Survey. An example of destruction of breeding habitat in the early 1970s is illustrated in Figure 2.3.

Other ways in which habitat became unsuitable also featured highly. Modification of habitat included ponds drying out, silting up, becoming overgrown and being

Figure 2.2 Different levels of impact from the creation of garden ponds: (a) the common frog has benefited greatly from their creation; (b) the common toad has gained less from their creation, generally preferring larger, deeper ponds (photo by Mihai Leu).

Figure 2.3 Loss of a frog breeding pond on the edge of the town of Chatteris, Cambridgeshire: (a) 1973, (b) 1974, (c) 1975.

cleaned out. Pollution of habitat included references, in order of number of replies, to rubbish, pesticides, fertiliser, detergent, oil/petrol and sewage. In contrast, collecting spawn or animals was believed to be a relatively minor factor, despite the 1960s being a decade when people with new garden ponds to stock were combing the countryside looking for spawn. Garden ponds were starting to become an extremely important refuge for frogs, and this was believed to outweigh any detrimental effect that collection might have had on frogs in the countryside.

Locality information will reflect suburban rather than rural views because replies from the Schools Survey outnumbered those from the Breeding Sites Survey by almost 10 to 1. More heavily urbanised counties tended to have the worst declines. There was a negative relationship between (1) the locality index of change during the late 1960s for the 21 best-recorded counties and (2) human population density for the county in 1969, as estimated by the Office of Population Censuses and Surveys (Cooke 1972a). Among the less densely populated counties where frogs had declined least were Cornwall, Norfolk, Suffolk and County Tyrone; while in more highly populated counties, such as Hertfordshire, Cheshire and Essex, frog declines had been severe. This highlighted problems on the edge of towns, but did seem to lack a similar focus on losses in the countryside, as was reflected in little emphasis being placed on loss of farm ponds. A contributing reason to this imbalance might have been that frogs had already suffered massive declines in some areas of the countryside by the 1960s. The agricultural sites that were reported on by participants in the Breeding Sites Survey (Table 2.2) may have tended to be those that had escaped earlier loss or modification and survived relatively well during the 1950s and 1960s.

The main conclusion from these two national enquiries was that both species had declined over much of the British Isles. Publication of the results in 1972 attracted considerable attention, and it is probably fair to say that it could be regarded as one of the first significant contributions to amphibian conservation in Britain, being closely followed by the publication of Trevor Beebee and Keith Corbett's similar work on the great crested newt and the natterjack toad (Beebee 1975, 1976, Corbett and Beebee 1975). At a personal level, the broad conclusions were apparent as early as 1970 when the responses began to flood in. It confirmed that I should push on with the pesticide studies, and that there were legitimate reasons for widening the scope of my investigations.

Monks Wood Experimental Station was set up by the Nature Conservancy in the early 1960s in the rural county of Huntingdonshire on the edge of the Fens, which was and still is one of the most intensively farmed regions in Britain. I was therefore ideally located to try to determine what had happened to frogs in the countryside, including in decades that pre-dated the 1950s – and this is addressed in the next two sections.

2.3 Changes in status of the frog and the toad in Huntingdonshire up to 1973

Huntingdonshire was one of the smallest counties in England, being only 948 km² in size. It was amalgamated with the Soke of Peterborough in 1965 to form the new county of Huntingdon and Peterborough (Figure 1.5). That entity, however, was short-lived, and it has been part of the expanded county of Cambridgeshire since 1974.

I began gathering information on the status of the common frog and common toad in the former county of Huntingdonshire in 1969. This exercise picked up pace during and following the national enquiries summarised above, and in 1973 Peter Ferguson and I completed a survey of virtually every village, town and fen area in Huntingdonshire, searching for frogs and toads and interviewing long-term residents about the status of the species (Cooke and Ferguson 1974). One aim was to gather information from this predominantly agricultural environment for further back into the past than had been achieved with the national surveys. Those surveys went back to the early 1950s; populations of both species were decreasing in England by then, but when did the declines begin, and could this be linked with pesticide usage in this agricultural county? By the early 1970s, its agriculture was very predominantly arable.

The people questioned usually showed keen interest and concern for their local wildlife, but doubtless not everyone could remember exact dates or was totally familiar with differences between the two species. The problem with dates was eased by the fact that many informants were able to relate to events around the Second World War. Most people believed that the toad showed similar changes in status to the frog, so only information for the frog is given here.

Comments on status were collected in 65 towns, villages and localities from 155 people, with between one and nine commenting in each place (Table 2.3). Information back to the 1930s was available for 45 (69%) of the 65 areas. There was a consistency about their observations, which are summarised in Table 2.3. As regards accuracy, I remember saying at the time that informants 'couldn't all be wrong'.

Table 2.3 Status and change in status of the common frog in towns, villages and other localities in Huntingdonshire, 1930–1973 (summarising data from Cooke and Ferguson 1974). Number of areas with information increased from 45 in the 1930s to 65 in the early 1970s.

	Percentage of the total number of areas with information				
Status or change	1930s	1940s	1950s	1960s	Early 1970s
Abundant	78	53	16	5	3
Increasing	0	0	0	0	6
Decreasing	4	29	51	36	5
Absent/rare	18	18	33	59	86

In the 1930s, the frog was reported as being abundant in 78% of areas, and absent or rare in 18%; but significant declines began in the 1940s. By the 1950s, it was absent or rare in a third of the areas and decreasing in most of the remainder. At the culmination of the study in the early 1970s, it was abundant in only 3% and absent or rare in 86% of areas, but the first glimmer of hope for a recovery arrived with several successful introductions.

Reasons suggested for declines (with number of times mentioned in brackets) were: use of agricultural chemicals (26), loss of ponds and ditches (18), habitat modification including improved drainage (11), loss of damp pasture (2), increased predation (2). The preponderance of views focusing on farm chemicals was very different to the opinions of participants in the national scheme and reflected the rural nature of the county. The role of pesticides in the decline of frog populations locally in the county and more generally in Britain is considered in Chapter 3.

By chance, we were able to tie in our information for one area with a survey undertaken by a colleague at Monks Wood. Judy Relton reported in 1972 that 35% of ponds disappeared between 1950 and 1969 in an area of roughly 20 km^2 around Kimbolton. This loss is consistent with the consensus view of six people interviewed in our survey that frogs were abundant in and around the village in the 1940s, but then declined to become rare by the early 1970s. One factor in this process was probably that as ponds were lost, so the distance between them increased, as did their isolation and the risk of extinction of their frog populations.

In 2008, another colleague at Monks Wood, the botanist Terry Wells, reviewed available agricultural census data on the amount of grassland in Huntingdon and the Soke of Peterborough. The area of permanent grass was fairly stable at about 40,000 ha from 1900 until 1939, but then decreased suddenly during and after the Second World War, declining by roughly 60% by 1945 and then by a further 20% by 1980 (Wells 2008). Arable land increased in area as the grass decreased. Rough grazing built up from less than 200 ha in 1900 to a peak of about 2,800 ha in 1939, before decreasing to below 1,200 ha by the end of the war. Terry Wells commented on how the severity of drainage and cultivation during the period 1950–1973 will have led to a decline in species of plants growing in wet grassland, but provided no data on drainage during the 1940s. However, these bald statistics suggest that loss of (damp) grassland will have had considerable detrimental effects on the habitat of local frogs

Figure 2.4 Examples of neglect and abuse of ponds in the early 1970s: (a) a pond left shaded and isolated in an arable field; (b) a roadside pond used as a refuse pit, full of unwanted containers and other rubbish.

and toads since 1940. With the conversion of grazed land to arable, ponds became redundant and were neglected, abused and eradicated (Figure 2.4).

In tandem with obtaining information on past changes of the frog and toad in Huntingdonshire, data on the status of the species were gradually collected between 1969 and 1973 (Cooke and Ferguson 1974). Our estimates of breeding adults in the county in 1973 were only 1,500 frogs and 4,000 toads. These were very low numbers for an entire county and equate to one or two frogs, and about four toads, per km². Old density data from anywhere in Britain are hard to come by, but I was informed by the Huddersfield Naturalists' Society that, in 1970, about 2,000 frogs were found within eight miles (13 km) of the town, which is a density of roughly four per km² (Cooke 1972a). These figures were very much lower than densities estimated by Savage (1961). Working in Hertfordshire, he proposed 3–4 adults per acre in an area of one square mile during the 1930s and 5–6 per acre for another similar-sized area in 1949; these estimates ranged roughly 750–1,500 per km². These localities were some of the best he knew for frogs, but 1 km² of Savage's study area in the 1930s and 1940s may have contained a comparable number of frogs to the whole of Huntingdonshire in 1973, thereby providing a graphic example of the catastrophic decline suffered by the species. Interestingly, when I met Maxwell Savage in 1974, he tended to blame deaths from road traffic for the losses of his local frogs. There is a comment in his book suggesting that in a survey he made in 1958, most sites still existed but the frogs had gone.

2.4 Changes in status of the frog and the toad on part of the Fens up to the early 1970s

In 1973, when Peter Ferguson and I were completing collection of the depressing status data from Huntingdonshire, we decided to find out what might have happened to the frogs and toads in an area of uniformly intensive arable farming on the Fens (Cooke and Ferguson 1976). The Fens is a flat, low-lying area that has been reclaimed over the centuries primarily for farming (Figures 2.5, 2.6 and 3.7a). Prior to the drainage, extensive marshy areas may have been an ideal wilderness for amphibians between periods of inundation with seawater.

We selected for convenience a 20 km square of peat fen (10 km grid squares TL 38, 39, 48 and 49; Figure 1.5), of which 15% was in Huntingdonshire and so was covered in the survey reported above. Although this fen area contains a number of settlements, with the town of Chatteris more or less in the centre, we concentrated entirely on the farmland. Information was gathered in 1973 on 85 localities by interviewing the following people either singly or in small groups: farmers, farmworkers, an eel catcher, a nature reserve warden and a drainage pump operator. In 16 of these areas, interviewees either admitted they could not distinguish frogs from toads or we considered that they could not do so; their views were omitted from this account. As we were undertaking this exercise, we appreciated the importance of drainage and held discussions with bodies responsible for drainage in the area.

Information received on status told a similar story to what we had learnt in Huntingdonshire, and again the frog and toad were said to have decreased in unison. Amazingly, we were provided with observations on seven localities back to Edwardian times and for 13 localities during the 1910s: back then, the frog was consistently described as abundant, as was the toad. Information for the frog is shown in Table 2.4 for the period 1920–1973. There was almost universal agreement that frogs and toads were abundant in the 1930s. I was told during the fen survey that in some areas people still referred to 'clearing off the spawn' when removing tea leaves from the surface of a cup of tea. In those days, tea bags had evidently not yet become widely used on the Fens.

The changes between the 1930s and the early 1970s were even more extreme than those reported in the Huntingdonshire survey: reports of frogs being abundant decreased from 94% of localities to zero, and reports of them being absent or rare increased from 6% to 97%.

For 15 of these 69 localities, interviewees did not express any opinion about why frog populations had declined. In the remaining 54, the principal reasons suggested were agricultural chemicals (mentioned in 78% of the interviews), improved drainage (37%), breeding sites destroyed (7%) and loss of pasture (4%). In 31 of the interviews in which agricultural chemicals were blamed, farmers and farmworkers were more specific, referring to herbicides in 29 (54%), insecticides in 22 (41%) and fertilisers in 3 (6%). The lethal effects of both of the pesticide groups were believed to be more important than their indirect effects on the frogs' habitat or food, but no one provided recollections of actual incidents.

Table 2.4 Status and change in status of the common frog in fenland localities, 1920–1973 (summarising data from Cooke and Ferguson 1976). Number of areas with information increased from 27 in the 1920s to 69 in the 1970s. No area reported increasing populations.

	Percentage of the total number of areas with information					
Status or change	**1920s**	**1930s**	**1940s**	**1950s**	**1960s**	**1970s**
Abundant	96	94	54	10	0	0
Decreasing	0	0	37	65	42	3
Absent/rare	4	6	9	25	58	97

One day in March 1974, I drove all round this fen area. I was looking for ditches close to the road that (1) were within the typical depth range of water where frog spawn is deposited (taken to be 8–35 cm), (2) had some semblance of freshwater communities of plants and invertebrates, and (3) had adequate suitable terrestrial habitat beside them. It quickly became apparent that such sites were virtually non-existent. One of the problems that farmers faced was boggy fields at times of cultivation; keeping ditches dry in spring helped the fields to be drier and easier to till (Cooke 1999). Eventually, I settled for finding sites of the right water depth – and discovered 11 spread through the area in four hours of searching. Depth was measured again on four occasions up to August 1974, and measurements were repeated in March and July 1975 (Figure 2.5). At least four sites and perhaps as many as six dried out before tadpoles would have metamorphosed. In the remaining five, depths increased to as much as 53 cm because of the need for crop irrigation in summer. None of the sites had water levels that remained within the range 8–35 cm. This small exercise hammered home the rarity of good frog habitat in the 1970s.

In this chapter, I am focusing mainly on the impact of habitat change, including drainage, and shelving most of the discussion of pesticides until Chapter 3. I will, however, discuss here whether the introduction of modern pesticides coincided with the initial decreases in frog and toad populations on the Fens. DDT came into use soon after the war ended, and my colleagues Derek Ratcliffe (1967, 1980) and Ian Newton (1979) documented the insecticide's unexpected effect of eggshell thinning in predatory birds, which began in 1947 and contributed to the reduction of raptor numbers in the UK in the 1950s. However, early instances of this phenomenon were probably caused by the predators taking racing pigeons dusted by their owners with DDT to control ectoparasites, rather than by widespread uses in agriculture, which began in 1948–1949 (Sheail 1998). Kenneth Mellanby (1967) recounted in his book *Pesticides and Pollution* how the cyclodiene insecticides, such as dieldrin, killed huge numbers of wild birds and mammals during 1956–1961, particularly in eastern England. So exactly when did frogs first decline on the Fens, and could these insecticides be implicated? The answer seems to be that the initial decreases in frog populations occurred during the Second World War, before these pesticides were widely used in British agriculture.

During the war, there was a drive for greater food production, both by farmers and by the public in gardens and allotments. A review of farming during the war years by J. R. Borchert in 1948 pointed out that there was a general decrease in the amount of cultivated land between 1870 and 1939 (apart from during the First World War), but by 1948 there was more tilled land in Britain than ever before. The cover of arable crops in the Fens by 1944 was as high as anywhere in the country (Martin 1993, in Oldham 1999a). The Fens saw a sudden switch away from livestock farming and uncultivated land to growing crops, accompanied by greatly improved drainage. Six farmers who were interviewed remembered the water table being lowered after the installation of new pumps, together with a decrease or total loss of frogs. Each pump had its installation date emblazoned on the wall, so helping to date the effect on frog populations. Although the level of the land eventually dropped due to oxidation, desiccation and erosion of the peat, the Fens are evidently much drier today than they were in the 1930s. Pumps driven by diesel and electric engines have enabled better control of water in the machine-manicured ditches, which has been good

Figure 2.5 Jon Hutton, who was assisting at the time, measuring the depth of a fenland ditch, March, 1975. Water level in the ditch would have been suitable for frogs on this occasion, but the terrestrial environment was too open and barren.

for farming but not for frogs – as I found out when trying to find sites that fulfilled criteria for spawning. The drive towards greater intensification of food production continued to grow for several decades. Those factors that affected amphibians in the 1940s still operated for many years. Jackson (1965) assessed changes in agricultural statistics between 1938 and 1962 for the Cambridge region, which included both this part of the Fens and Huntingdonshire. Among his conclusions were the following:

- The combined acreage of crops and grass remained fairly stable, but the area of barley increased nearly four-fold and the area of permanent grass more than halved.
- There were generally fewer grazing animals, with the number of horses kept for agricultural purposes being so small that it was no longer recorded in 1962.

Most people interviewed in our survey did not seem to appreciate the benefit to frogs of having pasture and scruffy areas of scrub rather than endless vistas of arable fields. Peter and I had several conversations when farmworkers described other species, such as grass snakes (now *Natrix helvetica*), disappearing at the same time as the frogs. When asked where they had seen such species, they might reply, 'In the fields with the cows.' 'OK,' we would say, 'can you see any cows now?' Realisation dawned. The bulk of arable land on the Fens became so hostile to creatures such as the frog that I doubt there would have been many, or any, left in the 1970s even if no farm chemicals had been used at all. Any direct poisoning effects of pesticides were likely to have been masked by the impact of changes in habitat. Both terrestrial and aquatic habitat became unsuitable for frogs, but toads hung on in the Fens, despite the views expressed by the farming community, by virtue of their ability to exploit deeper waterbodies and being more capable of surviving in dry, uninviting terrestrial habitat.

A small survey that my wife Rosemarie and I undertook in 2006 indicated that the toad still bred in the larger waterways in our local fen area (Cooke and Cooke 2008). The big rivers carry the water down to the Wash, and abutting them are district drains with a pump at the junction. These district drains are of middling size and connect the system of field ditches to the main river; district drains can be havens of freshwater life. Toads may have disappeared from field drains, but they still breed in some of the rivers and district drains. These, though, are unsuitable for frogs, being deep, steep-sided and often having good populations of fish as they never dry out. Even if frogs found them suitable, they would be unlikely to survive in the often desiccated arable environment adjacent to the waterways. Toads have been able to use these waterbodies because they can breed satisfactorily in the presence of fish and are capable of walking some distance from their main terrestrial haunts, which are likely to be patches of more diverse habitat.

I had been living on the edge of the Fens for more than 30 years before I appreciated how widely distributed toads were in this river system. The total of 4,000 in Huntingdonshire that Peter Ferguson and I estimated in 1973 was almost certainly a significant underestimate. In 2006, Rosemarie and I found them heading to breed in our three large fen rivers: the Forty Foot River (Figure 2.6), Bevill's Leam and the Old Course of the Nene (Cooke and Cooke 2008). Hopefully, one day someone will check the extent to which toads occur more generally in such habitat across the Fens. In our area at least, this could really only be done from a vehicle during suitable mild, wet nights at the peak of migration towards the breeding sites. The adjacent roads are generally unsuitable for searching on foot for road casualties. Some have little traffic, so there are unlikely to be many casualties; others are too busy and dangerous to walk along, and any signs of crushed toads are likely to disappear within a short time. Also, the rivers are too deep, wide and turbid to check for breeding toads during the day; and the water surface is often too far below the top of the bank. A few years ago, the local police dredged the Forty Foot River looking for a car belonging to someone who was missing. They failed to find that car, but unexpectedly found a different one. Interestingly, one of the bridges crossing this river is called Puddock Bridge – 'puddock' is an old name for a toad (Figure 2.6).

In summary, the frog declined almost to extinction on fen farmland between the 1930s and 1973, while the toad also suffered declines but continues to breed

Figure 2.6 An unexpected breeding site for common toads: the Forty Foot River in Cambridgeshire, viewed from Puddock Bridge, February 2012.

in some of the larger waterways. Improved drainage and a dramatic shift towards arable farming, together with loss of wet and damp habitats, were mainly responsible for the initial declines in the 1940s, and will also have played a major role in later decades.

2.5 Gathering information on the widespread amphibian species

2.5.1 *Changes in the 1970s and status in 1980*

In 1978 when I joined the NCC, the frog's future was looking more secure – at least in suburban gardens. Concern about newt species had been expressed a few years earlier (Beebee 1973, Prestt *et al.* 1974) but little evidence was available. This was soon to change, however, as Trevor Beebee, working on behalf of the Conservation Committee of the BHS, undertook an enquiry focusing on the great crested newt in the mid-1970s (Beebee 1975). He had useful responses from 142 sources with an interest in amphibians, and processed the information as I had done in the national frog and toad survey a few years earlier. His main conclusion was that crested newts had decreased dramatically, particularly in England. The species remained rare in Scotland and Wales. Suburban development followed by pollution effects, agricultural change and rubbish dumping were factors cited as being important in destroying habitat or rendering it unsuitable to breed or live in; collection and natural succession of breeding pond habitat were also deemed to be of significance. Its declines were more serious than for the other two newt species, possibly because the crested newt was less able to adapt to breeding in small waterbodies, which were often used by smooth and palmate newts.

This was a key piece of evidence in the parliamentary debate on the Bill which ultimately resulted in the appearance of the Wildlife and Countryside Act 1981. During the initial deliberations, I had collaborated with Henry Arnold to analyse distribution data held by the BRC at Monks Wood (Cooke and Arnold 1982). As an aside, throughout my career, I have always looked to maximise the usefulness of datasets and see if they might answer questions in addition to those at which they were originally aimed. Thus, Henry and I compared the number of 10 km squares in which each of the nine species of widespread amphibians and reptiles had been reported up to 1959 with records collected between 1960 and 1973. We divided the number for 1960–1973 by the number up to 1959, postulating that we might be able to compare this value between species to determine which had been most affected during 1960–1973. Results suggested we should be more concerned about the palmate newt. There were though obvious weaknesses in this approach, including that it told us nothing about abundance and it would be affected by national surveys for single species or by concentrated local distribution surveys. Because of doubts over the usefulness of this exercise, it seemed timely to undertake an enquiry with my colleague Al Scorgie, on all five of the widespread species of amphibians and also on the four widespread reptiles: common lizard (now *Zootoca vivipara*), slow-worm (*Anguis fragilis*), adder (*Vipera berus*) and grass snake (Cooke and Scorgie 1983). Information would be used to aid and inform local and national conservation strategies. Only amphibian data are presented and discussed here.

Questionnaires were circulated to organisations and individuals, such as county recorders for amphibians and reptiles, who might be able to provide information, if possible at a vice-county level. For each species, they were asked to describe status in 1980, changes in status since 1970, and likely reasons for those changes. Indices of status and change were calculated for the NCC's administrative regions and for countries. These differed from indices used in the earlier national survey for frogs and toads, but were similar in principle: rules were laid out for categorising (1) size of area (in relation to the size of a vice-county), and (2) comments on status and change; these categories were assigned numerical values and indices were calculated. Status indices varied from 0.00 (absent throughout a region or country) to 1.00 (abundant throughout); change indices varied from –1.00 (decrease throughout) to +1.00 (increase throughout), with 0.00 signifying no overall change. Total numbers of replies were: England 89–93 depending on species, Scotland 26–36, Wales 11–13. Indices are given for the frog and toad in Table 2.5 and for the newts in Table 2.6.

The frog and the toad were both widespread throughout Britain in 1980 and were considered to be generally abundant in many of NCC's regions, particularly in the west and the north. During the 1970s, the toad declined in about half of the regions, but nowhere were these changes as serious as in some areas in the 1960s (Section 2.2.2). Frog populations in England and Wales stabilised in the 1970s and even registered increases in the East Midlands and Southern England. These bald statements do though gloss over the fact that, in a number of regions, the frog had become an animal of suburbia rather than of the countryside.

None of the newt species was reported to be as generally widespread and abundant through Britain as were the frog and the toad. As is described elsewhere

Table 2.5 Information from the national enquiry on the widespread amphibians: national indices of status in 1980 and change during the 1970s for the common frog and the common toad (summarising data from Cooke and Scorgie 1983).

	Common frog		Common toad	
Country	Status index 1980	Change index 1970s	Status index 1980	Change index 1970s
England	0.80	−0.03	0.82	−0.14
Scotland	0.88	−0.21	0.77	−0.13
Wales	0.84	0.00	0.79	−0.16
Britain	0.83	−0.08	0.81	−0.13

Table 2.6 Information from the national enquiry on the widespread amphibians: national indices of status in 1980 and change during the 1970s for the three species of newt (summarising data from Cooke and Scorgie 1983). No index of change has been calculated for the crested newt in Wales because of its rarity and the small number of reports.

	Crested newt		Smooth newt		Palmate newt	
Country	Status index 1980	Change index 1970s	Status index 1980	Change index 1970s	Status index 1980	Change index 1970s
England	0.44	−0.41	0.83	−0.15	0.48	−0.12
Scotland	0.20	−0.14	0.43	−0.13	0.67	−0.15
Wales	0.13	—	0.29	−0.17	0.71	0.00
Britain	0.35	−0.37	0.67	−0.20	0.55	−0.12

(Section 8.3), crested and smooth newts tend to be associated with hard-water areas and the palmate newt with soft water. The hard-water pair of species tended to be rarer in Scotland, Wales and South West England, with the palmate newt rare in a band from East Anglia westwards across the Midlands. The survey indicated the crested newt to be rather sparsely distributed over most of England. In contrast, the smooth newt was common over much of the country. In Britain, the distribution of the palmate newt tended to be patchy, the species being widespread and even abundant in many regions, but virtually absent in others. Smooth and palmate newts suffered some population declines during the 1970s, but these were relatively slight, especially for the latter species. The crested newt, on the other hand, decreased in most regions where it was not absent or rare, and in some cases these declines were severe. Overall, this confirmed the conclusion of the earlier survey (Beebee 1975), although there were some differences in precisely where the worst declines occurred. Forty years on, this information may seem very basic and predictable, but this was the first attempt to go beyond distribution data and describe the status and changes in status for this group of species across Britain.

Loss of suitable habitat was considered to be the main reason for declines in the widespread amphibians during the 1970s, comprising 72–88% of reasons for the different species. Many correspondents referred to either urban or agricultural reasons (including pollution from farm chemicals) as explanations for losses. The ratio of agricultural to urban reasons varied from 1:1 for the crested newt, to 2:1 for the smooth newt and toad, and to 4:1 for the palmate newt and frog. This meant that the last two species were considered to be more affected by agriculture, while the crested newt was relatively more affected by developments unconnected with farming. Factors that were suggested on 5–10% of occasions were traffic deaths for toads, and collection for crested newts, toads and frogs.

As expected, creation of new wet habitat was the main reason given for increases in numbers, followed by colonisation of new habitat with or without help from people. The numbers of correspondents who mentioned garden ponds were, in ascending order: crested newt 6, palmate newt 7, common toad 18, smooth newt 23, common frog 38. In a study of gardens ponds in Brighton, Sussex, in the late 1970s, Trevor Beebee had similarly reported that the frog was best able to take advantage of this new breeding habitat, followed by the smooth newt and the toad (Beebee 1979). In the Brighton area, about 15% of gardens were estimated to have ponds, and amphibians bred in at least 50% of those ponds.

2.5.2 *Status information from the* Victoria County Histories

Following the publication of the enquiry of Cooke and Scorgie (1983), I decided to try to obtain more information on the historical status of the widespread amphibian species from the biological sections of the *Victoria County Histories* (VCHs). Compilation of these volumes was a project begun in 1899 and dedicated to Queen Victoria. For each county, a very wide range of topics was covered in detail, with natural history being a fairly minor component. For me, however, the sum of the information was of immense interest. Admittedly, I had to accept the statements at face value and could not query anything, but nowhere else was I going to find the breadth of information on species' status that was available in the pages of the VCHs. It is true to say, however, that a few accounts of the status of the five amphibian species were not very helpful or convincing. Those that simply listed species present without any qualifying comments were ignored. In total, volumes were located for 31 English counties; 27 were published during the first decade of the twentieth century, one appeared in the 1910s, one in the 1920s and two in the 1930s.

In this exercise, information was also gathered on the four widespread reptile species. Although I concentrate here on the amphibians, a significant difference between the accounts of the amphibians and reptiles should be mentioned. Statements in the VCHs focused on status, but there were a few comments on changes in abundance, which interestingly were much more numerous for the reptiles. The only references to declines of amphibians were for the frog in Lancashire and Oxfordshire, whereas decreases were mentioned for reptile species in 13 counties for reasons of habitat loss or persecution.

In order to calculate a status index for amphibians in the early years of the twentieth century, authors' statements were analysed in exactly the same way as was done during the national survey for 1980 (Cooke and Scorgie 1983). The status indices for the five amphibian species in England (with the values for 1980 in brackets) were:

common frog 0.98 (0.80), common toad 0.99 (0.82), crested newt 0.75 (0.44), smooth newt 0.93 (0.83) and palmate newt 0.50 (0.48).

Authors of the reports in the VCHs were in almost total agreement that the frog, the toad and the smooth newt were generally abundant in the last years of the nineteenth century (and the early years of the twentieth). However, the crested newt was reported as abundant in fewer counties, while the palmate newt was absent or rare in about half of them. Percentage changes in status index between the beginning of the twentieth century and 1980 were: palmate newt –4%, smooth newt –11%, toad –17%, frog –18% and crested newt –41%. While it would be wrong to treat these percentages as actual reductions in national population sizes, they are consistent with an overall view that the crested newt declined the most during that time, and the two smaller newt species declined the least. In other words, comparison of the two datasets may give a *relative* perspective on what happened to the five species during the intervening years. Note that the doubtful concern about the palmate newt indicated in the earlier analysis based on distribution records (Cooke and Arnold 1982; Section 2.5.1) was not supported by this comparison between information from the VCHs and the enquiry of Cooke and Scorgie (1983).

2.5.3 *The Leicester contribution*

Soon after the crested newt had become a protected species on Schedule 5 of the Wildlife and Countryside Act 1981, I turned mainly to Rob Oldham and his colleagues at Leicester Polytechnic to provide NCC with information on its natural history and breeding sites. I could also count on them to help with some of my local projects (Figure 2.7). This collaboration developed into a succession of contracts well into the 1990s, culminating in a lengthy one on the nine widespread species of amphibians and reptiles, the Herptile Sites project. Another contract was awarded to repeat the study of Cooke and Scorgie (1983) and collect and collate information on the status of the nine species in 1990 and how they had changed during the

Figure 2.7 Mary Swan and Rob Oldham of Leicester Polytechnic (now De Montfort University) using a seine net in Wood Pond at Shillow Hill in 1984 (see Section 10.2).

1980s. Indices of change for the widespread amphibians in the 1980s were: +0.16 for the common frog, between –0.02 and –0.10 for the smooth newt, palmate newt and common toad, but –0.24 for the crested newt (Hilton-Brown and Oldham 1991). The last species continued to give cause for concern. By 1990, the crested newt was much better known, but urban development, pond neglect and agricultural improvements were still eroding its national resource of suitable habitat.

In June 1992, Leicester Polytechnic became De Montfort University. The following year, the final report of the Herptile Sites project was published (Swan and Oldham 1993). This report contained some notable statistics. One element of the work was to determine pond occupancy for the five widespread amphibian species in Britain. Overall average occupancy for a large number of surveys was 52% for the frog, 30% for the toad, 22% for the smooth newt and 11% for both the crested newt and the palmate newt. The frog and the toad occurred most frequently in the west and north of Britain, with population densities mainly reflecting pond densities. Crested and smooth newts were found most frequently in central and eastern England, while the palmate newt tended to be found outside those areas. Overall pond loss was reported to be 17% from the 1950s to c.1990. Without the beneficial effect of some pond creation during the later years of that period, site loss would have been about 30%. In addition, 11% of sites were at an advanced stage of succession and 17% desiccated regularly.

A number of important papers emanated primarily from this work, such as Swan and Oldham (1994) and Oldham (1999a). These articles discussed issues such as the role of pond density in determining species status, because of the importance of metapopulation behaviour for most of our amphibian species (see Sections 8.6, 10.7.1 and 10.7.2); the likely exception was the common toad, for which isolated populations can often thrive in the longer term.

2.6 An overview of changes in status of the widespread amphibians during the twentieth century

2.6.1 *Introduction*

The enquiries that are described in this chapter gathered information on status and changes in status for different species for different periods of time in areas of varying size across the British Isles. In this section, I will attempt to explain in chronological order how and why changes occurred during the twentieth century. Two fundamental observations should be made. First, amphibian species in the countryside suffered declines during the twentieth century primarily because of loss or modification of habitat. This included destruction, contamination and lack of management of ponds, and unfavourable changes in terrestrial habitat. Therefore, I have tended to focus the review on changes in land use. Other factors are discussed elsewhere in this book. Second, amphibian populations in Huntingdonshire/Cambridgeshire, where I have resided for my entire working life, seem to have been affected as much as anywhere else, so the review tends to concentrate on this and nearby areas. Because the full spectrum of change of both freshwater and terrestrial habitat occurred in my area, changes elsewhere might have been more restricted and less severe. I also reflect in Section 2.6.3 on the standard of information gathered in the enquiries using questionnaires or interviews.

2.6.2 *A brief review of changes in the twentieth century*

Accounts gleaned from the VCHs indicated that the early decades of the twentieth century were probably a boom time for the widespread amphibian species in those parts of Britain where they were well adapted to survive and breed. During that period, British amphibians were likely to have enjoyed better environmental conditions than at any other time in recent centuries (Oldham 1999a), making full recovery from later declines extremely difficult and perhaps impossible to achieve. The palmate newt had the most restricted distribution, the crested newt was not as abundant as the smooth newt, but the common frog and common toad were widespread and truly 'common' animals. During the 1930s our agricultural depression was at its worst. Compared with today, much of the countryside verged on being a wilderness, as farmers desperately reduced any unnecessary expenditure. This will have been epitomised in the extent of poorly drained pasture and a general untidiness of the landscape. In contrast, house building was booming in the 1930s, and farmland was sometimes seen as a buffer against ribbon development, and against other industries that might take land, such as mineral extraction. The paradox of the extraction industry was that it could quickly destroy habitat, but provide extensive new aquatic and terrestrial habitat in the longer term (e.g. Figure 9.10).

It is important to derive as much information as possible on population changes that occurred in the 1940s. There was evidence of slight but accelerating decreases in frog populations overall in the British Isles during that decade (Section 2.2.2). This was reported to be particularly marked in agricultural areas of the Fens (Section 2.4). When I was writing this account, I was interested to determine whether the concomitant declines reported from Huntingdonshire (Section 2.3) occurred on the clay-lands as well as on fenland in the county; returning to the original data revealed that this was indeed the case. If these observations for the British Isles and for the Huntingdonshire clay-lands are both correct, it seems that such severe declines, as reported in my area of the country, were not especially widespread. Increases in the extent of arable land between 1939 and 1945 were striking in Huntingdonshire and nearby counties to the east and north (Wells 2008, Martin 1993 in Oldham 1999a), but less so elsewhere.

During the Second World War, the Ministry of Agriculture was reconstructed and charged with preserving and maintaining agricultural land *solely* for the production of food (Martin 2000). Tenancies might be terminated if land was neglected or badly managed, and measures were introduced to control pests such as rabbits (*Oryctolagus cuniculus*). Conservation of nature was not a subject to which people will have paid much attention. Intensification will not have just affected the frog and the toad: crested and smooth newts will have suffered sizeable losses too.

The fortunes of amphibian species did not improve in the post-war period. The Agriculture Act of 1947 aimed at encouraging a more stable and efficient industry through guaranteeing prices of main commodities (Martin 2000). In time, those farmers prospered who could exploit economies of scale to increase productivity, while small farmers struggled. Inevitably, amounts of arable land, field sizes and chemical use all increased. Agricultural output grew more quickly between 1945 and 1965 than at any other comparable period of time in the twentieth century (Brassley 2000). At the same time, housing building, industrial development and road construction did not slacken. One has only to compare old Ordnance Survey maps through the twentieth century to see how the countryside was eroded as villages, towns and cities spread and started to coalesce. My parents bought their new-build 'semi' on the extreme northern edge of Luton in Bedfordshire just before the Second World War. We were bordered by fields in the late 1940s, but 15 years later the house was buried deep in suburbia. My national surveys for the frog and toad (Section 2.2) indicated that the former species suffered widespread declines in England in the 1950s that worsened in the 1960s. For counties across the British Isles, the extent of declines for frogs in the late 1960s was negatively related to human population density: decreases in frog abundance were, for instance, more severe in the Home Counties than in more rural areas. This was akin to the large-scale relationship between the percentage of herpetofauna species in decline and human population density in 12 European nations (Beebee 2001).

Overall losses of amphibian breeding ponds continued for the rest of the century (e.g. Swan and Oldham 1993, Oldham 1999a). On the other hand, garden ponds became ever more fashionable, so frogs, smooth newts and, to a lesser extent, the other species were able to take advantage of them. But common toads and crested newts remained the species most at risk. People who grew up in the 1970s and 1980s, such my two sons, may well look back on that period with some nostalgia for the countryside, just as I did two or three decades earlier. In reality, however, much had already changed by 1980, and declines would continue for those two species (Carrier and Beebee 2003, Jehle *et al.* 2011). The impact of road traffic was of increasing concern, especially for common toads. Twenty-first-century studies that looked retrospectively at declines in these two species are discussed in more detail in Section 2.7. The palmate newt should have been less exposed to developmental and agricultural changes because of its distribution and habitat preferences. My few local breeding sites for this species tend to be protected – for example, ponds in the historic site of Bedford Purlieus and at Castor Hanglands NNR (Figure 2.8).

2.5.3 *Reflections on the reliability of information on status and change*

At this point, it may be worthwhile adding a few more comments on the reliability of enquiry information collected by asking people for their past observations and memories. While it is not an ideal way of gathering information on status and change, I could think of no other method when some knowledge of past changes was required (apart from consulting the VCHs). In Britain, we are fortunate in having a network of keen and often very knowledgeable amateur and professional naturalists who are willing and able to help with such schemes. Sometimes, however, I had to consult people without any specialist knowledge of amphibians. Nevertheless, the information that the surveys brought together does, I believe, provide a broadly accurate picture of what happened to our widespread amphibian species over the decades from the 1930s. The common frog is a fairly well known and conspicuous animal. This is probably less true of the common toad, and there are often difficulties with people not being able to distinguish between the two species. The three newt species tend to be less well known, and there are also identification problems with smooth and palmate newts; this particular issue should not have been a problem, however, as questionnaires which requested feedback on newts were targeted at people with specialist knowledge.

Figure 2.8 Palmate newt breeding sites close to Peterborough: (a) Bedford Purlieus, 1991, where several fire ponds held the species and were managed sympathetically by the Forestry Commission; (b) Castor Hanglands NNR, 1975, where a study was undertaken on the three species of newts breeding in this pond by the reserve warden, John Robinson.

For the frog in particular, there has often been broad agreement in what correspondents reported about status and changes in status over time, which I would suggest makes it more likely to be true than false. I would also suggest that overall generalities were more likely to be correct than some of the specific points that might be teased out of the data: for instance, national or regional changes were probably more accurate than those at a finer scale. To any doubters, I would point out that this information was provided over the years by large numbers of people; it is your choice whether you believe it not, but, in the main, conclusions are accepted by most herpetologists and other naturalists (Beebee *et al.* 2009).

Correspondents' opinions about reasons for change are likely to be less reliable, especially those associated with a lack of tangible evidence. This can also be true of more recent studies based on field surveys. If there is significant loss of breeding ponds in an area, then this is likely to be connected with reduced sightings of animals or spawn. But suggesting that losses due to collection, road traffic or pollution are translated into changes in population size may be fashionable, but proving or disproving cause and effect can be very difficult. In the case of pollution, statements are often made without any direct field evidence. This does not necessarily mean the statements are wrong, as some significant pollution effects can be subtle and difficult to detect, but there is a danger that wild speculation about causes can undermine the credibility of descriptions of changes in status. Young and Beebee (2004) investigated the validity of responses to a questionnaire enquiry on the status of common toads; they found that replies painted a picture of change that was broadly correct, but advised that closer scrutiny was necessary for accurate analysis.

2.7 Recent developments

In the early 1990s, my job changed and I was no longer officially responsible for issues concerning amphibian conservation at any level other than the very local. I continued to be involved with my personal conservation-based research, but I started no significant studies after the early 1990s, although I was an active recorder up until 2015. Consequently my participation in what was planned and undertaken to determine the status of the widespread amphibians was immeasurably less for the last 25 years than it was for the previous 25, and below is a quick sketch of the more important recent events relating to status and change.

The 1990s saw the advent of the UK's Biodiversity Action Plan (BAP). Under that umbrella, Species Action Plans (SAPs) were initiated for both the crested newt and the common toad, these being last revised in 2009. These are the two widespread species that have continued to give rise to the most concern, and targets for population, range and viability were set to help them recover. In addition, because the crested newt was listed on annexes of the Habitats Directive, the UK was legally obliged to report regularly to the EU on its conservation status.

A very detailed report (Wilkinson *et al.* 2011) evaluated previous attempts at assessing population status of crested newts and suggested new approaches. By that time, asking people for opinions was ancient history. Recommendations included: using data on occupied 10 km squares gathered by the National Biodiversity Network (NBN) to indicate range, and modelling numbers of occupied ponds to indicate status. The number of occupied ponds was estimated to be about 65,000 by these authors, but there had been debate over how many occupied ponds might exist, with a previous government figure of 75,000 being criticised by Tom Langton (2009) as a significant overestimate.

By that time, a National Amphibian and Reptile Recording Scheme (NARRS), organised by Amphibian and Reptile Conservation (ARC), was well under way, using volunteers to monitor and report on the status of British species (e.g. Beebee *et al.* 2009). Such an approach is referred to as 'citizen science'. Survey work was initiated on occupancy rates of the widespread species of amphibians and reptiles in random 1 km squares across the UK. It was the aim of NARRS to be a statistically rigorous approach to support the monitoring needed for the crested newt, common toad and the other species. A report by John Wilkinson and Andy Arnell in 2013 provided baseline results from the scheme for 2007–2012, against which future differences could be monitored. The five widespread amphibian species were assessed in 412 survey squares. Percentage occupancy was similar or higher for all species to occupancy figures provided by Mary Swan and Rob Oldham in 1993. Thus the occupancy rates for the crested newt were 11% (Swan and Oldham 1993) and 12% (Wilkinson and Arnell 2013) despite the pond losses reported in a number of local British surveys, as summarised by Jehle *et al.* (2011). It appeared that various initiatives to protect, manage and create habitat may have negated loss of former occupied ponds. Wilkinson and

Arnell (2013) suggested that NARRS surveyors may have been reluctant to submit negative results, but surely it was critical that such behaviour did not occur.

The intention with NARRS was to undertake a second survey during 2013–2018, but it was discovered that in order to meaningfully detect a change in the crested newt population, an unrealistically large number of squares would need to be surveyed in the second six-year period. Crested newts had been detected in only 48 squares during 2007–2012, and it was suggested that supplementary information would be needed. The same issue applied to several of the reptile species. The repeat survey was not undertaken and, in 2016, it was announced that random-square surveys had been suspended for the time being (Anon. 2016).

In 2021, ARC revealed that a National Amphibian Survey would build on and update the amphibian part of NARRS (ARC 2021b). This focused on both former NARRS sites and new ones suggested by the volunteers. A range of observations was requested, including counts of clumps of frog spawn and adult amphibians that would 'help ARC to improve its statistical analyses'. This would seem to bring our scheme more in line with a successful one in the Netherlands (Beebee *et al.* 2009).

The last report to the EU on the status of the crested newt (JNCC 2019) concluded that the population was stable during the period 2007–2018, based on the number of 1 km grid squares where the species occurred, but that area and quality of habitat had declined. Conclusions were said to be 'based mainly on extrapolation from a limited amount of data'. In addition, it was noted that England held 98% of the 'total UK resource', but about 80% of ponds were in poor or very poor condition. With the UK having left the EU, will similar official assessments be produced in the future?

As regards the common toad, Trevor Beebee (2012) undertook a long-term study on a population at Offham Marshes in Sussex that declined from the late 1980s. By 2001, he was sufficiently perturbed by this and other accounts of toad populations decreasing to instigate a large-scale enquiry (Carrier and Beebee 2003). Questionnaires about status over the period 1985–2000 were sent out by the charity Froglife to professional and amateur herpetologists, and roughly 100 replies were received. Information on the common frog was requested for comparison, and that species was generally reported to be doing well, including in rural areas. For the toad, on the other hand, serious declines were reported for eastern, southern and central England. Reasons for the decline remained unknown. Subsets of the replies were therefore investigated further in order to validate observations and attempt to establish causes (Young and Beebee 2004). This detailed investigation revealed that toad declines were indeed worse in this part of lowland England. Nevertheless, some of the reported declines could not be substantiated by closer inspection. Many of the populations were believed to be decreasing because of habitat change, but no single factor stood out. Concern was expressed by a number of correspondents and by the authors about the part played by traffic mortality.

Wilkinson and Arnell (2013) calculated a pond occupancy rate of 33% for the toad during the initial NARRS programme, compared with 30% reported previously by Swan and Oldham (1993). This apparent conclusion of little change was at odds with analysis of data held by Froglife from volunteers moving toads across roads, which revealed widespread declines across the UK between 1985 and 2014 (Petrovan and Schmidt 2016). Possibly this indicated the difference between population decline and total local extinction.

Citizen science will have a crucial part to play in any scheme to survey Britain's widespread amphibians in a statistically rigorous fashion. In order to achieve the necessary coverage, however, are there enough people sufficiently well dispersed across Britain willing and able to undertake surveys of amphibians? Some other animal groups have a much greater following than do amphibians. The considerably higher number of people with an interest in birds has already been exploited to help to record amphibians and reptiles, at least in gardens. Thus, Humphreys *et al.* (2011) had nearly 4,000 questionnaires returned on herpetofauna species in gardens. These returns were mainly from participants to the Garden BirdWatch project of the British Trust for Ornithology (BTO). Contributors provided information on species present, as well as characteristics of the gardens and surrounding areas. The project identified features that had positive or negative influences on whether species were recorded. For instance, large gardens were better than small ones, and rural gardens tended to be better than urban ones. Statistical examination of the data indicated that such an approach could detect changes in the national populations of the common frog, common toad and smooth newt in gardens. More recently, the similarly named Big Garden Birdwatch, which is run by the Royal Society for the Protection of Birds (RSPB), asked for information on frogs and toads in 2014 and 2018. Very recently, a research student has started a project at the University of Kent to examine a range of citizen science data to attempt to assess trends in amphibian and reptile species and then aid conservation strategies by forecasting future trends under different environmental scenarios.

CHAPTER 3

Pesticides and other hazards

3.1 Introduction to pesticide studies

This chapter primarily focuses on pesticides but later, in Section 3.6, ranges across a number of hazards: agents of disease, acid deposition, solar radiation, collection and fertilisers.

It may seem intuitive to many people that pesticides will have had a profound effect on frog populations. Pesticides kill animals and other organisms and are applied to places where frogs live, so frogs die and populations decline. And if they do not kill frogs directly, pesticides become concentrated up food chains and kill frogs that way. However, the situation is more complex. Many pesticides may be lethally toxic to frogs, but they have not necessarily been used in high concentrations where frogs occur. Active agents may be relatively non-toxic in the pure form, but their formulations may be more harmful. Some pesticides move up food chains, but most do not. Because of the frog's population dynamics, if mortality occurs, it may not be sufficient to affect numbers in a small area, let alone reduce numbers regionally or nationally. Moreover, readily observable habitat change, such as pond destruction, and other factors can affect frogs on a broad scale, so complicating the picture further. On the other hand, pesticides may affect frogs more subtly by impacting vegetation used as cover or invertebrates taken as food, and so reduce the density of frogs that an area can support.

In 1969, when I first began trying to find out whether pesticides might have affected frogs in Britain (Section 2.2), there was little hard information to mull over. A few knowledgeable individuals believed that the common frog was no longer so common. And my colleagues at Monks Wood were regularly discovering new facts about how plant and animal life was affected by pesticides. So what I needed most was some direction on how to start examining potential impacts. The two questions I was asking were (1) whether frog populations had been reduced by pesticides, and (2) whether impacts were still occurring.

The difficulties I faced were encapsulated in a short article by Colin Simms (1969) that was published just as I was beginning my research. During the period 1954–1967, he routinely netted adult amphibians each spring in a pond in Guisborough, North Yorkshire. The pond was fed by field drains from pasture and natural run-off. Some of the adjacent pasture was converted to other uses, including arable, during the course of the study, but Simms considered that 'modern' seed-dressings and insecticides had not been used in the catchment. The common frog and the other four widespread amphibian species were present in good numbers initially, but all were very rare or absent by 1967. Simms blamed hydrological changes and physical changes to the terrestrial habitat rather than collection or agricultural chemicals.

Nevertheless, contamination of breeding ponds seemed a likely route by which frogs might become affected in more intensively farmed areas such as the Fens. This could occur from over-spraying, drift, drainage or run-off from applications on adjacent land, from residues in rain, from general aerial fall-out or from the practice of dumping used pesticide containers in ponds. So I decided a good initial approach would be to dose tadpoles with a common chemical and record what happened. The easy part was deciding what pesticide to use. A new report by Manigold and Schulze (1969) had found DDT was the most common pesticide residue occurring in streams in the western United States. Earlier, Boyd *et al.* (1963) had found evidence of resistance to DDT in two species of cricket frog (*Acris crepitans* and *A. gryllus*) in areas of the United States where DDT had been used against pests in cotton. This insecticide had caused considerable harm to British wildlife after it came into widespread agricultural use in the late 1940s (Sheail 1998). DDT was one of a group of chemicals known as organochlorines. Other members of the group that were used as insecticides were the cyclodienes, such as dieldrin and aldrin. Among the properties of these chemicals were environmental persistence and the ability to move up food chains. In 1965, Monks Wood had hosted a landmark NATO-sponsored workshop involving scientists working on the environmental effects of pesticides. These scientists stressed their particular concern about the organochlorines (Moore 1966). The pesticide debate in Britain had been galvanised in 1963 by the publication of Rachel Carson's book, *Silent Spring*, and polarised into agricultural interests against those of wildlife conservation, the latter being led primarily by Nature Conservancy scientists and representatives of the voluntary bodies (Sheail 1998).

I should stress that there were two main drivers of my study. Herpetologists are likely to be primarily interested in whether DDT and other pesticides affected our frog populations. And deriving information to help conserve frogs was always one of my motivations. However, the principal driver was the need to increase the breadth of data on the impacts of DDT and other organochlorine pesticides on wildlife in Britain, in order to provide advice for the regulatory authorities that were making decisions about usage. Withdrawals of use of DDT began in Britain after the Advisory Committee on Pesticides and other Toxic Chemicals published its recommendations in 1969, but the compound was still in agricultural use into the 1980s. And DDT's environmental persistence meant it continued to have measurable effects on wildlife beyond that date.

It was during the Second World War that DDT was first used as an insecticide, especially to control body lice which carried diseases such as typhus. Kenneth Mellanby, the first Director at Monks Wood, worked on disease control during the war, and related in his book *Pesticides and Pollution* (1967) how volunteers wore underclothes for days that had been impregnated with DDT. The body lice died and did not re-infest for some time, while the volunteers suffered no ill effects. After the war,

Mellanby would often enjoy horrifying his lecture audiences by consuming small quantities of DDT. My father was a driver in the Royal Army Service Corps during the war, and was in the first detachment of British forces which liberated prisoners in the concentration camp at Bergen Belsen in 1945. He volunteered to shave prisoners and, like them, will have been liberally covered in DDT. Unfortunately, it did not prevent him from contracting typhus; he was sent away to convalesce and was demobbed in 1946.

When I started work at Monks Wood at the end of the 1960s, testing DDT against frog tadpoles seemed to make good sense, and over time I employed it almost as a model pollutant to try to understand how and why frogs might be affected by an example of the toxic and persistent organochlorine compounds. At the outset, I appreciated some of the problems that I would have with my investigation. For instance, it was very possible that frogs had been affected 10 or 20 years before, but that instances of poisoning were a thing of the past, so I might not find much or any contemporary evidence. What I was not expecting was to have difficulty in finding any tadpoles to dose: the decline of the frog in the old county of Huntingdonshire is described in Section 2.3. Initially, I had to go outside the county in order to find tadpoles. Another issue which I approached with some trepidation was being responsible for dosing animals. For several years, I had held an animal experimentation licence from the Home Office, but all I had done previously was modify chickens' diets to change the thickness of their eggshells. I justified treating tadpoles to myself on the basis of the fact that, on average, probably more than 99% of tadpoles never became adult frogs, and also because conservation was an ultimate underlying reason for the study.

Below, I discuss my work more or less in chronological order:

- Sections 3.2–3.4 present the various studies undertaken during the 10-year period up to 1978 that I was at Monks Wood trying to understand the part played by pesticides in the decline of the frog, and to some degree of other native amphibians, from the 1940s up to the 1970s.
- Section 3.5 discusses how my more general role on pesticide use and environmental safety during 1978–1998 was relevant to pesticide impacts on amphibians.
- Section 3.7 provides concluding remarks on pesticides, including new evidence on effects.

3.2 Dosing studies with pesticides in the laboratory and the field

3.2.1 Initial laboratory studies with DDT

The first of a series of laboratory studies was undertaken in 1969 (Cooke 1970). In that experiment, I dosed groups of tadpoles for one hour in water containing nominal levels of DDT ranging from 0.01 to 10 parts per million (ppm); then the tadpoles were kept in untreated water for several weeks. Treated tadpoles were compared with an untreated control group. These treatment concentrations were higher than would be expected in the field, but exposure time was brief. I wanted to ensure that if there was any chance of an effect in the field, then I would see it in this experiment.

The method was simple, but effects were dramatic. Tadpoles in the higher treatment groups became hyperactive with spasmodic tail lashing and body twisting.

This phase was followed either by recovery or by death. Reductions in body weights were seen in all treated groups, probably as a consequence of not feeding while hyperactive and also because of a snout deformity that developed later and will have inhibited feeding. Some froglets became hyperactive again; these died during tail resorption, probably because residues previously stored in fat reserves were mobilised. Tadpoles approaching the metamorphic climax cease feeding, and begin again after emergence. The second phase of mortality was a delayed poisoning effect. These froglets contained DDT levels in their bodies comparable to those sometimes found in other vertebrate wildlife. This experiment suggested that environmental levels could have been high enough in certain situations to have affected frog tadpoles, but in view of the very high loss of tadpoles naturally, this would not necessarily translate into an effect on the size of the population. Subsequently, I only ever witnessed hyperactive tadpoles or froglets in the field in experimental situations where the waterbody had been directly treated (Figure 3.1). And I am not aware of them ever being reported by other people in this country.

Hyperactivity meant a tadpole could be at greater risk of predation, which in turn might lead to the predator being selectively affected by focusing on poisoned prey. This was a topic that was attracting much scientific attention at the time. The phenomenon of concentration up a food chain was by then well known, but the role of selective predation in this process was unclear. I realised that if treated and untreated tadpoles were presented to newts, then selective predation might occur. Twelve adult great crested newts were acquired in 1970 and eight were used in the feeding tests. Single newts were simultaneously presented with one live treated tadpole and a similarly sized untreated tadpole, and I recorded what happened. Newts

Figure 3.1 Tadpoles dosed with DDT in the laboratory can become hyperactive and die as they turn into froglets, but this has never been reported from the field in this country.

appeared to locate their prey by sight and lunged at it – most lunges were unsuccessful, but eventually a tadpole was caught and consumed. Newts tended to focus on the hyperactive tadpoles, and significantly more lunges were made at them. However, success rate per lunge was unaffected by whether the tadpole was hyperactive or not. Although hyperactive tadpoles did not respond to external stimuli, this seemed counterbalanced by their persistent uncoordinated movements. By virtue of an increased rate of lunging, the newts caught the treated tadpoles during 90 of the 100 tests (Cooke 1971).

While these experiments were being undertaken, I was also organising two national enquiries into how the status of frogs and toads had changed over the previous two decades (Section 2.2). Responses indicated there had been significant declines in both species. Few people considered that pesticides had been responsible, but I was aware that these views tended to be from a suburban perspective rather than a farmland one.

3.2.2 *Broadening the studies in the laboratory and the field*

At that point, I had just started to understand the effect of a single pesticide on frog tadpoles. In the next dosing experiments (Cooke 1972b), I broadened my approach to include frog spawn and toad and smooth newt tadpoles, and also tested responses to the insecticide dieldrin, and the herbicide 2,4-D. Dieldrin was included because its widespread toxic effects on wildlife had been well documented; 2,4-D was tested as an example of a herbicide that could be used to control aquatic vegetation by spraying directly onto waterbodies. And for frogs, I examined the effects on spawn and tadpoles at different stages of development.

DDT evidently did not penetrate well-developed spawn, but treatment of freshly laid spawn resulted in frog tadpoles becoming hyperactive after external gills were lost – another delayed effect. Frog tadpoles were more susceptible than toad tadpoles to DDT at all stages of development and, unlike froglets, toadlets did not die during tail resorption. Toad tadpoles tended to acquire higher body residues than frogs from the same concentrations, and some tadpoles survived despite body concentrations of more than 100 ppm, the highest concentrations I ever found in live tadpoles. Newt tadpoles were tested with DDT, and body residues needed to be several parts per million before hyperactivity occurred, as was the case with frog tadpoles. At comparable concentrations, dieldrin caused lower mortality than DDT and had less effect on behaviour. Treatment with both chemicals resulted in a range of abnormalities, such as tail kinks and deflections, in both frogs and toads (Figure 3.2a). Some deformities affected swimming speed and could make affected tadpoles more likely to be predated in the wild. I began to wonder whether these physical abnormalities and the hyperactivity shown by tadpoles dosed with DDT and dieldrin could be useful in helping to identify sublethal poisoning incidents in the field. The aquatic herbicide 2,4-D was only tested on frog tadpoles: it had no obvious effect at concentrations that killed aquatic plant life.

By 1971, I had accumulated information from laboratory experiments on the effects of DDT on frog tadpoles. What I needed to know next was whether such effects ever happened in the countryside. I needed a worst-case situation to study. A colleague at Monks Wood, Mike Service, worked on mosquito control; I learned from him that some local health departments continued to spray DDT on the

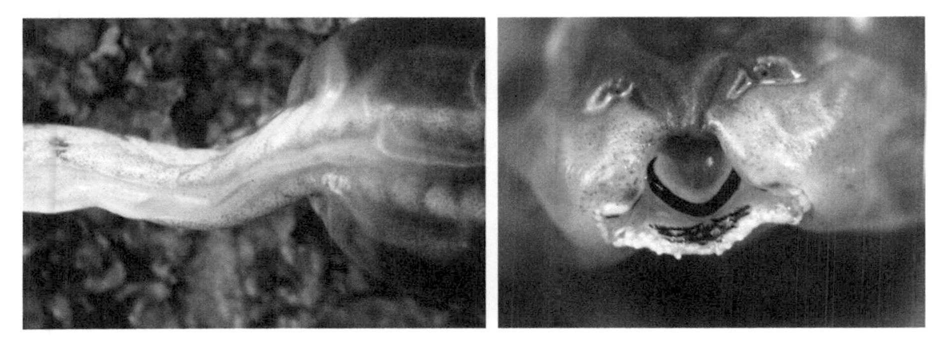

Figure 3.2 Examples of preserved abnormal tadpoles. (a) Spinal kink in a frog tadpole; this condition can be associated with environmental stress of various types, and is encountered in the field. Depending on the position of the kink, the resulting froglet's movement may or may not be inhibited; e.g. if the kink is confined to the tail, the froglet should appear normal. (b) DDT treatment in the laboratory can cause frog tadpoles to develop a hole in their snout with loss of the upper mandible. This condition has not been recorded in the field.

breeding sites of mosquito species that might be vectors of disease. This led me to set up a trial on marshes in the Isle of Grain in Kent with the collaboration of Strood Rural District Council Health Department (Cooke 1973a). I learnt in advance from local farmers and a schoolteacher that frogs were, at best, rare on the Isle of Grain, so I did not expect to be able to study local frogs. Instead I constructed open-top mesh cages that were fixed to posts in the waterbodies to be sprayed and took frog tadpoles to keep in the cages. I was greatly helped in the study by Paul Harding (Figure 3.3). With the assistance of the Health Department official who undertook the spraying, we selected five sites in the marshland and a control site in a nearby nature reserve; and then positioned a pair of cages with tadpoles in each site. The insecticide was sprayed at the usual rate, and we monitored effects on the tadpoles and took samples of water, tadpoles and other aquatic life for analysis. DDT was found in the water and in the tadpoles, but concentrations varied according to the size of the site, with the highest levels in the smallest site. In the three smallest waterbodies, some tadpoles became hyperactive, some developed abnormalities and some died, just as had occurred in the laboratory. This showed that DDT spraying could, in a worst-case situation, detrimentally affect frog tadpoles in observable ways.

Amazingly, however, two days after treatment, three free-living tadpoles were netted under a mat of vegetation in the most contaminated site – and these tadpoles contained no DDT residues. DDT was found in snails and in a dead three-spined stickleback (*Gasterosteus aculeatus*) in the same ditch, but the local tadpoles demonstrated that animals might avoid harm even in acutely contaminated sites. Mike Service told me that DDT was used annually to control mosquitoes in a large number of pools and ditches, mainly in Essex, Surrey and Kent. DDT spraying might, therefore, have been a contributing factor in the declines reported from those counties by contributors to my enquiries in 1970 (Cooke 1972a). A more recent review reported that 11 local authorities were still spraying insecticides to control mosquito populations (Medlock *et al.* 2012). Some sites were sewage works and water treatment works, but others were on coastal wetlands where there were concerns

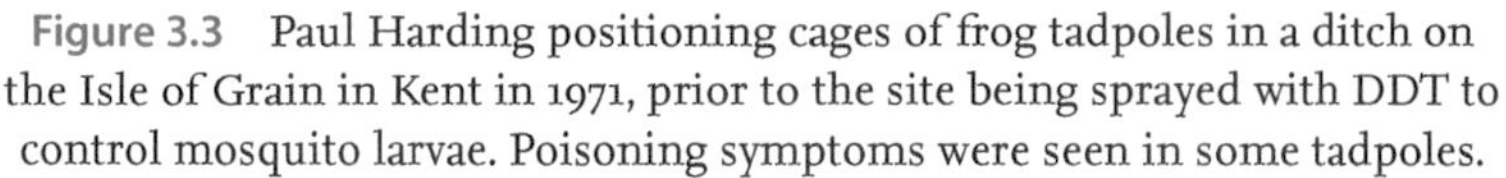
Figure 3.3 Paul Harding positioning cages of frog tadpoles in a ditch on the Isle of Grain in Kent in 1971, prior to the site being sprayed with DDT to control mosquito larvae. Poisoning symptoms were seen in some tadpoles.

for nature conservation interests, but presumably not for frogs. Several of the local authorities were known to be using the much safer microbial larvicide *Bacillus thuringiensis israelensis*. All uses of DDT have been banned in the UK since the 1980s. A study of frogs and toads in parks and reserves beside Lake Erie in Ontario, Canada, found local extinctions occurred in areas where DDT had been frequently applied for mosquito control until the late 1960s (Russell and Hecnar 1996); more than 25 years after the last applications of DDT, elevated residues still occurred in the fat of amphibian species that had survived.

Later in the 1970s, I undertook five more studies to understand better how DDT affected tadpoles and frogs. First, I demonstrated that frog tadpoles exposed to chronic levels of DDT (i.e. low levels for longer periods of time) were more tolerant than those exposed acutely in the earlier dosing trials (Cooke 1973b). Second, frog tadpoles were reared at two different densities to produce siblings of two sizes, considered to be at either end of the 'normal' size range (Cooke 1979). When exposed to an acute dose of DDT, the large tadpoles displayed no adverse effects, whereas the small ones became hyperactive and many developed a hole in their snouts. These results demonstrated that the largest, fittest tadpoles in a waterbody will be the most likely survivors of this type of pollution incident. This is the same as the probable outcome in the absence of pollution. In other words, 'moderate' contamination may tend to kill those tadpoles destined not to metamorphose anyway (see also Section 3.6.3).

Just before I left Monks Wood, I teamed up with my colleagues Dan Osborn and Sue Freestone to undertake a histological study of the snout abnormality (Osborn *et al.* 1981). It appeared that DDT disrupted development of skin glands above the upper mandible, and when the tadpole became hyperactive it caused the lower mandible to repeatedly strike the inner surface above the upper mandible. In some cases, this was sufficient to cause loss of the upper mandible (Figure 3.2b). Although some animals recovered, others developed into froglets with blunt snouts and deformed brains. As froglets with deformed snouts tended to be smaller than normal, they were doubly disadvantaged and unlikely to survive for long in a field situation. This characteristic deformity was never seen in wild tadpoles in the field.

In 1974, I collaborated with J. A. Zoro of Bristol University to test the toxicity of frog tadpoles to a compound known as DDCN, which was formed when DDT was incubated with biologically active sewage sludge (Cooke and Zoro 1975). The compound had the potential to be widespread in mud and sediment in the field, but was of reassuring low toxicity to tadpoles.

The remaining experiment was with adult frogs, which I had been reluctant to kill unless it was essential. However, previous studies in other countries with DDT and adult frogs were concerned with calculating a precise figure for toxicity or finding the chemical's site of action. I wanted to know how DDT affected the frog's behaviour and physiology. A small number of frogs were dosed, and symptoms displayed included: a marked tendency to hide less than normal; a colour change, becoming paler or greener; adoption of abnormal postures; increased croaking and sluggishness (Cooke 1974a). No difference in jumping ability was noted prior to these other symptoms being seen. There were no significant differences between the groups in food consumption, faecal sac excretion, breathing rate, liver or heart weight, and fat body or gonad size. All of the frogs in the top-dose group contained measurable residues of insecticide, as did several in the medium-dose group. Frogs with some of these symptoms would be expected to be more vulnerable to predation in the field, and also more liable to be seen and reported by people, but no incidents have been reported. Scientists at Monks Wood demonstrated that earthworms and other invertebrates in heavily treated areas such as arable fields or orchards could contain significant residues of organochlorine insecticides although these were considered to be insufficient to cause acute lethal poisoning in birds (Davis 1968, Davis and French 1969). Nevertheless, the potential existed for poisoning to occur in frogs, especially by eating slugs (Cooke *et al.* 1992).

3.2.3 *Effects of herbicides*

I mentioned above the laboratory study with 2,4-D in which no effects were noted in frog tadpoles. During the 1970s, several herbicides were cleared for use in water to control excessive amounts of vegetation. Not surprisingly, these chemicals should have been previously tested on a range of aquatic animals, such as fish and invertebrates. They probably had not been tested on amphibians – and certainly would not have been formally tested in the field to elucidate any indirect effects on amphibians. So I was interested to study the impacts of other herbicides that were applied directly to water, and in particular to examine what happened in field situations. In 1972, my colleague Chris Newbold was undertaking a detailed study of the effects of two aquatic herbicides, diquat and dichlobenil, on plants and invertebrates in a set of experimental ponds in the nearby wetland nature reserve at Woodwalton Fen (Newbold 1975). I was therefore fortunate to be able to take advantage of Chris's study

Figure 3.4 Chris Newbold treating one of the experimental ponds at Woodwalton Fen with a granular formulation of dichlobenil, May 1972.

to investigate whether these herbicides had effects on the food supply and cover required by amphibians (Cooke 1977).

Chris treated two ponds with diquat and two with dichlobenil at normal field rates (Figure 3.4). I had a cage of frog tadpoles and a cage of toad tadpoles in each pond, as well as cages in two untreated control ponds. His wife, Pam, helped me to make the observations. Neither herbicide caused death, delays in development or changes in activity. Tadpoles of both species in ponds treated with diquat weighed less than controls after 4 days, but those in both types of treated ponds were heavier than control tadpoles after 18 and 32 days. The increases in weight were attributed to the treatments causing a proliferation of algae that the tadpoles found highly acceptable – the herbicides were designed to kill 'aquatic weeds' rather than algae. Thus the treatments had no obvious detrimental effects of the tadpoles and may have been beneficial in leading to larger tadpoles as they reached the climax of metamorphosis.

In 1970, an article was published in the *British Journal of Herpetology*, which, as far as I am aware, is the most detailed account of a non-experimental incident in this country where frogs appear to have been affected by a pesticide. The author was Ellen Hazelwood, a schoolteacher in Bolton, Lancashire. With her students, she had studied the frog population in the school pond for a number of years. In 1965 and 1966, spawn hatched and developed normally, but no spawn hatched in 1967. Therefore, the situation was observed even more closely in 1968. That year, spawn failed again in the pond, but a pair of frogs brought into the school produced spawn which disintegrated, but from which a small number of tadpoles hatched; these grew abnormally quickly and reached a large size at metamorphosis, but developed spinal kinks. In the summer of 1968 the pond dried out, leaving a grey deposit in the muddy centre. The sediment in the base of the pond was removed and retained in bags. Some of this mud was analysed for organochlorine residues at Monks Wood, but no residues were detected.

The pond was 9–10 m below the level of a railway line. In the spring of 1967, a notice near the local station stated that weed-killer trials were being undertaken along the line. That summer, a white froth had been noticed where field drains entered the pond. Mrs Hazelwood was informed by British Railways that the line had been sprayed annually for some years with a product containing the herbicide atrazine, and that an 'improved' product with a higher content of atrazine had been used after 1966. The manufacturers analysed the mud samples from 1968 for Mrs Hazelwood, but failed to detect atrazine residues. In 1969, Mrs Hazelwood carried out dosing trials in the school and found spawn failed to hatch in spring water containing the pond mud or added atrazine product, but hatched normally in uncontaminated water. She concluded that 'a substance possibly originally including atrazine' had destroyed two years of frog reproduction.

Naturally, I found this account very interesting, and I corresponded for some time with Mrs Hazelwood. One point worth making is that it was only because she had been studying the pond for several years that any problem was noticed. I did not follow up on this report with further experiments because I had my hands full looking at potential effects of the organochlorines, and also because atrazine was a herbicide, a class of compounds which generally had low toxicity to animals, and it was confined to use on land. What I did not know was that by virtue of its mobility, persistence and toxicity, it would be banned in this country and in the rest of the EU more than 30 years in the future. Its use in this country was never especially extensive, but it is one of the most widely used pesticides in the world and is the most commonly detected pesticide contaminant of ground, surface and drinking water. Tyrone Hayes from the University of California and his various collaborators have reported endocrine effects on frogs down to concentrations of 0.1 parts per billion (Hayes *et al.* 2003, 2010). Interference with the endocrine system can have a number of effects, including changes in sexual development leading to a preponderance of one sex over the other. There is, however, a range of conflicting evidence refuting that atrazine is so toxic to frogs (Mann *et al.* 2009).

Daniel Pickford of Brunel University has been interested in the possible effects of endocrine disruptors on toads for many years. Once, I took him out onto my patch of the Fens and showed him where toads bred in the large rivers and district drains adjacent to arable fields. Subsequently, he and his colleagues reported on a study in which nine breeding sites of toads in England and Wales were screened for activity of endocrine disruptors and the development of caged tadpoles was monitored (Pickford *et al.* 2015). Results indicated some signs of endocrine-disrupting chemicals at low level, but these were contaminants 'other than plant protection products'.

Atrazine has remained at the back of my mind since Ellen Hazelwood published her study. It is one of a class of chemicals known as triazines. With my colleague Al Scorgie, I studied the effects of a new triazine, cyanatryn, on invertebrates and tadpoles (Scorgie and Cooke 1979). In the laboratory, juvenile pond snails (now *Radix peregra*) were unaffected by levels of cyanatryn recommended for use in the field, whereas frog tadpoles and adult common water fleas (*Daphnia pulex*) were sublethally affected. The frog tadpoles fed less and weighed about half that of control animals after 11 days of dosing with granules of product. In a field trial with caged tadpoles exposed to one-tenth of this concentration, no adverse effects were noted. So cyanatryn also raised some concerns, but was withdrawn from use many years ago.

3.3 Analysis of organochlorine residues in field samples

From 1969 to 1972, I took opportunities to analyse amphibian samples for organochlorine residues to aid understanding of exposure in the field (Table 3.1). When possible, analysis was restricted to remains of predated adults or animals found dead for unknown reasons, but adult and juvenile amphibians were also collected alive, as were all of the tadpole samples. Tadpoles and juvenile smooth newts were composite samples of several individuals.

In total, 102 samples were analysed and 17 contained detectable residues of DDT and/or its metabolites. Limit of determination was 0.05 ppm. The highest residues were in a crested newt in 1970 from a field pond in a Huntingdonshire village; this newt and three others with detectable residues were collected alive. A local source of contamination was indicated, but these levels had no visible effect on the newts. Three crested newts and three smooth newts were analysed from a roadside pond in the same village, but no residues were found. The frog with detectable residues and two of the smooth newts with residues came from a garden in Manchester; they were found dead and were desiccated, which will have increased the concentration of the residues. Seven of the toads came from St Neots Common in 1971 and 1972, and three of these contained detectable residues; the toad population was flourishing at that time (Section 6.2). None of the 19 tadpole samples had detectable residues; from the laboratory experiments, tadpoles with residues below about 2 ppm would

Table 3.1 Total residues of DDT and its metabolites in amphibians, 1969–1972.

Species	Life stage	Sample	No. of samples	No. with residues	Maximum residue (ppm)
Common frog	Adult, juvenile	Whole body	7	0	—
		Liver	19	1	1.8
	Tadpole	Whole body	12	0	—
Common toad	Adult, juvenile	Whole body	2	0	—
		Liver	10	5	1.7
	Tadpole	Whole body	2	0	—
Smooth newt	Adult, juvenile	Whole body	34	6	0.9
	Tadpole	Whole body	5	0	—
Crested newt	Adult	Whole body	11	5	4.3

not be expected to show any symptoms of poisoning. Adult frogs which displayed symptoms and died in the laboratory test had liver levels higher than 8 ppm (Cooke 1974a). Sample sizes are small and conclusions are necessarily tentative: although residues were detected in a number of samples, there was no evidence to suggest that any of the animals died of DDT poisoning or were displaying sublethal effects when collected.

All samples were analysed for cyclodiene residues: each sample was clear apart from 0.23 ppm of the insecticide aldrin being found in an adult smooth newt. Cyclodienes used as cereal seed dressings had caused considerable mortality of granivorous and predatory birds and mammals during the late 1950s and early 1960s prior to being banned from such use (Sheail 1998). Other uses, however, continued up to the mid-1970s. The general absence of cyclodiene residues in the amphibian samples indicated very limited exposure to these persistent chemicals.

At this stage, it is probably salient to mention analysis of residues in grey herons (*Ardea cinerea*), which were found dead by members of the public and sent to Monks Wood. Ian Prestt (1970) concluded that herons fed largely on slightly contaminated fish, but concentrated residues in their tissues. The last task I had at Monks Wood in 1978 before leaving for my new post in the NCC was to draft a review on pollutant residues in predatory birds with Tony Bell and Margaret Haas (Cooke *et al.* 1982). In heron livers, geometric mean residues of the main metabolite of DDT were fairly stable at about 8 ppm from 1963 to 1971, but fell to 1.5 ppm by 1977. Residues were found in all 108 samples analysed in the period 1963–1975, with some livers containing more than 100 ppm. The main effect of DDT and its metabolites on the herons appeared to be failed reproduction by causing eggshell thinning leading to breakage prior to hatching (Prestt 1970, Cooke 1975b). Amphibians do of course form part of herons' diets. Whether DDT in adult frogs might have caused spawn failure is a moot point. It was not a topic that I considered investigating. By that time, frogs were very rare in my local arable landscapes and environmental levels of DDT were falling. Some toads at St Neots Common were contaminated in the early 1970s, but there was no evidence of large-scale and unexplained spawn failure either for them or for the frog population at the site (Sections 5.2.2 and 6.2.2).

3.4 Field surveys to detect pollution events

3.4.1 *Mill Pond, Ramsey*

I was continually looking out for situations where frogs might be exposed to pesticides, but in my area frogs did not seem to live where pesticides might be used. A potentially polluted pond occurred on the edge of Ramsey, the town to which I moved in 1973, although frogs did not breed in the pond and any contamination was not via pesticide application. This pond was close to the toad pond beside Field Road (Section 6.3.1) and was on land owned by a mill. Among other types of business, the mill treated grain with pesticides, prior to the grain being drilled into farm fields. Used boxes and containers that had once held pesticides and other chemicals were piled up close to the pond and periodically set on fire; although if it had been windy, it was not unusual for cardboard boxes to blow into the pond before the mill workers had a bonfire. The containers had held organomercurial fungicides and gamma-HCH, an organochlorine insecticide. Analysis of mud from Mill Pond for mercury showed levels were roughly 10 times higher than in Field Road Pond.

In 1974, the foreman at the site told me that Mill Pond had teemed with frogs when he was a boy, 20–30 years before, but it was at least 10 years since he had seen frog spawn in the pond. I visited Mill Pond regularly in the spring and summer from 1974 until 1977. Small numbers of toads and crested newts bred there sporadically, and sticklebacks were occasionally abundant. The foreman told me that the water in the pond became cloudy each summer. I recorded dead fish in 1975 and 1976, and caged frog tadpoles failed to survive in 1976, when a milky sediment rose slowly to the surface; oxygen levels were then extremely low in the pond. I last checked Mill Pond in the spring of 1977: toads were seen occasionally, but there was no sign of spawn, tadpoles or fish. The pond was filled in without warning during the winter of 1979/80.

In conclusion, this was a pond with a seasonal problem, which worsened from 1974 to 1976, possibly being exacerbated by the exceptionally hot summer weather. The relevance of any pesticide contamination was unresolved. Whether it was significantly contaminated was in the end irrelevant, as it went the way of so many other ponds.

3.4.2 *Frogs breeding in potentially polluted sites*

In 1975, I decided to be more proactive in my search for polluted situations and organised a survey of frogs breeding in potentially polluted sites and in sites that were unlikely to be polluted. In this survey, I was interested in pollution more broadly, not just pesticides. The aims were to provide further information on the

Figure 3.5 Trevor Beebee in a pasture pond at Ash Green, Hampshire, during the 1975 survey. Cattle drank the tadpole sample in the orange bowl prior to examination and holed the author's pond net, but frogs bred successfully in the pond.

part played by pollution in the decline of the frog and to determine whether it might be a suitable organism for indicating the impact of local pollution. Although I visited nearly all of the sites, I was helped by a team of volunteers including specialists on frog ecology such as Maxwell Savage (Figure 1.1) and Trevor Beebee (Figures 3.5 and 12.1). For each site, information was collected on dead adults, fate of spawn, survival of tadpoles and abnormalities, and whether any froglets emerged in the summer.

There were 20 sites in the potentially polluted group. These were in or next to arable land or pasture, and/or beside roads. They were selected on the basis of surrounding land use, not because of any knowledge of their likelihood of being polluted. Of the 13 ponds in unpolluted situations, all but two were in gardens of various types or in nature reserves. Sites were in the south of England or the Midlands.

Dead frogs were reported from a total of 11 sites from both groups. Motor traffic was responsible for most of the deaths. At one site, where about 50 clumps of spawn were laid, 26 dead frogs were noted, suggesting that traffic had killed a significant proportion of the adult frogs. Because adult mortality was blamed on road traffic, the survey mainly concentrated on unexplained failure of spawn, presence of abnormalities among tadpoles, and absence of froglets (Table 3.2).

Out of a total of about 1,500 clumps laid in all sites, nearly 1,000 were observed to hatch normally (here taken to mean more than 95% of ova in a clump became tadpoles), about 200 were assumed to have been taken by people (from 12 sites), and about 300 failed to hatch normally (in 19 sites). Most of the clumps that failed were in a single unpolluted site where 247 had been laid, but only about 20 hatched – an event that was blamed on freezing conditions. Cold weather was also blamed for spawn failure at three other sites. At four sites, between one and six clumps failed totally and were labelled 'infertile'. In all, some spawn failed at 19 sites, but the reasons remained unknown at most of them. If an arbitrary level for unusual spawn failure is set at 10% or more of the clumps, then six or seven sites in the potentially polluted group had an unusual amount of spawn failure, as did two in the unpolluted group (Table 3.2). Colonisation by fungus was noted on ova or embryos in spawn in nine sites. Most of these clumps had low or zero hatching rates. Many of the clumps that failed to hatch normally, however, were not colonised by fungus.

Some of the volunteers found difficulty in identifying tadpole abnormalities. This probably stemmed both from lack of experience and from failure to examine every tadpole closely. I routinely studied tadpoles individually when they were confined in a glass tube, as in Figure 3.6. When available, samples of toad tadpoles were also examined, as were tadpoles raised in captivity from spawn laid in the sites. One of the most common abnormalities was a bitten tail on a well-grown tadpole. This

Table 3.2 Number of sites in the 1975 survey of common frog breeding sites with unusual or unexplained events.

	Potentially polluted group	Unpolluted group
Number of sites	20	13
Number with an unusual level of spawn failure	6–7	2
Number with an unusual level of tadpole abnormalities	4	1–2
Number with an unexplained absence of froglets	6	1

Figure 3.6 Sue Freestone examining a tadpole for abnormalities in a glass tube to which is attached a small rubber bulb for suction.

will have resulted from the tadpole being strong enough to tear itself away from a predator, probably a newt. Smaller, less vigorous tadpoles cannot free themselves and are eaten whole (see Section 4.3). Morphological deformities in the early stages of development consisted mainly of lateral or vertical curvature of the tail or body and tail: these were found at nine sites across both groups. Abnormalities on older tadpoles tended to be tail or body kinks (three sites; Figure 3.2a) or distended bodies (two sites). I personally checked for abnormalities in 20 sites: 6 out of 11 in the potentially polluted group had morphological abnormalities (i.e. excluding bitten tails), as did 5 out of 9 in the unpolluted group. Thus abnormalities were quite widespread in both samples. I took an incidence of at least 5% of abnormal tadpoles as being unusual; such levels were found in four potentially polluted sites and one or two unpolluted sites (Table 3.2). Many years later in a study in Sussex, Hitchings and Beebee (1997) found higher levels of tadpole abnormalities in populations with low genetic diversity such as occurred in gardens and parks.

Froglets were seen in 22 sites. One other site was very difficult to search, but at the remaining 10, no tadpoles or froglets were found despite thorough searches. Water persisted at 7 of these 10 sites; 6 of these were in the potentially polluted group.

This survey focused on searching for and defining unusual and inexplicable events: unexplained spawn failure of at least 10% of clumps, unexplained frequency of tadpole abnormalities in samples of at least 5%, and an unexplained total lack of froglets (Table 3.2). Sixteen of the 33 sites had at least one unexplained event, demonstrating that such events were in fact not unusual. Twelve of these sites were in the potentially polluted group. Four sites had two or three unexplained events. There was some evidence to suggest that unusual events were more prevalent among the potentially polluted group. This project was never formally published but I did circulate a report to inform and thank the volunteers. In 1976, I looked at tadpole hatching and development in a number of sites, focusing mainly on those sites where there might have been pesticide effects. Some of the sites were those with problems in 1975, but results were generally unrevealing.

3.4.3 *Frog tadpoles caged on arable fenland*

In 1977, I undertook a field study using caged tadpoles in an area of 76 ha of typical high-intensity arable land on the Cambridgeshire fens adjacent to Woodwalton Fen NNR (Cooke 1981a; Figures 1.5 and 3.7). The aims of this trial were two-fold. First, I intended to expose frog tadpoles in a commercial arable situation to determine whether any problems could be detected. Second, I had become interested in whether monitoring captive tadpoles closely in such situations might be a simple and economical method of detecting possible effects in potentially polluted situations, prior to those effects being investigated in more detail.

Cages of tadpoles were in place for up to 11 weeks during April to June. The drainage system of the study area was separate from that of the surrounding farmland. Most of the fields had tile drains; the remainder were adequately drained without tile systems. Ditches occurred between all of the fields, but only some held water during that summer. No free-living amphibians of any species were encountered during the trial. The principal crops that year were potatoes, barley and sugar beet.

Sixteen cages of tadpoles were used in the trial, arranged so that eight were next to each of the main crops (some cages were between two fields with different crops). I recorded survival, development stage, weight and abnormalities through the trial, and each cage was removed when the first tadpole acquired front legs. Next to potato fields, vertical curvature deformities were 12% on average, while tail-tip defects were 15%; away from potato fields, these averages were 2% and 4% respectively. During the trial, I was in constant touch with the farm manager over cropping and pesticide application. If the significantly higher incidence of abnormalities next to potato fields was due to pesticide use, the chemical most likely to be implicated was oxamyl. This is a granular product for use against insects and nematodes, which is drilled 3–4 days before potatoes are planted. On the study area, it was known to have been drilled during the second half of April. Oxamyl is not very persistent, but is water-soluble, so could conceivably have run off into the ditches. It was tested against tadpoles in the laboratory: many had similar, but not identical, deformities to those seen in the field. The evidence did not prove oxamyl was responsible, but it did suggest that of the 19 pesticides applied to the study area that summer, it warranted further consideration. However, that was to be my last summer of fieldwork at Monks Wood. In 1978, I joined the NCC.

When I published this study, I pointed out the possible usefulness of caged frog tadpoles as 'bioassays' to identify potential problem chemicals in the field (Cooke 1981a). Russell Hall and Paula Henry (1992) discussed how pesticide registration might be improved for amphibians, and signalled the value of this novel method. I also noted that tadpole cultures of different origins often have very different percentages of morphological abnormalities, and that the main deformity can vary between cultures. Thus, when undertaking such a trial with tadpoles in the field, it may be better to utilise a culture showing some susceptibility to producing a certain type of abnormality. More than two decades later, Joseph Kiesecker (2002) demonstrated how exposure to pesticides can increase the incidence of an abnormality primarily caused by a different agent (Section 3.7.2).

Figure 3.7 The arable pesticide study at Woodwalton Fen in 1977: (a) part of the area, with a tractor spraying; (b) Sue Freestone recording oxygen levels in a ditch beside one of the tadpole cages.

3.5 Pesticides: the later years, 1978–1998

I joined the NCC to become its adviser for Britain on the effects of toxic chemicals on fauna and flora, and the conservation of reptiles and amphibians. At first glance, this might suggest that I would continue to be actively involved in the subject of pesticides and amphibians, but that was not the case. Although pesticides remained a major part of my work on toxic chemicals, amphibian input was mainly on issues pertaining to the conservation of crested newts and natterjack toads and their habitat. I continued to have various conservation-based research projects in my spare time, but it's hard to do practical pollution work in isolation. Although I had been ploughing a fairly lonely furrow when I investigated the potential effects of pesticides on the common frog and other amphibian species during 1968–1978, I was a member of a team of scientists working on pesticide issues in laboratory surroundings and had the support of analytical chemists and statisticians.

While I was reasonably sure that pesticides had been a relatively minor contributor to amphibian declines, I never forgot that pesticides and other forms of pollution had the potential to cause problems for amphibians. This was always at the back of my mind when tackling significant issues of the time, such as the impacts of acid rain (Section 3.6.3) or the environmental levels of radioactivity emanating from Sellafield or from fallout from the Chernobyl explosion (Section 7.2). I was able to include some of these issues when I was asked to convene the session on pollution impacts at the First World Congress of Herpetology at Canterbury in Kent in 1989 (Figure 3.8). Having provisionally agreed a programme, I brought in a speaker at a late stage to address the subject of global declines in amphibians – which were just starting to be noticed. One thing I remember from that session was a European

Figure 3.8 Delegates at the First World Congress of Herpetology, held at the University of Kent in 1989 . Photo courtesy of Richard Griffiths.

delegate standing up at the beginning of the final period set aside for questions and complaining that having an extra speaker had reduced time for questions. I explained the importance and relevance of hearing up-to-date information on the worldwide declines – and asked what question did he have? He said he did not have one and sat down. It is impossible to please or convince all of the people all of the time! Similarly, many people in this country remain convinced that pesticides must have caused, and are still causing, rarity of frogs in the countryside.

Throughout the 20-year period 1978–1998, I was actively involved with advising on pesticide use in Britain, including the safe application of pesticides on nature reserves (Figure 7.3b). I never told reserve wardens or managers that they should use pesticides, but I pointed out both how any risks could be minimised and what non-pesticide options were available. Sometimes it was a choice between using a pesticide and doing nothing, but seeing the conservation value of some asset decline. Much of my time was devoted to trying to ensure the environmental safety of pesticides was not compromised, via input to the departmental Advisory Committee on Pesticides, its Scientific Sub-Committee and the Environmental Panel. For eight years in the 1990s, I also served as a member of the Veterinary Products Committee, being the first person appointed to represent environmental safety on that committee.

I cannot recall any consideration ever being given to amphibian species during the assessment process, but potential impacts on freshwater life were carefully considered, using information on toxicity to fish and aquatic invertebrates; and impacts on terrestrial life forms generally were tacitly taken into account by attempts to ensure the safety of higher vertebrates and beneficial invertebrates. The problem has always been that it is neither possible nor desirable to test every new chemical against every species that might be exposed. Wagner *et al.* (2013) drew attention to the particularly vulnerable position in which amphibians are placed by virtue of inhabiting both land and water and having permeable skins. Generally, acceptable levels of risk are achieved by applying safety factors to worst-case situations. Hall and Henry (1992) reviewed, mainly from a North American perspective, the problems and needs of the assessment process in order to ensure the protection of amphibian species. This paper stemmed from a presentation I invited Russell Hall to give at the World Congress in 1989. These authors suggested that research was needed (1) to identify amphibian taxa and types of pesticide for which prediction of effects across species was most difficult, and (2) to understand how to minimise numbers of tests and harm to test animals in those situations where dosing amphibians was deemed essential. This subject was reviewed in 1998 by scientists from the UK's Central Science Laboratory working under contract to Defra. Their conclusions were that toxicity data on fish could be used to safeguard tadpoles, but that data on birds did not satisfactorily protect other terrestrial life forms. Instead, dietary exposure should be estimated for adult frogs, and simple calculations of dermal exposure made to test whether animals were at risk under credible worst-case conditions.

A subject that had been little researched, even by the mid-1980s, was the impact of pesticide spray drift. I decided that if no one else was going to do the research, then we had to do it. So, from 1986, I commissioned research on behalf of the NCC and later English Nature to understand better the principles of drift and buffer zones, and to determine the impacts of applying pesticides at a range of distances away from higher plants, lichens and invertebrates. Tadpoles were not studied in bioassay trials, but aquatic invertebrates were, so some of the work was highly relevant for conserving freshwater life in general. A meeting was organised in 1992 to publicise the results and conclusions to the conservation and agricultural communities, and I edited a report that was published the following year (Cooke 1993b). The synthetic pyrethroids were probably the group of insecticides of the greatest concern for freshwater life at that time. Clive Pinder and his colleagues from the Institute of Freshwater Ecology carried out the work on aquatic invertebrates, and calculated the distance at which no mortality occurred after spraying different pyrethroids: the longest 'safe distances' were 28 m for ground spraying and 200 m for aerial spraying. NCC/English Nature should have been consulted over any aerial spraying within three-quarters of a nautical mile of a Site of Special Scientific Interest (SSSI), so could influence what spraying was done; but aerial spraying operations became rare during the 1990s. During the approval procedure, statutory buffer zones were sometimes applied to new pesticides to reduce the risk from over-spraying or from drift, so it was possible for me to request, if necessary, a buffer at that stage.

In July 1995, the Ministry of Agriculture, Fisheries and Food (MAFF) organised the first meeting of an Expert Working Group on Buffer Zones. The idea was to tighten up protection for freshwater life in particular. At that meeting, it was revealed that very little was known about farm ditches, such as how big they tended to be and whether they generally held water. As a member of the group, I volunteered to obtain such data from the Fens and also to measure how close to the ditch cropping currently occurred. Eventually, this exercise resulted in a standard 6 m buffer being introduced. It clearly would not protect every species in every situation, but it was designed to help while not hugely inconveniencing farmers. Now, if you drive through an arable landscape, you may notice a bigger uncropped edge to arable fields next to ditches. Sadly, there seems to have been little or no research performed to determine whether such buffers make a significant difference (Freshwater Habitats Trust 2013).

One of the last major pieces of work I commissioned before leaving English Nature was on the indirect effects of pesticides on birds. A summary document, edited by Lennox Campbell of the RSPB and myself, pointed out that many farmland bird species had declined in numbers in Britain in the previous two decades (Campbell and Cooke 1997). The report's conclusion was that there was evidence of pesticide use contributing to the decline in abundance of many types of invertebrates and plants on which the birds fed. Moreover, there were temporal associations between increases in the frequency of pesticide use and the periods of rapid decline of some bird species. A package of measures was recommended to help to reverse the declines:

- Encourage the use of integrated crop management (ICM), narrow-spectrum pesticides and similar techniques to reduce the overall use and impact of pesticides.
- Introduce an integrated package of prescriptions, including non-pesticide techniques, in an agri-environment scheme.
- Introduce a successor to set-aside designed to be sympathetic to farmland birds.
- Encourage a switch to organic farming.

Although this work focused on birds, it could be inferred that the same principle of pesticide-induced indirect effects might also affect non-avian farmland species (Mann *et al.* 2009). And as use of pesticides is not confined to agricultural land, a cautious approach is needed wherever these chemicals are applied, to minimise the risk of direct and indirect effects on amphibians and other wildlife (Cooke 1997a). It should though be remembered that some indirect effects may be beneficial (e.g. Cooke 1977; Section 3.2.3).

Surveillance of pesticides in use may help compensate for predictive failures at the approval stage (Hall and Henry 1992). Throughout my time on the pesticide committees, the agricultural authorities ran a Wildlife Incident Investigation Scheme (WIIS); it later became the responsibility of the Food and Environment Research Agency (now Fera Science). The scheme exists to investigate incidents that may have been caused by pesticides. Only very rarely were incidents involving amphibians reported by WIIS. I still have four annual reports to hand to check, because during my time as chairman of the Environmental Panel of the Advisory Committee on Pesticides, it was one of my tasks to sign off the WIIS annual report. During those four years, 1994–1997, two such incidents were investigated. In 1994, toads were suspected of having been poisoned, but died from an unspecified disease (Fletcher *et al.* 1995). In 1996, a frog was found to have died from pasteurellosis, not pesticides (Fletcher *et al.* 1997). There had, however, been two incidents investigated by the scheme in 1989 and 1993, in which frogs died from metaldehyde poisoning, possibly as a result of secondary poisoning after the chemical had been used to kill slugs (Cooke 1997a).

Secondary poisoning, including in gardens, was a topic I would like to have followed up, but I tried unsuccessfully to commission work when I was with NCC. Interestingly, some information on the use of pesticides in gardens was obtained from a citizen science project which reported on whether amphibian and reptile species occurred in nearly 4,000 gardens (Humphreys *et al.* 2011). Correspondents reported on a range of characteristics and features of gardens, including whether slug pellets or other pesticides were used. Counterintuitively, the use of slug pellets was positively associated with the occurrence of slow-worms, but no relationship was found with any of the amphibian species. However, the use of herbicides was positively associated with the presence of the common toad and smooth newt. The authors summarised these observations as there being no evidence of pesticides having negative effects. While I would not argue with this interpretation, if there had been evidence of a negative effect, this would probably have been reported as an observation rather than a 'non-observation'. This illustrates the problem with correlations and associations – they are not necessarily linking cause and effect (see also Section 10.6.4).

3.6 Other environmental hazards

3.6.1 Introduction

This seems a good point to break away from pesticides and mention some other potentially serious hazards facing amphibians that are not covered elsewhere in the book. The eclectic collection of hazards introduced below have all created widespread concern, but, with the exception of ranavirus, do not appear to have had dramatic effects in Britain.

3.6.2 *Disease*

Threats of various infectious diseases to European amphibians and reptiles have recently been reviewed by Allain and Duffus (2019). A virus or group of viruses of the genus *Ranavirus* has been proven by Andrew Cunningham and his colleagues at the Institute of Zoology in London to be responsible for widespread and serious die-offs of adult frogs in a number of areas of England since the 1980s (Cunningham *et al.* 1996, Teacher *et al.* 2010). Infections can be transient or persistent with recurring phases of mortality. Incidents are most often reported from garden ponds, where they are capable of causing long-term reductions in frog numbers. Often a number of thin frogs are found dead in a small area around the same time. If affected frogs are still alive they are likely to be lethargic and their skin may be red in colour and ulcerated (Figure 3.9). I have often thought that infected frogs were, sadly, victims of circumstance. It is trite, but true, to say that if we had not created garden ponds, then high suburban densities of frogs would not have formed and there might have been no epidemic. But it goes further than that: in the heat of summer, frogs cluster back in garden ponds to try to keep cool and wet, so presumably being more likely to infect or be infected. More recently a large team of scientists from the Institute of Zoology, University College London and Queen Mary University demonstrated that higher temperatures increase ranavirus propagation, and very hot weather in summer increases the risk of a ranavirus outbreak (Price *et al.* 2019). And tadpoles are affected too. The scientists predicted that the situation will continue to worsen because of global warming. I am typing this account while in isolation because of my age and underlying health conditions during the coronavirus epidemic. Frogs do not know about self-isolation. In order for them to have more options in hot

Figure 3.9 A dead frog that tested positive for ranavirus, showing some reddening of the legs and an ulcer on its 'knee'. Photo by Jim Foster.

weather, the authors recommended people create more shady and moist patches in gardens as well as deepening pools to keep frogs cool.

Another threat to amphibian populations on a global scale is cutaneous chytridiomycosis, caused by chytrid fungus (Cunningham *et al.* 2005). The disease it causes is highly infectious and has affected populations in warmer climes. In the UK, volunteers from local amphibian and reptile groups collaborated with the Institute of Zoology in 2008 and 2011 to swab several thousand amphibians from across the country for chytrid (ARG UK 2013). Some positive results were found. Its effects are most obvious when it kills froglets, toadlets and the efts of newts just after metamorphosis (Froglife 2013). The susceptibility of species in Europe is not fully understood, but there is both a wide geographic range and a wide host range within the continent (Allain and Duffus 2019). This is a situation that requires careful monitoring and control. Advice on biosecurity to reduce the risk of introducing and spreading amphibian pathogens is available (ARG UK 2017). This document includes recommendations such as 'handle amphibians only when necessary' and 'in general translocation should be avoided'.

3.6.3 *Acid deposition*

Problems caused by 'acid rain' came into prominence in the early 1980s. Gary Fry and I reviewed the generalities of the issue for the NCC (Fry and Cooke 1984). We attempted to identify areas of conservation importance which might be at risk from continuing high levels of acid deposition or from changing patterns. Vulnerable areas included south-west and central Scotland, the Lake District and the Pennines. Fish and invertebrates in freshwater habitats on granite or other slow-weathering rocks were at higher risk, but we also drew attention to the risk to amphibians living in sensitive areas. Experiments with frog tadpoles led Clive Cummins (1986) to conclude that tadpoles would be adversely affected in certain types of acidic situation such as he had observed in Scotland. His experiments increased in complexity and culminated in a trial with 54 replicated pools examining the implications of varying acidity, tadpole density and other factors (Cummins 1989, 1990), work which I part-funded on behalf of the NCC. The later tests examined whether density effects could overcome those of acidity. Simplifying the results, the essence of the issue centred on the fact that at high density there is competition between tadpoles, causing a hierarchy to develop in which a few tadpoles out-compete the remainder. Clive found that in acidic pools mortality tended to occur early in the experiment, allowing survivors to grow more quickly than tadpoles in otherwise comparable neutral pools. In neutral pools with high tadpole density, there was little mortality in the early stages, but growth rates were slow and mortality increased later in the experiment. These results demonstrated how difficult it can be to determine the outcome of events in a frog pond and how, even under adverse conditions, a few tadpoles may ultimately metamorphose at a large size. Similarly, I remember the poor recruitment in the plastic pond in my third garden during 2002–2007 (Section 5.3.3): despite almost total failure of spawn and tadpoles, there were usually a few fat, smug tadpoles metamorphosing at the end of the season. In a later American review, Benjamin Pierce (1993) concluded that there was little direct evidence of acid precipitation causing widespread population decreases in amphibians.

However, this is only part of the story in this country. Acid conditions in a breeding site below pH 6 are not well tolerated by natterjacks, with spawn and small tadpoles being especially vulnerable (Beebee and Griffin 1977, Beebee 1986). Poorly buffered heathland pools will have been sensitive to acidification, and some pools used by the species acidified during the twentieth century (Beebee *et al.* 1990). There were several weekend occasions about 40 years ago when Rosemarie and I stopped off at Roydon Common when we were on our way to the north Norfolk coast for some sunshine and a swim. Natterjacks died out on the common in the 1970s when available water was too acidic to support them. I hunted for ground water of at least pH 6: it did occur, but only in a small area that was overgrown with sallow bushes (*Salix* spp.). As far as I know, natterjacks were never seen again on the common and have not been reintroduced.

On a brighter note, through the NCC, I funded examination of sediment cores from two ponds at Woolmer in Hampshire where natterjacks had previously bred: analysis of diatoms, heavy metals and soot particles revealed that atmospheric pollution had caused acidification down to about pH 4 (Beebee *et al.* 1990). Pollution control in recent decades has led to improvements in water quality, as has direct management of natterjack ponds, such as removal of acidified sediments or addition of lime.

3.6.4 *Increased exposure to UV-B*

A concern in the 1990s was that anthropogenic destruction of the ozone layer allowed higher amounts of solar radiation at mid-wave ultraviolet (UV-B) wavelengths to reach ground level, and this additional radiation might be harmful to amphibians breeding in shallow water. UV-B is associated with skin burning in humans. Clive Cummins was also involved with this issue (Cummins *et al.* 1999). Newly fertilised frog eggs were exposed to sunlight with supplemental UV-B which approximated to increases of UV-B of up to 25%, but this did not increase mortality or result in a high incidence of abnormalities.

3.6.5 *Collection of adult frogs*

Collection of frogs for the purposes of teaching and research has occurred in parts of the world for many years. This is different to other human actions that might be detrimental at a population level, in that frogs are deliberately removed and eventually killed rather than being incidentally affected. As I never had a biology lesson beyond the age of 14, I escaped having to dissect a pickled frog at school. Events did though catch up with me when I was a research student at Reading University in 1966. One of my duties was to be a 'demonstrator' for agriculture students, showing them how to undertake various experiments and explaining what was going on. One unpleasant process I remember clearly after an interval of more than half a century was killing frogs and showing how the heart continued to beat.

Dorcas O'Rourke (2007) outlined the usefulness of amphibians generally in this respect from an American perspective, but cautioned that 'society should not tolerate the irresponsible removal of animals from the wild' because of concern about global declines in amphibian species. In the UK we have been worried about the impact of collection of frogs from the field for much longer, perhaps because our populations began decreasing several decades earlier. Smith (1953) and Taylor (1963) shared this concern, with the former author estimating the demand for frogs in laboratories to be about 150,000 per annum.

In the early 1970s, when I came to consider why frog populations had declined, it seemed that the demand was as high as it had been 20 years before. The biggest supplier of frogs in Britain told me that they sold about 85,000 in 1970 (Cooke 1972a), while the figure for those collected by schools, universities and research establishments was suggested to be as high as 100,000 (Kelly and Wray 1971). By the mid-1980s, dissection of frogs had been largely removed from examination syllabuses; and under the Wildlife and Countryside Act of 1981, dealers had to be licensed by the Department of the Environment and send in returns every six months stating how many they had sold. At that time, most frogs sold in Britain had been collected from either the Republic of Ireland or Cornwall. My information about the status of frogs in Cornwall abstracted from previous national enquiries suggested that the county's population had changed relatively little in previous decades, but worries were expressed by some local people about the activities of collectors, so I decided in 1985 to obtain further information.

Questionnaires were distributed to about 50 individuals and schools in Cornwall that had helped with previous enquiries. Correspondents generally considered that frog populations had not changed markedly, but some were aware that collection was still occurring (Cooke 1985b). In addition to these replies, Jim Wright of the Cornwall Trust for Nature Conservation appealed via the media for information on my behalf; responses to this request differed in producing a number of reports of recent population decreases, probably mainly for reasons other than collection.

Five years later, I reappraised the situation, this time working with my NCC colleague David Morgan and also with Mary Swan from Leicester Polytechnic (Cooke *et al.* 1990). The Department of the Environment supplied trade information which revealed that the number of frogs collected in Cornwall decreased from about 14,000 in 1983 to 6,500 in 1988, but some of that decrease might have been due to the death of a well-known local collector. Because of Mary's work on pond density and population size, we were able to estimate that the total collected might represent about 3% of the frogs living in Cornwall outside gardens. This, we considered, would not be enough to cause population decreases at county level. We also approached, for their opinions, Jim Wright, who had helped in 1985, and Stella Turk of the Cornish Biological Records Unit. Both replied there had been no appreciable recent change in the county's frog populations, and neither was aware of collection having any local effects. However, whether there might have been local or regional effects on frog populations elsewhere in the British Isles during the 1950s, 1960s or 1970s remains unknown.

3.6.6 Fertilisers

In the mid-1990s, Rob Oldham approached me with a concern about the toxicity of agricultural fertilisers to frogs. A spatial association had been noticed between (1) the regional index of change in status for the frog during the 1980s (Hilton-Brown and Oldham 1991; Section 2.5.3), and (2) the percentage of farmland receiving inorganic nitrogen fertiliser. Frogs had done less well in areas receiving more fertiliser. There had been various changes in fertiliser use over the years, such as a trend for application to occur more frequently in spring. These observations raised the possibility that fertilisers might affect frogs during spring migration. On behalf of English Nature, Alastair Burn and I commissioned Rob Oldham and his colleagues to undertake laboratory and field studies to determine the risks to frogs (Oldham *et al.* 1997).

In the laboratory, frogs were exposed on soil or moist chromatography paper to fractions of a field dose of granular ammonium nitrate fertiliser. They were removed and washed with water when their breathing rate indicated a toxic response that would have led to death had the trial not been stopped. A dose of roughly one-tenth of the recommended local application rate was enough to cause a significant level of response. Trials were then carried out on arable fields with captive frogs. Realistic field rates of fertiliser caused a toxic effect; however, further field trials demonstrated that the toxicity disappeared after one hour as the granules dissolved. We concluded that, because fertilisers would be applied in daylight and frogs would tend to move at night, high levels of mortality were probably fortuitously avoided. We did though propose other avenues for investigation, such as defining the potentially increased risk of granules which dissolve more slowly.

Rob Oldham collaborated with other researchers to document the effects of ammonium nitrate on frog tadpoles (de Wijer *et al.* 2003), having earlier reviewed the topic (Oldham 1999a). In a replicated pond experiment, there was some increase in mortality and a delay in development, but weight at metamorphosis was increased, probably because of increased growth of periphyton and phytoplankton on which the tadpoles fed. In my various field trials with caged tadpoles, growth was almost invariably good in any eutrophic ditch with substantial amounts of algae. However, in ditches with clear water and little or no plant and animal life, growth was poor. This observation also applied to ditches and pools away from farm fields. After conducting a 15-year study on a toad population that had declined on Offham Marshes in Sussex, Trevor Beebee (2012) found no evidence of mortality that might have been caused by applications of fertilisers or pesticides.

3.7 New evidence and conclusions on pesticides

3.7.1 *Pesticide effects in Britain*

Having just briefly reviewed information on other potential hazards, it is striking (and fortunate) how the worst concerns about impacts on frog populations from chytrid fungus, acid deposition and UV-B have not (yet) been realised in this country. This is similar to my experiences with frogs and pesticides: when I began at the end of the 1960s, I expected to unearth evidence of considerable direct effects, but that did not happen. With ranavirus, however, there is ample evidence of high mortality, suggesting that if there are widespread effects resulting from an environmental pollutant or another problem, some incidents are likely to be discovered and reported, even if they occur in sites that are less well observed than garden ponds. And in terms of its habitat preferences, behaviour and susceptibilities, the frog is the most likely amphibian species to be poisoned by pesticides and the event to be reported. The frog is the species on which I have mainly focused.

Nevertheless, I demonstrated how DDT could be harmful to frog tadpoles and adults in the laboratory and occasionally in the field, and that a proportion of animals contained residues in their tissues. The organochlorine era may seem to be in the remote past now, at least in Britain, but DDT was still in agricultural and horticultural use throughout the period that I worked on it and beyond into the 1980s (Cooke *et al.* 1982). In 2009 and 2010, the main vertebrate metabolite of DDT was detected in frogs in remote habitats in the Sierra Nevada mountains in California (Smalling *et al.* 2013) demonstrating both the persistence of such residues and their ability to travel considerable distances.

My conclusions for Britain are that:

- The initial population declines of frogs and toads in the 1940s occurred before DDT, the first of the organochlorine insecticides, came into widespread agricultural use in Britain (Section 2.4). Up until then pesticides should not have been a significant problem for our amphibians.
- Pesticides will have contributed to some degree to the documented declines of the common frog and common toad in Britain since the 1950s, although it has proved difficult to find evidence of effects in the field. The 1950s and 1960s were peak times for organochlorine usage.
- By the 1980s and 1990s, population declines to which pesticides might have contributed directly were probably largely over, at least for the frog (Section 2.5), which had become rarer in intensive arable areas primarily because of habitat loss. Pesticides may though have continued to suppress recovery to some extent, including by interacting with one another and other agents such as disease, or indirectly via reductions of food and cover on exposed land.

Loss of the only breeding site(s) to which a population has access will, in the space of a few years, lead to extinction of that population. However, it may need just one successful year of breeding every few years for a frog or toad population to survive. Thus, intermittent breeding problems due to pesticides in the water are less likely to cause extinction. In addition, contamination of a breeding pond could cause mortality amongst tadpoles, but the fittest or luckiest may survive. It could be argued that death of adults via topical exposure or consumption of residues in prey would be more harmful to a population – and that is generally true. Had mortality incidents been known to occur, a question that might have been asked was whether they were killed in greater numbers than died on the roads. All amphibian species are at risk crossing roads (see e.g. Sections 3.4.2, 7.2 and 9.3.2). As discussed elsewhere in the book, I have evidence of road mortality of toads making a significant contribution to the virtual extinction of my local suburban populations (Sections 6.2 and 6.3). My point here is that in the last 80 years there have been obvious and widespread factors killing our amphibians and affecting their ability to breed properly, which would have had massive effects on status even if chemical pesticides had never been invented and used.

It remains difficult to find evidence of pesticide effects on frogs or other amphibians occurring in our countryside. With the benefit of hindsight, if I could now be transported back to the late 1940s I would know what I had to do to study whether pesticides were affecting frogs. I would find farms where frogs still existed in good numbers in the newly created arable areas of the Fens, and I would settle down for as many years as possible to look at specific populations, recording availability of suitable breeding sites, dead frogs, spawning success, tadpole abnormalities and froglet emergence in relation to (changes in) the landscape and pesticide usage. As it was long ago, I probably would not be able to analyse pesticide residues in water or wildlife, so would need to judge whether frogs and tadpoles were being poisoned by, for instance, their behaviour and likely levels of exposure. And I could expose captive animals. If I found effects were occurring, I would then need to discover how widespread and severe they were. I would have a number of years to continue my research before frogs became too rare to study. And I should have the opportunity to see what happened when the cyclodiene insecticides were introduced in the late 1950s. Dieldrin, for example, was used on crop fields and in sheep dips but was progressively withdrawn from use from 1963. These compounds were implicated in the decline of the otter (*Lutra lutra*), mainly via poisoning of adults (Chanin and Jefferies 1978, Jefferies and Woodroffe 2008). However, only one amphibian sample out of more than 100 that I analysed had cyclodiene residues (Section 3.3).

Life might have been frustrating and sad. The landscape would be changing around me as farming practices continued to modernise. Frogs had declined on the Fens during the war years primarily because of the switch from undrained pasture to well-drained arable, and such a change would continue to cause decreases in frog populations and other fauna and flora for another couple of decades or more. Predominantly arable landscapes were said by Mary Swan and Rob Oldham (1993) to be 'particularly inimical' to frogs. Any pesticide effects would need to be teased out from the influence of these broad landscape changes. But if I had been able to do that, would the farming authorities have been interested? The ethos just after the war finished was, not surprisingly, that food production must continue to increase by making farming more efficient. If frogs suffered as a result, then 'so be it' would almost certainly have been the official reaction. Our small country was too crowded and too hungry for frogs to flourish everywhere. Gradually, however, attitudes and interests changed and increasing numbers of people found time and space for animals like frogs, even if it was only in their gardens.

I could not leave this account without saying something about the neonicotinoid group of pesticides, which has often been in the news in recent years and has attained a position of some notoriety mainly because of harmful effects on bees and other pollinators (e.g. Goulson 2014). Honey bees (*Apis mellifera*) are, sadly, unintentional barometers of the effects of environmental contamination. It is almost like having a network of bioassay stations with beekeepers assiduously looking after their charges and quickly becoming aware should anything be amiss. As John Sheail (1998) reported, it was beekeepers in Fife who first drew attention in 1957 to the effects of organochlorines because of their losses caused by indiscriminate spraying of charlock in flower. And, during the 1980s, I worked through the similar scene of destruction with bees being poisoned by the spraying of the insecticide triazophos on flowering oilseed rape (Greig-Smith *et al.* 1994).

As regards aquatic life, in a review of the environmental risks of neonicotinoids, Wood and Goulson (2017) concluded there was 'widespread contamination of waterbodies of all kinds with levels of neonicotinoids known to be harmful to sensitive aquatic invertebrates'. This group of chemicals is both quite stable in soil and soluble in water, a dangerous combination. Nevertheless, North American studies have indicated minimal detrimental effects on frogs at environmentally relevant concentrations in the absence of additional stressors such as prior exposure to disease (e.g. Pochini and Hoverman 2017, Robinson *et al.* 2017, 2019).

3.7.2 More recent evidence from around the world

Since 1998, I have been an armchair spectator, occasionally reading about evidence of impacts from areas of the world that have received liberal treatment with pesticides. My impression is that, despite an undercurrent of concern and some pertinent observations, there has not been an avalanche of damning material focusing on amphibians from other countries.

The rest of the world caught up with us by having their widespread declines of amphibian species from the late 1980s. The year 2000 saw the publication of a book of 900 pages entitled *Ecotoxicology of Amphibians and Reptiles*. It was edited by Donald Sparling, Greg Linder and Christine Bishop and had grown from a seed sown at the First World Congress of Herpetology and other meetings in the 1980s that discussed the declining status of amphibian populations around the world. Despite being a hefty tome, one of the recurring admissions on its pages was that scientists did not have enough information to understand the role of environmental contamination. Paul Stephen Corn reviewed hypotheses for why the declines had occurred. He considered that, despite much speculation, there was little direct evidence that contaminants had caused declines in widespread species. Another paper by Deborah Cowman and Laura Manzanti focused on the potential role of the newer pesticides in recent population declines. Generally there has been a move since the 1970s to approve the use of some extremely toxic new pesticides, but which were relatively unstable when compared to the old organochlorines. These authors concluded that certain new products might harm or kill individual amphibians or even populations under certain conditions. However, toxicity would vary because of a wide range of disparate factors, and any contributions to the worldwide declines were more likely to be via cumulative effects of different pesticides in different situations.

There has long been a concern that pesticides might act with other factors to harm amphibians, and this was elegantly illustrated in a study by Joseph Kiesecker (2002). Some limb deformities have been shown to be caused by trematodes. Kiesecker undertook a field experiment in which tadpoles of the wood frog (now *Lithobates sylvaticus*) were maintained in enclosures exposed to or away from run-off from farm fields and exposed or not exposed to trematode infection. Rates of limb deformity in froglets varied from zero in those shielded from trematode infection to 29% in those exposed to trematodes beside farm fields. Complementary laboratory experiments were run in which tadpoles were treated first with different doses of atrazine, malathion or esfenvalerate, and then exposed to trematodes. Pre-exposure to pesticides depressed the tadpoles' immune response and led to higher levels of trematode infestation. Thus, the pesticides alone did not produce limb deformities, but pesticide exposure predisposed tadpoles to being more affected.

Various methods have been employed to help understand better the current and potential effects of pesticides. Thus in the Western Ghats in India, organophosphates, such as methyl parathion and malathion, have been aerially sprayed in agricultural areas; Vasudev *et al.* (2007) set up mesocosms mimicking field conditions which revealed various lethal and sublethal effects on tadpoles and adult Indian cricket frogs (*Fejervarya limnocharis*). Tadpoles also displayed behavioural and morphological abnormalities similar to some of those that I witnessed in my experiments with DDT (Cooke 1981a).

Organophosphates and carbamates are among pesticides which inhibit the enzyme cholinesterase. This enzyme controls hydrolysis of the neurotransmitter acetylcholine, thereby allowing a cholinergic neuron to return to its resting state. Inhibition of the enzyme can lead to a variety of sublethal symptoms including twitching. Peltzer *et al.* (2007) looked for changes in enzyme activity in three species of toads (*Chaunus* spp.) living in an area of intensive agriculture in Argentina: they found that cholinesterase levels in the blood of adult toads caught on soybean fields were significantly lower than in toads in control sites. They also reported that 6% of the toads from soybean fields had morphological abnormalities. Davidson (2004) investigated the connection between the spatial patterns of decline of five species of Californian anurans and historical pesticide use, 1974–1991. Amount of pesticide applied upwind was 'a strong, significant variable in logistic-regression models' for four of the species. And use of organophosphates and carbamates was more significantly associated with amphibian declines than was the use of other classes of pesticides. This and a following paper (Davidson and Knapp 2007) described historical, spatial approaches that sidestepped the usual problem of determining whether a pesticide effect might be translated into a population decline, but still did not *prove* cause and effect (Mann *et al.* 2009).

In Germany, Wagner *et al.* (2013) and Berger *et al.* (2018) reviewed effects of the very commonly used herbicide glyphosate. Products containing this active ingredient often bear the name Roundup. Some formulations have been found to be toxic to tadpoles and juvenile amphibians, with an additive being mainly to blame. Although severe effects have been noted in laboratory studies, evidence of toxicity under field conditions was described as 'rare', despite the herbicide having been used on a huge scale for a number of decades. The only examples of pesticide incidents involving amphibians that I have noticed in literature published this century were four in Sri Lanka (de Silva 2011): one of these concerned caecilians (*Ichthyophis glutinosus*) being found with skin burns a few hours after Roundup had been sprayed. So, with this paucity of incidents in the field, it comes as no surprise that it is difficult to be sure of what happened before I or anyone else started to work on population declines of the frog, that is during the 20-year period from the late 1940s to the late 1960s.

Research and debate into the effects of pesticides and other toxic chemicals continues today, and I am sure will continue for as long as such chemicals occur in the environment.

CHAPTER 4

Frogs and toads: collation, predation and translocation

4.1 Introduction

Quite a number of my studies have been undertaken simultaneously on the common frog and common toad. Many of the surveys in Chapter 2, for example, were on the two species, often to try to ensure that contributors knew the difference between them. In other instances, I have worked with both species because it has been interesting and relevant to their conservation to compare and contrast them. This short chapter brings together and summarises three deliberately very different projects when both species have been studied. First, I run through a desk exercise in which information on spawn site selection and colony size was collated from many sources. Second, I summarise trials with captive animals recording levels of predation by newts on frog and toad tadpoles. Finally, a large-scale introduction in Peterborough is described where the objectives were to conserve significant populations of amphibians in a purpose-built reserve and to learn more about how to do it. This collection illustrates the variety of work undertaken and also how it evolved from the first two straightforward solo studies in the early 1970s to the more complex and joint fieldwork involved in the introduction during the late 1980s and early 1990s.

The simple processes that culminated in these studies were very different to the thought, justification, planning and search for funding that has typically been required in more recent times. The first two pieces of work were begun when I was in the Toxic Chemicals and Wildlife Section of the Nature Conservancy. An immediate reaction might be that they were nothing to do with toxic chemicals, but they grew out of the need to obtain more background information on the two species. My initial enquiries into national changes in frog and toad populations (Cooke 1972a; Section 2.2) had been essential, and provided extra information on breeding sites, as well as contacts that might be able to contribute further help if needed. So it was an opportunity to bring together many strands of information relatively easily to provide useful data not just to underpin an investigation of pesticide effects but to aid conservation of frogs and toads more generally. By the early 1970s, the enquiries indicated that both species had declined considerably over much of Britain.

The study on newt predation of frog and toad tadpoles developed from the investigation into whether newts selectively preyed on poisoned tadpoles (Cooke 1971; Section 3.2.1). Conservation authorities were not yet concerned about crested newts being kept in captivity (although the Home Office needed to sanction any experiments that involved exposing amphibians to toxins). At the same time, I was gathering information which demonstrated the increasing importance of garden ponds as refuges for amphibians. I wondered what might be happening in those ponds when tadpoles came face to face with predators such as newts. It was quick and easy to determine the outcome in a worst-case situation.

In 1984, I had to terminate my monitoring of frogs and toads on St Neots Common (Sections 5.2 and 6.2) because the NCC office moved from Huntingdon to Peterborough and the extra distance meant that I was no longer able to squeeze in lunchtime excursions to carry out fieldwork. Amazingly, however, authorities in Peterborough constructed a new reserve for amphibians the following year. This was beside the River Nene not far from the city centre and was part of a development consisting of footpaths, cycle ways, a rowing course and Ferry Meadows Country Park. Several years before, because of my interest on the effects of disturbance on wildfowl, I had been asked for advice in a personal capacity on how the wetland habitat might be partitioned to cater for various conservation and leisure pursuits in the country park. However, I knew nothing about the amphibian reserve until after its construction, when the NCC was consulted about stocking it. Not only was it an opportunity to learn more about how to translocate animals rescued from development, it was also close enough to visit at lunchtime. The word 'opportunity' often crops up when I reflect on what I did in the distant past.

This chapter is followed by two others in which I describe work on the frog and the toad individually. The separation into three chapters may be rather artificial, but one intention is to show the variety of work in this chapter and then present two chapters in which the stories are linked by common themes.

4.2 Spawn site selection and colony size

4.2.1 Introduction

By the early 1970s, some information existed on the types of site in which frogs and toads spawned, but it tended to be local in nature. Thus, Maxwell Savage (1961) reported that frogs rarely bred in ponds on the London clay in his study area on the border between Hertfordshire and Middlesex, while Graham Bell (1970) gave an account of the distribution of amphibians in Leicestershire. Malcolm Smith's New Naturalist volume on amphibians and reptiles mentioned that the toad selected ponds, canals and ditches where the water was 'deepish'. He continued that shallow waters were avoided, and if the two species bred in the same pond, toads chose the deeper parts and the frogs spawned in the shallows – the two species seldom mixed.

In his section about the frog, I only learnt that they spawned 'in a greater variety of ponds than toads'. Other questions came to mind:

- What type of sites did they prefer in different situations?
- Exactly how deep was the water where they spawned?
- How many individuals constituted a typical breeding colony?

4.2.2 *Types of breeding site*

I realised that replies that I had received from the Breeding Sites Survey undertaken in 1970 gave me some of the answers. This enquiry (described in Section 2.2) indicated, for instance, that garden ponds had increased considerably in importance in Britain between the early 1950s and late 1960s. Information was considerably bolstered by unpublished data from Peter Tinning on breeding sites in London; and, in 1973, Mary Crichton of An Foras Forbartha (the Irish state environment agency) distributed questionnaires for me to schools in the Republic of Ireland asking for information on types of breeding site and spawn depth.

Results for type of site used for breeding in the early 1970s are shown in Table 4.1. The Breeding Sites Survey provided a general picture for Britain, probably with something of a rural flavour. At that time, 35% of frog breeding sites were in the following (sub-) urban situations: parks, commons, wasteland, schools and private gardens. The London data were collected by members of the London Natural History Society within a 20-mile (32 km) radius of St Paul's Cathedral; not surprisingly, 68% of frog sites were in such places. In contrast, Irish observations showed the importance of sites on agricultural land, marshland and moorland for the frog in that country. Garden ponds rarely featured as refuges for frogs in Ireland in that survey.

Table 4.1 Occurrence of frogs and toads in different types of breeding site in the early 1970s, as recorded in surveys for Britain, London and the Republic of Ireland (summarising data from Cooke 1975a). The toad is not native to Ireland.

	Common frog (% of reports)			Common toad (% of reports)	
Nature of site or surrounding land	Britain	London	Republic of Ireland	Britain	London
Garden ponds	20	26	4	16	20
Schools, large houses	1	10	Less than 1	3	7
Parks, commons, wasteland	14	32	2	16	47
Agricultural land	36	7	27	22	0
Woodland	5	3	3	4	2
Marshland, moorland	11	2	39	16	0
Roadside waterbodies	2	8	4	1	2
Streams, lakes	4	4	6	11	0
Artificial pits	5	4	3	9	18
Other sites	2	4	12	2	4
Number of sites	268	138	358	94	45

When compared with frog breeding sites in Britain, rather more toad sites were in marshland, streams, lakes and artificial pits, while garden ponds and waterbodies on agricultural land were reported to be relatively less frequently used. In London, 85% of toad populations were in ponds in parks, commons, wasteland and gardens or in artificial pits. While this was very basic information, it appears to have been the first time that data on where the species bred had been presented to show differences between frogs and toads and differences between situations.

4.2.3 *Depth where spawn was laid*

I was provided with data on water depths at spawn sites by Deryk Frazer, who kindly passed on to me a large amount of information collected by herpetologists during the period 1952–1959. Some of it was published in a series of papers in the *British Journal of Herpetology* in the 1950s, the last being Frazer (1956). The original depth data were provided in inches; the ranges are also shown in centimetres in Table 4.2. Schools in Ireland supplied information on frog breeding depths, which did not differ significantly from the British data, so are omitted from the table. Nearly half of frog spawning took place in water up to six inches (16 cm) deep, whereas more than half of toad spawning occurred in water in the range 7–18 inches (17–46 cm). For photographs of spawn sites of the two species see Chapters 2, 5 and 6.

Nearly 40 years after Deryk Frazer collected the original data on water depth, an analysis of habitat characteristics was published for about 3,000 British ponds, of which roughly half were in gardens (Swan and Oldham 1993). This was part of the programme on the widespread amphibians and reptiles in Britain commissioned with Rob Oldham by NCC and English Nature. One conclusion was that frogs liked all sizes of ponds, including small and shallow garden ponds: they were found in 85% of garden ponds. They did not mind ponds drying out and tolerated the presence of fish in garden ponds. Toads preferred larger, deeper ponds than were

Table 4.2 Information on depth of water where frog spawn and toad spawn was laid in Britain and offshore islands, 1952–1959 (summarising data from Cooke 1975a).

Water depth range		Percentage of observations	
Inches	Centimetres	Common frog	Common toad
Up to 6	Up to 16	48	13
7–12	17–31	23	32
13–18	32–46	13	27
19–24	47–61	7	13
25–30	62–77	6	6
31–36	78–92	2	5
37–42	93–107	0	1
43–48	108–123	1	1
More than 48	More than 123	1	1
Number of sites		189	82

normally found in gardens, especially those which did not usually dry out. They often occurred with fish. So differences were again noted between the two species.

Why do frogs select such shallow water for spawning? One explanation could be that shallow water tends to be warmer. Therefore, spawn laid there is likely to hatch faster and tadpoles will develop more quickly. In two of my garden ponds, frogs spawned consistently earlier and in greater quantity in the shallower, warmer pond (Section 5.3.3). Another reason could be connected with what eats what in a pond – and this is where fish enter the equation. Predatory fish, such as goldfish (*Carassius auratus*), tend to eat frog tadpoles rather than toad tadpoles (Cooke 1975a). If frogs breed in shallow water, this increases the risk of catastrophic loss of tadpoles if the pond dries out too early in the summer, but reduces the likelihood of a pond containing fish if it is prone to dry out occasionally. Based on a worst-case study, I calculated that a couple of goldfish weighing 50 g each would be capable of consuming all of the tadpoles from five clumps of frog spawn (Cooke 1975a). In a pond situation, however, at least a few tadpoles would probably survive in places where the fish found them difficult or impossible to catch (Beebee 1979). Looking at it from the point of view of toads, their tadpoles are distasteful to fish, but are liable to be eaten by tadpoles of the earlier-breeding frogs (Heusser 1970, Beebee 1979). By breeding in deeper water, toads may tend to avoid frog tadpoles eating their progeny even if frogs have spawned in the pond. Furthermore, by selecting a pond with deep water they are more likely to share the space with fish, which should reduce numbers of frog tadpoles. On the other hand, invertebrate predators are likely to be present at higher densities in more permanent waterbodies. The predator–prey web is complex, but hopefully Section 4.3 on newt predation and information in later chapters will shed more light on it.

4.2.4 *Colony size*

Turning to colony size, I defined a colony as animals breeding in one waterbody, unless there was likely to be a high degree of exchange of adults between two or more spawning sites. Thus, frogs spawning in a range of areas within a few hundred metres on St Neots Common (Section 5.2) would be regarded as a single colony. Information from streams and ditch systems were excluded. For the frog, colony size was calculated by multiplying the number of spawn clumps laid by two. Maxwell Savage may have been the first person to use this simple method in Britain in the 1930s and 1940s (Savage 1961). I appreciated the method's inherent uncertainties, but it was a start, and I was only looking to assign colony sizes to very broad categories. For toads, it was more difficult. Multiplying the maximum count of males by two was often employed, on the basis of missing some males but assuming males outnumbered females (e.g. see Gittins 1983a, Halley *et al.* 1996). Years later, I found this simple calculation worked reasonably well for number of toads in the water on St Neots Common (Section 6.2.2) but not at the Boardwalks Reserve at Peterborough, where it would have seriously underestimated numbers in the population (Section 4.4). Clearly, a reasonably high level of detectability is needed for the method to have any chance of providing a meaningful estimate. As is discussed later for newts at Shillow Hill (Section 10.3), the accuracy of counting depends on many factors: proportion of animals in the water, their behaviour, amount of vegetation, turbidity, time of day or night etc. In short, it was assumed that colony sizes for toads were probably underestimated in many cases.

Information on colony size for Britain was collated from replies to the Breeding Sites Survey, from information held by the national BRC, from my own records, and from other literature, correspondents and colleagues. Data were also available from the London Natural History Society and from the survey of Irish schools organised by An Foras Forbartha (Table 4.3). The only clear difference between reported colony sizes in different types of site was that colonies in garden ponds tended to be smaller, a fact indicated in Table 4.3. Because of the skewed nature of the datasets, mean colony sizes were consistently higher than their respective medians. For example:

- For the frog in garden ponds in Britain, mean colony size was 30 individuals and median size was about 10.
- For the frog in other sites, mean size was 100 and median size was 40.
- For the toad in garden ponds in Britain, mean and median sizes were similar to those for the frog, while for other sites mean size was 600 and median size was 60.

Despite the problems associated with estimating numbers of toads, 23% of their populations in Britain were reported to comprise more than 1,000 individuals, compared with only 2% for the frog. Maxwell Savage told me that colonies of more than 1,000 frogs occurred more frequently prior to 1950. The largest spawning in a single pond reported in his book amounted to 3,000 clumps in a site in 1948 (Savage 1961). Sites in London had fewer populations of greater than 100 frogs or toads than were found in the sample of sites in Britain. Derek Yalden (1967) commented that the largest known colony of frogs in London was composed of about 100 pairs at Ruislip Local Nature Reserve. Since the 1970s, there has been much further work on toad populations and, because they can be found in relatively large waterbodies, populations exceeding 10,000 animals have been recorded (Beebee and Griffiths 2000). Halley *et al.* (1996) commented that toads rarely seemed to colonise sites where the carrying capacity was less than 300 adults.

Table 4.3 Percentages of colonies of the frog and toad of different sizes reported from Britain, London and the Republic of Ireland (summarising data from Cooke 1975a). Statistics for garden ponds are given in brackets. The toad is not native to Ireland.

	% of common frog colonies (garden ponds)			% of common toad colonies (garden ponds)	
Colony size (estimated number of adults)	Britain	London	Republic of Ireland	Britain	London
Less than 10	30 (54)	41 (60)	34 (38)	35 (50)	31 (86)
10s	39 (35)	47 (40)	57 (54)	25 (38)	41 (14)
100s	29 (11)	12 (0)	9 (8)	18 (13)	28 (0)
1,000s	2 (0)	0 (0)	0 (0)	23 (0)	0 (0)
Number of colonies	174 (26)	74 (15)	326 (13)	57 (8)	29 (7)

4.3 Newt predation on frog and toad tadpoles

If you happen to be a small, tasty creature in a pond, you really should not wriggle in front of a newt. Yet when I put frog tadpoles in a tank with a crested newt, this is exactly what they did. They even swam through the newt's legs. The serious side of this small experiment was that I was assessing how many tadpoles a newt could eat, and I wanted to check if tadpoles took any evasive action. Their behaviour appeared unaffected: they ignored the newt apart from occasionally rasping at skin on its back. Newts are voracious predators, both on the developing embryos in frog spawn and later on tadpoles. In Switzerland, Hans Heusser (1971) had found that crested newts ate frog and toad tadpoles, whereas smooth newts took frog tadpoles but refused those of the toad. I decided to extend Heusser's study by setting up large plastic tanks outside containing water to a depth of about 10 cm, and with two terracotta flowerpots on their sides to act as refuges (Cooke 1974b; Figure 4.1). A single adult newt and 25 tadpoles of defined species and weights were added, and tanks were covered with plastic mesh. Trials lasted 48 hours, but tanks were checked during that time to ensure that no cannibalism occurred. There were five trials for each combination of newt species, tadpole species and tadpole weight.

For both smooth newts and crested newts, numbers of frog tadpoles eaten decreased as the tadpoles grew. When tadpoles were losing their external gills and weighing about 20 mg, crested newts ate an average of 10 in two days, while smooth newts managed a creditable nine. However, by the time tadpoles weighed 180 mg, crested newts took an average of four, but smooth newts, being smaller, only caught an average of 0.8. Crested newts could catch small numbers of larger frog tadpoles,

Figure 4.1 A predation trial with a female crested newt and frog tadpoles.

even up to the point of the metamorphic climax. The weight at which frog tadpoles became safe from smooth newts was about 260 mg. However, some smaller tadpoles had bitten tails showing they had been caught by newts, but had torn themselves free. What I did not do was test whether tadpoles with bitten tails were then more prone to being predated by smooth newts or other predators because their swimming ability had been affected. Later, in the United States, Raymond Semlitsch (1990) demonstrated that tail injury put tree frog (*Hyla chrysoscelis*) tadpoles at greater risk of predation by dragonfly larvae.

Toad tadpoles were almost totally rejected by smooth newts: single tadpoles were eaten by two newts out of the five tested with small toad tadpoles (averaging 14 mg in weight). With larger toad tadpoles, averaging 370 mg, none was taken in five trials. Crested newts took some toad tadpoles, but numbers were reduced when compared with the tests with frog tadpoles: an average of three toad tadpoles weighing 25 mg were eaten, decreasing to an average of 1.2 with larger tadpoles weighing 370 mg. No toad tadpoles were found that had been chewed, so they had not been caught and then spat out. The newts had recognised them as being distasteful either visually or by chemoreception.

I calculated that 12 crested newts could theoretically account for every tadpole from one clump of frog spawn. This did not include predation at the spawn/hatching stages. In reality, total loss would not necessarily result from such predation, just as when fish are present (Figure 4.2; Section 4.2.3), but frog population declines have been noted in gardens when high densities of smooth newts occur, largely it seems from predation at the earlier stages (e.g. Beebee 2007; Figure 4.3). Outcomes in a breeding pond are a delicate balance between species present and local conditions.

Figure 4.2 Risk of being caught in a pond by crested newts and other predators is reduced for frog tadpoles if the number of hiding places is high. Photo by Mihai Leu.

Figure 4.3 Predation by smooth newts can have a major effect on the hatching rate of frog spawn, especially in garden ponds. Photo by Mihai Leu. Figures 4.2 and 4.3 illustrate a potential point of confusion for an uninformed observer: female crested newts do not have a crest, whereas male smooth newts do.

Fish and some of the larger invertebrates prey on newt tadpoles and, in some cases, take adult newts. Thus, the extensive survey of Swan and Oldham (1993) on the requirements of widespread amphibian species concluded that crested newts preferred ponds without fish. Similarly, Beebee and Griffiths (2000) stressed the smooth newt's preference for fish-free pools. As will be become clear in later chapters, predatory fish often cause problems for crested newts (e.g. Sections 8.4.4, 9.3, 9.4 and 9.5).

As a result of undertaking these experiments on predation, and also the trial with poisoned, hyperactive tadpoles (Cooke 1971; Section 3.2.1), I understood better how newts hunted. Although this chapter focuses primarily on frogs and toads, this is a convenient point to digress and elaborate further. Newts appeared to locate a tadpole by sight, moved towards it and then directed a fairly inefficient snapping lunge. Only about one lunge in eight was successful. One of the crested newts collected for the trials seemed to be blind, so it was not used (Cooke and Fulford 1971). Its eyes were blue and opaque instead of brown and clear, and it did not respond to movement of a small object against the outside of its glass tank – such movement induced lunging in other newts. It ignored tadpoles unless they came within 1 cm, when it would lunge. Its success rate was one in three, but it seemed to have more difficulty than the sighted newts at catching large tadpoles. Malcolm Smith (1969) believed newts hunted by sight and smell, but if the blind newt was hunting by smell, it needed the tadpole to come very close. It seemed that this newt might be reacting to being touched by the tadpole or because it sensed turbulence in the water.

4.4 Establishment of frogs and toads in a newly created reserve

4.4.1 *Site creation, introduction and monitoring*

The charity Froglife has undertaken management and survey since 2015 at the Boardwalks Local Nature Reserve, which it has described as 'a hidden gem in the heart of Peterborough' (Figure 1.5). This site was said to contain populations of frogs, toads, smooth newts and a small number of crested newts, as well as grass snakes. It was no accident that this community of species occurred there, as in the 1980s large-scale introductions were made to the site, part of which was specifically designed for amphibians (Cooke and Oldham 1995).

The reserve was created in 1985 by the former Peterborough Development Corporation (PDC) beside the River Nene. It was 6–7 ha in size, with the western end having several shallow pools intended to be attractive to waterfowl and wading birds. The remaining land comprised the amphibian reserve: 16 pools were excavated, with the largest measuring 125 × 15 m and the smallest being 2 m across (Figure 4.4). In winter, the river sometimes inundated the reserve (Figure 4.5a), so many of the pools intermittently held fish. Much tree planting and seeding of the ponds and their edges was done by the PDC, and management was undertaken initially by the Peterborough Wildlife Group.

NCC headquarters had transferred to Peterborough just before the reserve was constructed, and, as I recall, I was asked for advice on establishing animals in the reserve. When I saw its size and the number of pools that had been created, I realised that it had enormous potential as an introduction site. In the 1980s there was a general problem of knowing what to do with animals rescued from developments. Releasing them into areas where the species already occurred might have caused numbers to exceed the upper limit the site could support, and areas where the species did not occur might be unsuitable. But here was a custom-made core site of several hectares with extensive reasonable habitat further afield that could support large numbers of amphibians.

Figure 4.4 The Boardwalks Local Nature Reserve, Peterborough, April 1986: (a) the eastern end, showing part of the boardwalk; (b) the western end, including the River Nene in the upper left of the photograph.

Figure 4.5 The Boardwalks reserve in the early 1990s: (a) flooding of the eastern end, December 1992; (b) growth of emergent vegetation in part of the sinuous waterbody seen in Figure 4.4b, September 1991.

I contacted Rob Oldham at Leicester Polytechnic to ask if he knew of animals being rescued. Fortuitously, he happened to be working where a large number of toads were due to be moved because planning permission had been given for development that would destroy their breeding site. The timetable fitted in perfectly with creation of the Boardwalks reserve. Toad spawn would be available in 1986, and adults would be collected and translocated in 1987. Had it been 10 years earlier I would have been unable to find much frog spawn to translocate, but by the late 1980s it was easier to find 'surplus' spawn in garden ponds. And I managed to rustle up a few smooth newts too. A decision was made not to attempt to introduce crested newts, because of the fish that might be present from time to time. No crested newts were seen during the monitoring period of 1986–1993, so the small number that have turned up recently have either colonised naturally or have been introduced. It was always a terrific place for grass snakes, so it is good to know that the population survived for at least 35 years.

Amounts of spawn and number of adult toads introduced to the site were: 200 clumps of frog spawn and toad spawn strings totalling approximately 500,000 ova in 1986; and 150 clumps of frog spawn and 5,911 adult toads (5,116 males and 795 females) in 1987. In addition, the following numbers of frog spawn clumps appeared to have been left at the site by the public: 8 in 1990, 4 in 1991 and 14 in 1993 (see Section 5.2.2 for the type of evidence that indicated spawn was introduced by other people). Adult toads introduced in 1987 were marked by toe-clipping.

Our intention was to monitor the site until either population levels stabilised or the introductions failed. We hoped to establish strong populations of both species, in other words a frog population of more than 100 adults and a toad population of more than 1,000 (Table 4.3). Rob Oldham was also interested in determining whether it was better to use spawn or adult toads as a means to establish a new colony. I monitored the frog population by counting and mapping spawn clumps, but monitoring toads was more complicated. Time did not permit mark–recapture estimates of population size except in 1988 and 1989. As a means of monitoring in the longer term, I counted adult males and pairs in daylight regularly through the breeding season, and we used peak counts as indicators of population change. No night counting of toads was undertaken, in part because of the commitment to evening work on other species and projects. In the early years, the entire edge of every pond could be readily accessed and water clarity was consistently good. With the passing years, though, counting became progressively more difficult because of the site and its vegetation maturing (Figure 4.5). Floodwater following river inundation was checked by wading (Figure 4.5a). I also introduced a system of 'counting' toad spawn. I used a cane to lift spawn strings to the surface to see them more clearly, and then estimated by eye the number of strings that might be in each mass, based on what a single string looked like. The technique was intended to make comparisons, not to be an absolute measure. Between 1988 and 1990, samples of 50–100 adult toads were caught at night, measured and checked for clipped toes. Those already marked had a dye-spot added to their skin; unmarked toads had a toe clipped to indicate a specific year. Toe tips removed in 1990 were examined to reveal the ages of the toads using a method pioneered by Agnes Hemelaar and Jan van Gelder (1980). Monitoring also informed conservation action, such as moving spawn to avoid desiccation or providing suggestions on how to manage the habitat. Because of other commitments, fieldwork stopped in spring 1993. Frogs were monitored that year, but toads were not. By then populations were judged to be reasonably stable.

4.4.2 *The frog population*

Froglet emergence was good in 1986, the first year of introducing spawn. Many sub-adult frogs were subsequently seen on the site in 1987, but there were fewer froglets from the introduced spawn that year. The first naturally laid spawn was in 1988, when 92 clumps were found (Table 4.4). A small amount disappeared, presumably having been taken by people. Evidently not everyone in Peterborough had enough spawn in their gardens. The observations were interpreted as indicating that many of the frogs from the spawn introduced in 1986 had matured and spawned in 1988. This outcome was expected (Section 5.3.2), but even so it was gratifying when it happened. If there were roughly 300,000 frog eggs added in 1986 and about 200

Table 4.4 Numbers of clumps of frog spawn introduced or laid at the Boardwalks, and the fate of naturally laid clumps, 1986–1993 (adapted from Cooke and Oldham 1995).

			Reasons for losses of clumps laid naturally			
Year	Clumps introduced	Clumps laid	Drying out	Taken by people	Failed to hatch	Total loss
1986	200	0	—	—	—	—
1987	150	0	—	—	—	—
1988	0	92	0	8	0	8
1989	0	162	0	6	0	6
1990	8*	121	12	10	0	22
1991	4*	147	13–16	12	0	25–28
1992	0	117	0	3	1	4
1993	14*	112	1	0	2	3

* Introduced by members of the public

adult frogs turned up to breed in 1988, this indicated a survival to maturity at two years of age of approximately 0.07%. Some animals may have moved away from the site, while others had not yet matured. The spawn count peaked at 162 in 1989, and changed relatively little during 1990–1993. Collection and desiccation were the main reasons why spawn failed to hatch (Table 4.4). Floodwater beyond the confines of the ponds occurred in 1988 and 1990 – and frogs used it for spawning. Had I not moved the spawn, losses in 1988 could have amounted to 88%. 1993 was the first year that no spawn was collected from the site – and 14 clumps were considered to have been deposited there by people.

4.4.3 *The toad population*

No adult toads were seen in 1986 (Table 4.5), so there was no direct evidence of any being on the site prior to the translocation. Some spawning occurred in 1987 after the adults had been introduced. This spawn and the spawn introduced in 1986 resulted in good numbers of toadlets being recorded in both years. In 1988, 64% of a sample of males caught at the Boardwalks had been marked in Leicestershire, as had 89% of a sample of females. We assumed that unmarked toads had developed from the spawn introduced in 1986. Body lengths of unmarked males were typically in the range 51–56 mm, while the marked ones, which were at least three years old, were mainly in the range 58–62 mm, so there was relatively little overlap between the two groups.

Another sample was caught in 1989: just 21% of males were marked toads from Leicestershire, which were aged at least four years; 2% of the sample had been marked at the Boardwalks in 1988; and the remaining 77% were unmarked. Some of the unmarked toads were 65–70 mm in length, and Rob wondered whether their large size indicated that local toads from beyond the Boardwalks had found the site. By 1990, 15% of the males were marked toads from the original translocation three years before. Age estimation from bone rings in clipped toes indicated that eight original males from Leicestershire were 5–7 years old. Previously unclipped

Table 4.5 Monitoring details for toads at the Boardwalks, 1986–1992 (adapted from Cooke and Oldham 1995). Spawn was introduced in 1986 and adults in 1987. Population estimates for males in 1988 and 1989 used a mark, release and recapture method. Tentative estimates for 1990–1992 were derived from peak counts.

Year	Peak count	Peak pair count	Index count for spawn laid	Estimated population size of males
1986	0	0	0	—
1987	No counts	No counts	30–50	5,116
1988	127	6	22–27	2,890 ± 1,420
1989	311	7	29–34	5,620 ± 4,140
1990	181	10	58–63	3,180–4,100
1991	328	34	100–105	5,430–7,010
1992	306	54	107–114	4,660–6,010

males were mainly aged 3–5 years. The two unmarked toads that were five years old confirmed that some indeed were neither from Leicestershire nor from spawn laid at the Boardwalks.

Counts of males appeared to have stabilised by 1992, but counts of pairs were still increasing. Spawn string index may have been stabilising by that year. This seemed to be the first time that anyone had attempted to quantify all losses of toad spawn based on an estimation of amount laid. Overall, spawn losses were similar to those of frog spawn, although none was lost in three of the six years 1987–1992 (Cooke and Oldham 1995). Roughly one-quarter was lost during that period to, in descending order of importance: desiccation, removal by people and failure to hatch normally (being covered by fungus).

Population estimates for male toads were derived by Rob Oldham and are shown in Table 4.5. His estimate for loss until 1988 of adult males translocated in 1987 was 64%, which included those that may have left the site as well as those that had died. Loss of all males between 1988 and 1989 was 39%, and it was 42% during the following year. Thus, after the first year, mortality decreased and/or site fidelity increased. New adult males formed 36% of the male population in 1988, 79% in 1989 and 85% in 1990. Peak counts of males in the ponds in 1988 and 1989 represented about 5% of the estimated totals.

The total number of toads exceeded our criterion for success of 1,000 adults. These seemed to forage over a considerable distance, with dead ones, thought to be from the Boardwalks, being found up to 400 m away. In view of the loss of 64% of the translocated toads in the first year, establishment was considered to be primarily due to recruitment from the spawn brought from Leicestershire in 1986 and the spawn laid at the site in 1987. Then, as females from these cohorts matured, the population stabilised. Rob suggested that introducing spawn to a site on three successive years should be a good method of establishing a new colony.

4.4.4 *Postscript*

Silviu Petrovan told me that when Froglife took over management of the Boardwalks site in 2015, ponds were neglected and few toads remained. Excessive growth of terrestrial and emergent vegetation shaded and choked the ponds. Within a couple of years, however, targeted clearance and re-excavation of two ponds had resulted in good recruitment of toads and the future looked considerably better for the amphibians. In addition, three large ponds were created a short distance away with toads in mind. When I visited the reserve area in December 2020, it was unrecognisable from how it had been some 30 years before. Growth of trees and scrub in the main pond area was remarkable (Figure 4.6a) and many of the original pools were heavily colonised by common reed (*Phragmites australis*), but still retained water. Other pools looked more promising (Figure 4.6b). Conserving important populations of frogs and toads requires regular monitoring and the ability to undertake whatever management is necessary. Assuming populations can look after themselves for long periods of time is a dangerous approach. Fortunately, there is such a range of waterbodies that some should continue to be suitable breeding sites, but clearly without regular management the site will not fulfil its potential. When the reserve was designed and constructed, insufficient thought was given to its very long-term

Figure 4.6 The Boardwalks reserve, December 2020: (a) a similar view to those in Figures 4.4a and 4.5a after the boardwalk had been renovated (photo by Rosemarie Cooke); (b) an open pool towards the western end of the site.

management. For instance, zealous tree planting in the 1980s is causing problems in the early decades of the twenty-first century.

4.5 Conclusions

These three investigations were very much a product of their time. The first two were from the early 1970s when little was known about many aspects of the natural history of our amphibians, and even very straightforward and/or brief investigations could provide useful information. Later work helped to demonstrate just how complex life in a pond could be in terms of the balance between species. Below are a couple of examples of studies in the later decades of the twentieth century that revealed what to my mind were astonishing facts. The second example is about crested newts, rather than frogs or toads, but this is a species whose numbers help shape events in frog ponds.

The first story concerns intra- and interspecific crowding effects of frog and toad tadpoles, a phenomenon which had been observed for many years without the causative agent being fully characterised, although algal-like cells seemed to be implicated (Heusser 1972). The agent and the mechanism were elucidated about 20 years later (Beebee 1991, Beebee and Wong 1992). Tadpoles that fed on detritus rather than by filter feeding were affected by the unicellular alga *Prototheca richardsi*. At times of food shortage large tadpoles produced greater amounts of these cells in their faeces. When small tadpoles ingested the algal cells, they were diverted away from food of high quality and concentrated on feeding on faeces. In this way, large tadpoles were able to retain more of the better food at times of low availability and grow at the expense of smaller animals.

The second focuses on the survival of crested newt embryos (Horner and Macgregor 1985, Macgregor 1995). Newt cells carry 12 pairs of chromosomes. Millions of years ago, an exchange of genetic material occurred between a pair of chromosomes, with the result that they became unequal. Subsequently, only embryos with the two different forms of the chromosome developed normally. The complementary nature of one of each type was needed to make a viable newt. Those embryos with identical versions of the chromosome died during the first few days of development: in one case each chromosome of the pair was deficient, while in the other it had duplications. The overall effect of this condition was to eliminate half of the reproductive output of the species. It is remarkable that this defect evolved and then persisted for so long, especially as the species usually breeds in waterbodies that also contain related species of newts whose potential for reproductive success is 100%.

As regards population ecology, a range of detailed studies were under way by the 1980s, some lasting a number of years and usually focusing on a single species, particularly common toads. Our study at the Boardwalks reserve (Cooke and Oldham 1995) was unusual in that it tackled two species simultaneously and was oriented more towards conservation than towards basic ecology. Despite being neglected for a number of years, the site still supports an important community of amphibians and reptiles and provides an interesting wild corner in the city of Peterborough. I should add that translocation is no longer recommended, unless it is absolutely necessary, because of the risk of spreading disease or invasive alien plants (e.g. see Section 5.5). Such concerns were not considered to be significant in the 1980s, but perhaps they should have been.

CHAPTER 5

Spawn story

5.1 Introduction

This chapter on frogs deals with three related topics. It begins at St Neots Common, which must now be the 'ancestral' site of numerous garden populations of frogs in the former county of Huntingdonshire. Like many other people, I collected frog spawn from the common in the distant past, and although most of it was used for research of various kinds, St Neots genes will have been to the fore in the populations of frogs that I established in gardens of the houses where Rosemarie and I lived in Sawtry, Ramsey and Bury (Figure 1.5) – and this is the subject of the second part of the story. I also moved spawn around, mainly to ponds in other people's gardens. That was before the risks of spreading disease were understood (Section 3.6.2), and such actions are no longer recommended. The final part is a description of frogs colonising two of the local toad ponds: Bury Pond and Horse Pond in Ramsey. I was deeply interested in our local toads over a period of about four decades (Section 6.3), and, while observing them, I was able to monitor the changing fortunes of frogs. The three parts illustrate what has occurred more widely, with frogs in traditional sites declining, but sometimes doing well in gardens, and in some cases spilling out or being moved out to colonise larger suburban ponds. Life, though, is still not easy for frogs.

5.2 St Neots Common

5.2.1 The site and surveillance methods

It must have been in 1970 that someone at Monks Wood drew my attention to St Neots Common and its frogs. By then, I was well into studies on the possible effects of pesticides on frogs, but I was still having difficulty finding many, or indeed any, frog populations nearby. St Neots is 15–20 km south down the A1 from Monks Wood (Figure 1.5), so was not especially close, but the common was an example of a more traditional landscape than was evident in 1970 in most of the county – and it certainly had a good frog population. Initially, it was somewhere to go for spawn. But I soon realised that it was also an interesting place to undertake low-key surveillance with no designated end point (Cooke 1985a).

The common is an SSSI and the citation was revised in 1971; it stated that the site was one of the last traditionally grazed alluvial grasslands in the area, having a diverse habitat with ponds, ditches and sallow carr. The citation mentioned that flourishing populations of both frogs and toads occurred on the common, and that frogs were locally rare. In short, it was what I was looking for to demonstrate and learn how frogs lived in such a landscape. This section focuses entirely on frogs, and the fortunes of the toads are discussed in the next chapter, in Section 6.2.

A review of the history, management and conservation interest of the common was published by Peter Walker (2008) on behalf of the local Fauna and Flora Society. The common is 33 ha in size and is divided into two parts by a hedge (Figure 5.1). The area in the west beside the River Great Ouse is Lammas Meadow and the part to the east, between the hedge and Common Road is Islands Common. On a facsimile Ordnance Survey map from the 1830s, published by Cassini, Lammas Meadow was called Island Meadow and Islands Common was Island Common, which then extended further to the east beyond the road.

As regards amphibian breeding habitat, a ditch runs through the northern part of Islands Common and then down Lammas Meadow parallel to the river. A lagoon occurs in the north of Islands Common and this often floods seasonally, as does an area to the west of the ditch on Lammas Meadow (Figures 5.2 and 5.3). Frogs spawned in all of these waterbodies.

My objectives were simply to get to know frogs better and record the fortunes of the population for as long as possible. Current researchers probably struggle to understand the flexibility we enjoyed 50 years ago! The relaxed approached was, however, more constrained after 1973 when I worked for ITE and again from 1978 with NCC. The study was long term in that it started in 1971 and continued until spring 1984; it only stopped because the NCC office in which I was then based moved from Huntingdon to Peterborough, which meant it was more difficult for me to escape to the common for an hour or two. It was not an intensive or particularly scientific study. Islands Common was very disturbed by the public, so equipment could not safely be left there. Most of the fieldwork was undertaken during the spring, assessing amount and survival of spawn, particularly in relation to collection by people.

I must have been one of the first people in this country to use counts of spawn clumps as a measure of the changing number of frogs breeding at a site in a long-term study. Ashby (1969) counted spawn clumps over four years for a captive population in a garden. Savage (1961) undoubtedly counted spawn in his study area on the Middlesex/Hertfordshire border, as he mentioned this as a means of estimating frog density. In his book, however, he presented an impressive table showing whether or not frogs bred in each of his 92 study ponds during 1929–1938 without providing spawn counts.

Each spring, the common was typically visited twice per week from before the first spawn was laid until after spawn had hatched. Thereafter, visits were timed to check whether spawning sites dried out before froglet emergence. Because of other commitments, little attempt could be made to estimate emergence each year, but froglets and juveniles were noted when seen Clumps which disappeared between visits were usually assumed to have been collected – an observation often confirmed by finding human footprints on the bank, fragments of spawn in the grass or disturbed spawn left in the water.

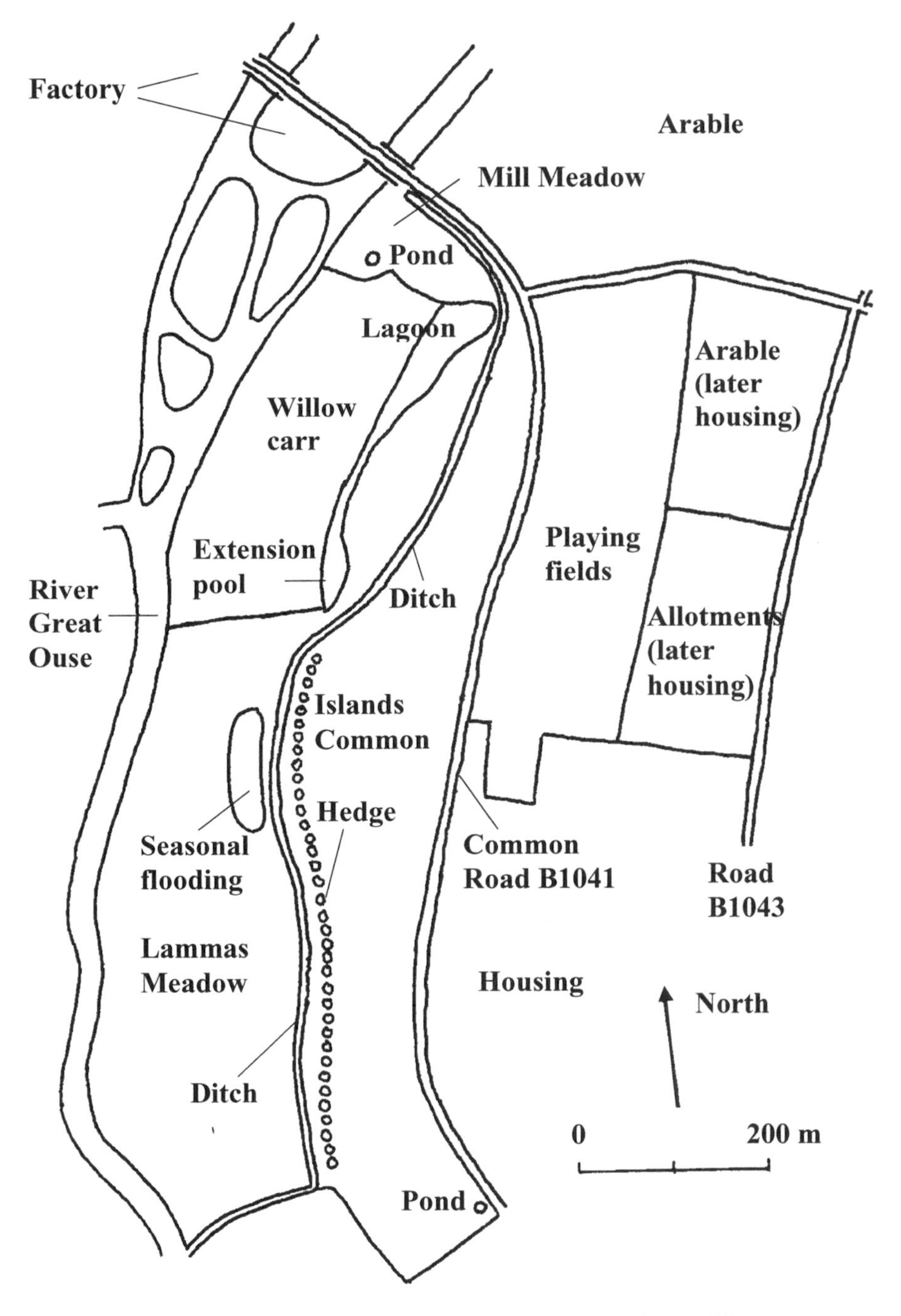

Figure 5.1 A sketch map of St Neots Common and part of the surrounding area, as it was during the study in 1971–1984.

Figure 5.2 The northern end of Islands Common, St Neots, showing typical wet spring conditions during the 1970s. The reed at the rear of the lagoon can be seen in the distance on the left. The waterbody in the foreground is the seasonally flooded extension pool of the lagoon. This photograph can be compared with Figure 6.5, which shows the same view in 2005.

Figure 5.3 The northern end of Lammas Meadow, St Neots, spring 2005. The main ditch is to the right, with seasonal floodwater to the left. Boats on the river can be seen far left, with the wet woodland in the distance.

In some of the later years of the study, spawn was found that was considered to have been laid elsewhere and then brought to the common. When initially found, this was often part-developed, covered with silt, broken up, unattached to vegetation and in areas that frogs did not usually select for spawning. This presumably occurred because some people felt they had too much spawn in their garden ponds, possibly as a result of taking spawn from the common several years before.

5.2.2 *Surveillance data*

Numbers of spawn clumps laid on the common are shown in Figure 5.4. I did not realise that frogs spawned on Lammas Meadow until the summer of 1972, so there were no counts for the meadow in 1971 or 1972. In addition, there was an isolated pond at the extreme southern tip of Islands Common where a few clumps were occasionally found. Frogs also spawned in a pond in pasture called Mill Meadow just to the north of Islands Common. This field was fenced off throughout the study, and it was only in 1981 that I discovered frogs spawned in the pond, with totals building up to a maximum of 109 clumps in 1983. Spawn in these sites is not included in totals for the common in Figure 5.4.

Average numbers of clumps produced annually were 136 on Islands Common and 58 on Lammas Meadow. The total number of clumps counted on the common during the entire study was 2,607. What happened to the spawn is shown in Table 5.1, expressed as percentages of clumps remaining after I had collected a total of 52 (2.0%) for research of various kinds. Amounts taken by people were higher on Islands Common, which is closer to the road and the car park (Figure 5.5). Collection tended to be worse when fine weather coincided with a weekend. On one such weekend in 1974, part-way through the season, 57% of clumps on Islands Common disappeared. In percentage terms, amounts of frog spawn taken by people decreased

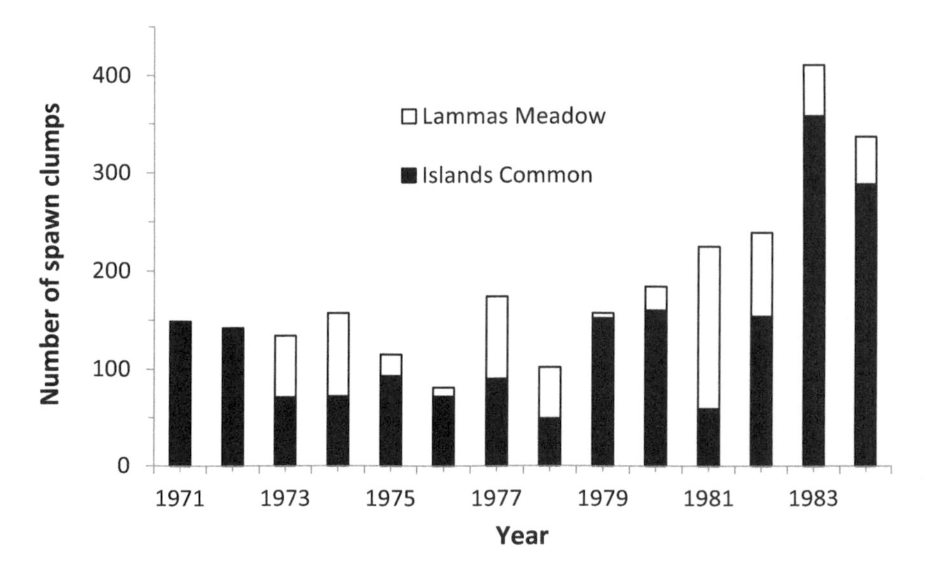

Figure 5.4 Total number of spawn clumps laid on the two parts of St Neots Common, 1971–1984 (updating and adapting Cooke 1985a). Spawn on Lammas Meadow was not counted in 1971 or 1972.

Table 5.1 Average percentage of clumps lost during the spawn phase on the two parts of St Neots Common up to 1984. Also shown are percentages of clumps which hatched normally in areas that subsequently did or did not desiccate during the tadpole phase (averages of data up until 1983).

	Percentage of spawn clumps	
	Islands Common 1971–1984	Lammas Meadow 1973–1984
Taken or destroyed by people	19.0	4.8
Lost to desiccation, flood, animals	1.6	13.1
Failed to hatch normally	1.2	5.4
Hatched normally	78.2	76.6
Desiccated during tadpole stage*	13.6	28.8
Hatched in suitable site*	55.6	47.3

*Data up to 1983

Figure 5.5 The ditch on Islands Common was one of the main spawning sites for frogs and was an accessible location where spawn was vulnerable to collection.

steadily after 1974 (Figure 5.6). In 1981 and 1983, more spawn was deposited in waterbodies on the common than was taken.

The percentage of spawn lost to desiccation, flood and animals was higher on Lammas Meadow. In 1973 and 1974, considerable numbers of clumps were laid in receding floodwater and then stranded. I moved as much spawn as possible into the ditch, but have recorded it here as having desiccated. In 1979, the ditch on Lammas

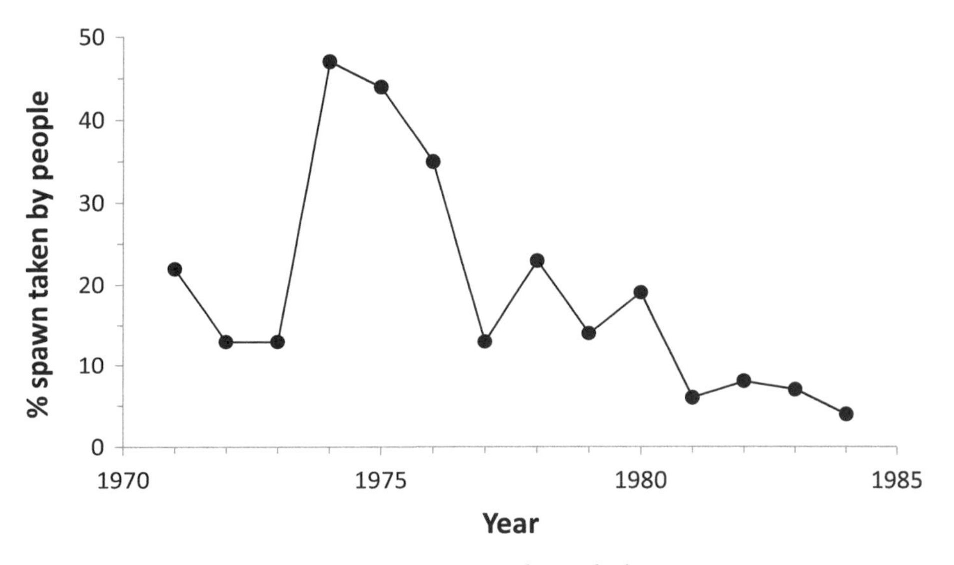

Figure 5.6 The percentage of frog spawn clumps laid on St Neots Common that was recorded as taken or destroyed by people, 1971–1984.

Meadow was dredged, and a low line of spoil prevented rainwater or floodwater from running off into the ditch, so presenting frogs with what may have seemed to be an ideal shallow pool during the breeding seasons of 1980–1983. Spawn laid there survived to hatch but the pool dried out during the three tadpole seasons of 1980–1982, resulting in large losses. In 1979, spawn was carried away by floodwater and most was never seen again. Occasionally spawn was broken up by cattle wading through shallow water or by mallard (*Anas platyrhynchos*) dabbling. Some cattle-affected spawn was taken back to Monks Wood for study, but less than 50% hatched. At St Neots, I defined a hatch rate of less than 80% as being unusually low for a spawn clump. Failure rate was higher on Lammas Meadow, but some of the loss referred to spawn that had partially desiccated in receding floodwater or because of rapid growth of vegetation lifting spawn out of the water. Ova in such spawn were often covered with fungus. Overall, percentage of spawn hatching satisfactorily was similar on the two parts of the common, although the importance of factors for losses varied.

Data on desiccation losses were not recorded in 1984, as the study was terminated after the spawning season. Losses during the tadpole phase were higher in Lammas Meadow because of drying out of the shallow pool separated from the permanently wet ditch. For the whole period, it was estimated that 57% of the spawn on Islands Common and 47% of the spawn on Lammas Meadow hatched satisfactorily in sites that remained wet throughout the larval stage. The only year when there were no desiccation losses at the tadpole stage was 1983. In broad terms, roughly a quarter of the spawn failed to hatch, and a quarter hatched in places that dried out before froglets could emerge.

5.2.3 *Interpreting the data*

In trying to understand why the population was changing, the elephant in the room was of course what happened to tadpoles that hatched in places that did not dry out – and in particular how many were predated. I have mentioned mallards already; and populations of smooth newts and predatory fish, such as ten-spined sticklebacks (*Pungitius pungitius*), were sometimes very large. And there was presumably some contribution from predatory invertebrates.

In order to try to obtain information on levels of predation, one day in April 1976 I released 1,000 frog tadpoles that had just lost their external gills into two areas of the ditch, one on Islands Common and one on Lammas Meadow. These were more than 50 m away from spawn sites in 1976, but were close to places where frogs had spawned in previous years. No tadpoles could be caught at the introduction sites prior to the releases. After 30 days, I returned to try to catch tadpoles to carry out a capture–mark–recapture exercise. The aim was to examine how well tadpoles survived in two of the main spawning sites on the common, in which water generally persisted through the summer. On Islands Common, only four tadpoles were netted in 30 minutes and three of those had bitten tails. The numerous netted sticklebacks were presumed to be responsible for the dearth of tadpoles and for the state of their tails. It was not worthwhile marking the tadpoles, as survival was clearly very low; by that time, tadpoles had small hind legs, but the ditch dried out before completion of metamorphosis. In the ditch on Lammas Meadow, 34 tadpoles were caught in one hour within 20 m of the release point, including 31 within 10 m. Again many sticklebacks were caught. The netted tadpoles were temporarily turned a pale red colour by immersing them in a dye called neutral red, using a method I published later (Cooke 1978); they were then released back into the ditch where they had been caught. The following day, tadpoles were recaptured and the proportion of dyed ones indicated a population of about 140 close to the original release site, a survival of 14%. Sadly, this area also soon dried out in the long, hot summer of 1976, so none of the introduced tadpoles survived to metamorphosis.

That year was the worst for desiccation losses throughout the entire study: only three clumps hatched in sites where water persisted through to metamorphosis. Total spawn production in 1976 was the lowest of the study period at 81 clumps. The previous year had also been dry. In contrast, good emergence was likely to occur in the wetter years when the more ephemeral breeding sites retained water through the summer; then tadpoles might be able to enjoy less pressure from predators, particularly fish. Such conditions occurred in 1983 but that was too late to have affected numbers of clumps counted during the study.

Spawn production on the common indicated that the population level changed little during the study from 1973 until the late 1970s. However, a decrease may have occurred prior to 1973 because amounts of spawn on Islands Common in 1971 and 1972 were comparable to totals on both parts of the common in the rest of the 1970s. There were large numbers of small frogs sheltering in the ditch on Lammas Meadow in the summer of 1972, suggesting good recruitment in 1971 and/or 1972. There was, however, no immediate recovery; and totals remained low, perhaps restricted by collection of spawn and by the hot, dry years in the middle of the decade. Numbers of spawn clumps then increased from the late 1970s and had more than doubled by 1983; that year, an additional 109 clumps were counted in the pond in Mill Meadow to the north. The breeding population may then have been in the region of 1,000. Unlike toads (Table 6.1), no dead frogs were recorded on Common Road (although smooth newts were found dead in three years), indicating little or no migration to and from the east. Unless frogs crossed the river, they appeared to be more or less confined to the common at a density of approaching 30 per hectare.

The increase from the late 1970s will have resulted from good recruitment in certain years. Several years stand out as likely to have been unusually productive. First, 1977 followed the drought in 1976, which led to the drying out of nearly all of the waterbodies on the common and probably the widespread reduction in abundance of predatory fish and invertebrates. In 1977, most of the spawn was laid in the ditch on Lammas Meadow, and this was likely to be the most productive year on that part of the common. This process was repeated the following year, 1978, but less spawn was laid in the ditch and predator numbers were presumably building up again. Another potentially good year was 1981, when people dumped 51 clumps on the common in areas which did not desiccate. If frogs matured from these summers in two or three years then this could explain much of the increase noted in Figure 5.4. Around that time, St Neots Common was well known for its frog population regionally, not just locally. In April 1982, when the respected naturalist Ted Ellis wanted to film an item on frogs for his weekly slot on regional television, he approached me about visiting the common – I met him there and showed him the various spawning sites (Figure 5.7). It may be a reflection of the rarity of frog-rich sites like St Neots Common that Ted Ellis and the crew ventured over from Norfolk to film the item.

5.2.4 *Postscript*

It is now about 40 years since I was a regular visitor to St Neots Common. I returned there on five occasions in spring 2005 to check how the frog population was faring,

Figure 5.7 Ted Ellis being filmed for television in April 1982, pointing out frog spawn in part of the lagoon on St Neots Common.

but managed to find only 22 clumps of spawn in the usual sites; I thought perhaps the proliferation of aquatic vegetation (e.g. compare Figures 5.7 and 6.4) made the site less suitable and any spawn harder to see. In addition, there were 12 clumps in the pond at the southern end of Islands Common, but the pond in Mill Meadow was dry. In 2005, the first spawn was seen on 22 March. Most recently, I visited twice in 2021, on 17 March and 7 April. No evidence was seen of frogs or their spawn in the usual sites: the pond in the south held no spawn, and the pond in Mill Meadow seemed to have disappeared. In 2021, the common looked much as I remembered it in 2005. Ominously, there was no mention of frogs or toads on English Nature's notice board at the main entrance to the site. Roger and Sarah Orbell have produced annual reports on amphibians and reptiles for the Huntingdonshire Fauna and Flora Society during the period 2009–2020: no records were quoted for frogs on St Neots Common during those 11 years. I am left with the conclusion, therefore, that the frog population died out during the early years of this century. Possible reasons are discussed in Section 5.5.

5.3 Garden ponds

5.3.1 *Introduction to garden ponds*

During the 55 years that Rosemarie and I have been married, we have lived in two flats and four houses. The flats were in Reading in Berkshire, whereas the houses were all in the Huntingdon–Peterborough area. Each house had a garden, and in each garden I constructed a pond or two with frogs in mind. I have been fascinated by garden ponds and their inhabitants for as long as I can remember. So, as soon as we owned a garden, some sort of pond was an almost essential addition, bringing an extra element of interest and delight into everyday life, which was enhanced further by my professional curiosity about amphibians.

This account is primarily about ponds in the second and third houses, because we have had only very small pre-formed plastic pools in the other two. Our son Steven has referred to the current pond as a 'puddle', but it has had frog spawn occasionally and serves as a refuge for adults and juveniles in the summer and the winter. The main reason I have never installed a larger one here is that the garden is surrounded by mature trees and there is a continuous fall of twigs, branches, leaves, acorns, maple seeds, bird droppings and unidentifiable debris, and I need a token pond that is simple to manage.

The very first pond, in Sawtry, was a similar size, but that was designed to fit in with our tiny garden, most of which was taken up by an over-ambitious rockery. In the early 1970s I introduced frog tadpoles, smooth newts and even a pair of great crested newts, knowing very little about the requirements of the last species at the time. The smooth newts bred successfully despite the small size of the pond. A few frog tadpoles metamorphosed and there was some dispersal of frogs to other gardens in the estate by the time we moved out.

5.3.2 *Ponds in the second garden*

Our second house was a new-build in a small estate in Ramsey on the edge of a wartime airfield that was gradually being reclaimed for agriculture (Figure 5.11). When we moved there in 1973, our sons were three years and 18 months old, and we decided a sandpit was probably more appropriate and safer than a pond. By 1978, the boys had outgrown the sandpit and were both interested in wildlife, so the sandpit was converted into a pond (Figure 5.8a) and an even smaller version was added close to it. Spawn from St Neots was introduced in 1978, and froglets were seen through the autumn. Subadults were around in 1979 and frogs returned to breed in 1980. These were the first adult frogs I had seen within 1 km of our garden in the seven years we had lived there. That year, Trevor Beebee had a letter published in the *British Journal of Herpetology* about how long it takes frogs to reach sexual maturity. He recounted that, in four instances of spawn being introduced, adults returned to breed for the first time two years later (Beebee 1980a). I realised that I had an opportunity to know my frogs individually and follow their growth over a year or two if they survived. Because of the extreme rarity of frogs locally, I could be reasonably certain that they were all two years old.

Frogs were caught at night in the main pond. I concentrated on males, as any females tended to be in amplexus. They were weighed, length was measured and they received individual toe-clips. I also drew their spot patterns and found I could recognise them in the pond without catching them. Silviu Petrovan has told me recently that he has modernised this technique by photographing back patterns and processing the results on his phone. As far back as 1932, Maxwell Savage (1961) used paper labels with numbers to identify individual frogs in a breeding pond at Mill Hill in north London. He described standing among them, having a roll call and writing down the behaviour of specific frogs as they swam about.

Figure 5.8 Three of our garden ponds in Ramsey: (a) the larger pre-formed plastic pond in the garden of our second house; (b) the concrete pond in the garden of our third house; (c) the plastic pond in the garden of our third house.

Observations up to spring 1981 were included in my letter to the *Journal* in response to Trevor's original one (Cooke 1981b). Between 1980 and 1981, average weight increased from 30 g to 40 g, and average length increased from 59 mm to 67 mm. Average weight loss of the eight males between February and May 1980 was 28% (range 24–38%); the breeding season is a time when they do not feed.

Just because Trevor Beebee and I had found that spawn introduced to our gardens resulted in breeding after two years, it does not follow that all frogs in a population attempt to breed at two, neither does it mean that this will happen in an established population where there will be competition from older frogs. As summarised by Trevor and Richard Griffiths in their New Naturalist volume, frogs first breed in this country when they are two or three years old and are at least 50 mm in length (Beebee and Griffiths 2000).

Our final full year in that house was in 1982. I decided it would be interesting to estimate tadpole survival in the larger pond in the presence of smooth newts, which had colonised. Just before metamorphosis, I caught a sample of tadpoles and dyed them with neutral red (as described in Section 5.2.3). They were replaced in the pond and given time to mingle with the undyed tadpoles. Another sample was caught and, from the proportion of red ones, I estimated the population to be about 320. A total of seven smallish clumps of spawn had been laid in the spring, containing perhaps about 7,000 ova. Survival from spawn to large tadpoles was therefore calculated to be in the region of 5%. There was a high incidence of bitten tails, suggesting considerable smooth newt predation earlier in the season. Also blackbirds (*Turdus merula*) had fished tadpoles from the shallows, presumably to feed to their young.

5.3.3 *Ponds in the third garden*

On a personal and financial basis, 1982 and 1983 were complicated years. My father had died suddenly in 1981 and we had decided to pool our resources with my mother's so all of us would move into a larger house, in which she could live independently. For a time in 1982 and 1983, we had two houses about 500 m apart, and we did not all reside in the new one until later in 1983.

The new property in Ramsey came with a traditional circular concrete pond in a shady part of the garden: 2.1 m in diameter with a uniform depth of 45 cm (Figure 5.8b). There was no sign of frogs in the garden in 1982, so in March 1983 I constructed a submerged shelf in the pond to try to make it more attractive to frogs should there be any around. There were at least two, because a single clump of spawn had appeared by the following day. Early in 1984, I constructed a new pond with a plastic liner 65 m from the concrete pond. Its dimensions were intended to be irresistible to frogs: it was 2.9 × 1.8 m with a maximum depth of only 25 cm and a water volume of 500–600 litres (Figure 5.8c). Initially, it was in a sunny position, but it became progressively more shaded with the growth of newly planted trees and shrubs. Five clumps of frog spawn, taken from the garden of our previous house (i.e. St Neots stock), were introduced to this pond in the spring of 1984. That year, no clumps were laid naturally in either pond; but in 1985 five clumps of spawn were laid in the plastic pond and one in the concrete pond.

I later discovered there were ponds in two of the adjacent gardens. There was no evidence of frogs being introduced to either and their populations remained very small. One of these ponds was destroyed in the winter of 1992/93. A third pond was created in another neighbouring garden in 1999: the maximum amount of spawn recorded was 25 clumps, but tadpole survival was poor because many large koi carp (*Cyprinus rubrofuscus*) were kept there. I have often wondered about the origin of the frogs that spawned naturally in our third garden in 1983 and 1985. Other people in the town will have been introducing frogs to their gardens, and it is likely that a small number of adults from such sources were associated with the two ponds in neighbours' gardens, and that a pair of these frogs produced the single clump in the concrete pond in 1983. On the other hand, there was a slight chance that the single clump in 1983 and the six clumps laid in 1985 could have been produced by frogs or froglets migrating 500 m from our second garden, where there had been successful reproduction since 1978. An alternative explanation for the spawn laid in 1985 was that adults had developed from the single clump laid in the concrete pond in 1983, but no emergence had been recorded.

My principal involvement with the two ponds in our garden was counting spawn production in the spring (Figure 5.9) and following the tadpoles through to metamorphosis. Again, there was no defined end point – I wanted to follow how and why the frog population changed over the years. From 1983 until 2007, when we moved again, the concrete pond had spawn laid in it in 19 years. Amounts varied up to a maximum of 23 clumps in 1992, when the plastic sheet had to be replaced in the other pond and the frogs appeared not to appreciate the change. That was the only year in which the concrete pond came close to having the same amount of spawn as the plastic pond. As is discussed later, the concrete pond was cooler and deeper than the plastic pond (Section 11.5), and this evidently rendered it less attractive to breeding frogs. No froglets were observed emerging from the concrete pond during the 25-year observation period. Tadpoles could survive in the water, but it is likely

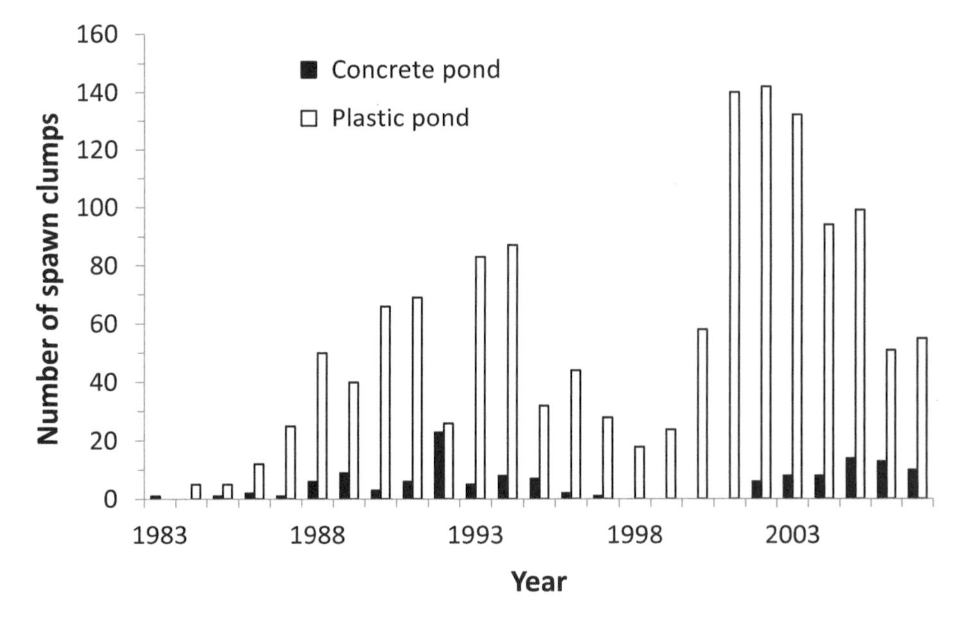

Figure 5.9 Numbers of frog spawn clumps in the concrete and plastic garden ponds, 1983–2007 (updating and adapting Cooke 2003).

that cool water slowed development, leading to greater predation of embryos and tadpoles by newts, and by the goldfish that were periodically kept there.

Numbers of clumps laid in the plastic pond varied cyclically, depending at least in part on past froglet emergence. In some years very large numbers of clumps were produced for such a small pool. Not only was the pond itself attractive to breeding individuals, but it seems frogs are also drawn to suitable sites where others have already spawned (Cooke 1985a; Figure 5.10). Froglet production in the plastic pond and changes in population size can be summarised as follows:

- During 1984–1986, emergence was good each year, allowing the population to build up.
- During 1987–1996, emergence generally failed or was sometimes poor, and the population declined from 1994 to 1998.
- During 1997–2001, emergence was good each year and the population recovered quickly to 2002.
- During 2002–2007, emergence was poor each year and so the population decreased again.

Why then was there such poor recruitment in the plastic pond during 1987–1996 and again in 2002–2007? There appeared to be no single reason. In 1990 and 1991, the pond had a slow leak and dried out in both summers. I blamed our sons, accusing them of throwing homemade spears into the pond, but was forced to apologise when I realised that the pairs of holes in the liner were probably caused by a heron trying to catch frogs.

Failure of more than 10% of clumps to hatch normally, often with a covering of fungus, was recorded in 1994, 1995, 1996, 2002, 2003 and 2007. In each of these years, more than 30 clumps were laid in the plastic pond, with an average of 87. Dead and dying tadpoles were seen in 1989, 2002, 2003 and 2005. If distressed tadpoles were transferred into water taken from the concrete pond, they sometimes recovered. Tadpoles completely disappeared in 1987, 1988, 1992 and 1993. Newt predation was not thought to have been as severe as in the concrete pond; in 2002, for instance, newt breeding failed in the plastic pond, perhaps for the same reasons that afflicted frog spawn and tadpoles.

Figure 5.10 Frogs seem attracted to breeding sites where others have already spawned. Photo by Mihai Leu.

Many of the problems may have been due to the shallow nature of the pond (1) leading to a greater likelihood of spawn being frozen at or above the surface and (2) later in the summer exacerbating the effects of high temperatures and overcrowding on the tadpoles (Cooke 2003; Figure 11.2). Excluding the two years when the plastic pond leaked, recruitment was good in 6 out of 11 years when fewer than 30 clumps of spawn hatched; whereas emergence was good in only 2 out of 11 years when more than 30 clumps hatched. So there was some evidence of the effects of overcrowding. While the pond may have been attractive to breeding frogs, it may not have been conducive to safeguarding the spawn and tadpoles after there had been relatively large gatherings of adults. Because of the concern about overcrowding, 223 clumps of spawn were moved from the plastic pond to sites elsewhere during 1990–1994. In addition, 17 clumps were translocated from the concrete pond in 1992–1993. I was repeating the actions of those people who had taken spawn to St Neots Common during the early 1980s.

It was amazing that the population survived the, at best, poor recruitment for the 10-year period 1987–1996. Indeed, spawn counts in 1993 and 1994 were relatively high. Survival may have been helped by immigration from neighbouring ponds. No significant die-offs of adults were recorded or reported by neighbours. Very few dead frogs were seen within the garden, and I cannot recall any road casualties. One dead frog was seen in 2003 that may have been a victim of ranavirus (Section 3.6.2). These observations hammered home again that so much can and does affect frog recruitment that it is difficult to determine cause and effect even when it is occurring in your own garden.

Although garden ponds were known to be increasingly important refuges for frogs, not much in the way of detail had been published about them before 1980 apart from articles by K. R. Ashby (1969) and Trevor Beebee (1979). Trevor was later to write several articles about the aggregation of ponds and amphibians in his garden in Brighton, where he lived for many years. When he looked back on his first 30 years, he described how the frog population had first boomed, but then declined (Beebee 2007). He blamed predation of smooth newts on frog embryos in spawn in the main frog pond. In 1991, he introduced three-spined sticklebacks to the pond; they apparently predated the newt tadpoles rather than the frogs, so the newts declined, allowing the frog population to stabilise and then increase slightly.

5.4 Frogs in toad ponds

5.4.1 Frog colonisation of Bury Pond

Bury Pond was for a time a local Wildlife Trust reserve because of its toad population. The pond started out life as a lagoon at a vegetable washing and processing plant. Toads bred there in the 1980s. When the plant and the surrounding land were due to be cleared for housing, part of the lagoon and the fringing land were retained as a small reserve for the toads (Figure 6.9a). The full story of the toads is related in the next chapter. The locations of the principal toad ponds, and also of our two gardens which featured in the previous section, are shown in Figure 5.11.

I visited the site several times while the plant was still in operation and usually saw evidence of toads, but never recorded frogs. Bury is a small village which was once a kilometre or so from the town of Ramsey. Now both settlements have expanded and there is no longer any obvious physical separation. When we moved to Ramsey in 1973, toads were numerous but frogs, as far as I was aware, only lived in one garden in the centre of the town. I started regular spring monitoring at Bury Pond in 1988, and the plant was pulled down that May, with the pond being reduced in size. Monitoring continued until 2011, by which time it was clear that the toad population had declined to just a few individuals. The pond contained significant populations of smooth newts and sticklebacks.

The first two male frogs were recorded in the pond on the night of 31 March 1989; on the same occasion, 252 adult toads were counted. Those two were the only frogs seen at the site in 1989 and no spawn was laid that year. The following spring, on 21 February, 10 frogs were recorded, including two mixed pairs with male frogs on female toads. By 25 March, 12 clumps of frog spawn had been laid. The population built up to a peak of 283 clumps in 2002, and frogs produced 68 clumps when annual monitoring stopped in 2011 (Figure 5.12). On an isolated visit on 15 March 2021, 26 clumps were counted including fresh spawn, suggesting there might have been more to come.

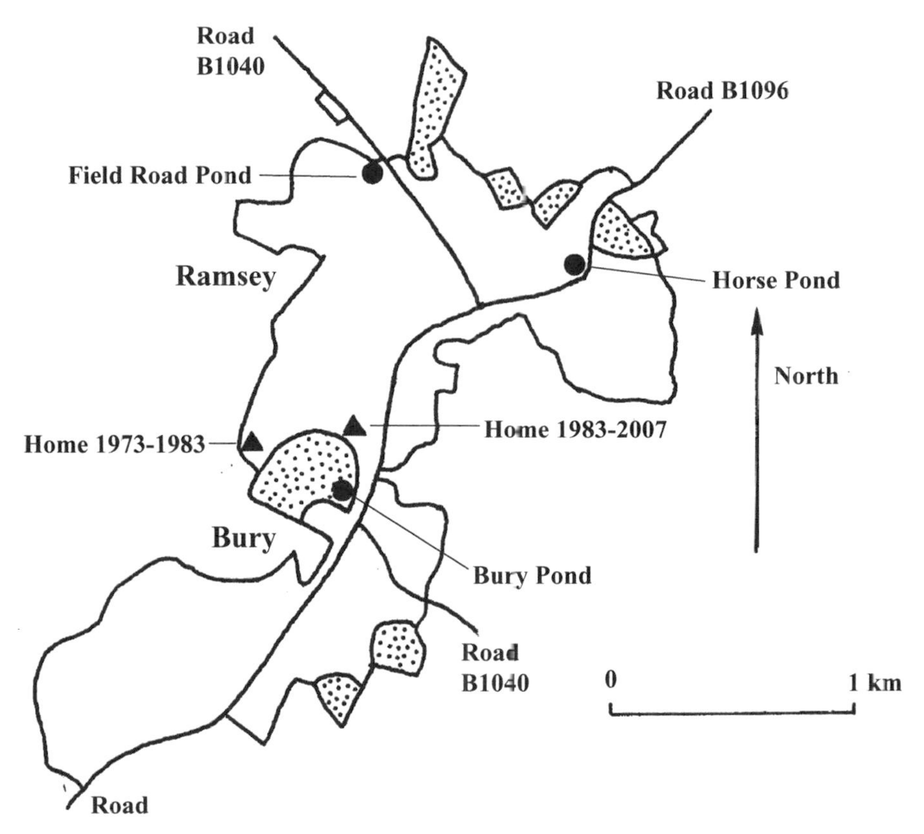

Figure 5.11 A sketch map of Ramsey and Bury, showing the locations of our second and third houses (black triangles) and the three principal toad ponds (black dots). Unshaded areas indicate development up to 1975, and stippling shows new development up to 2000.

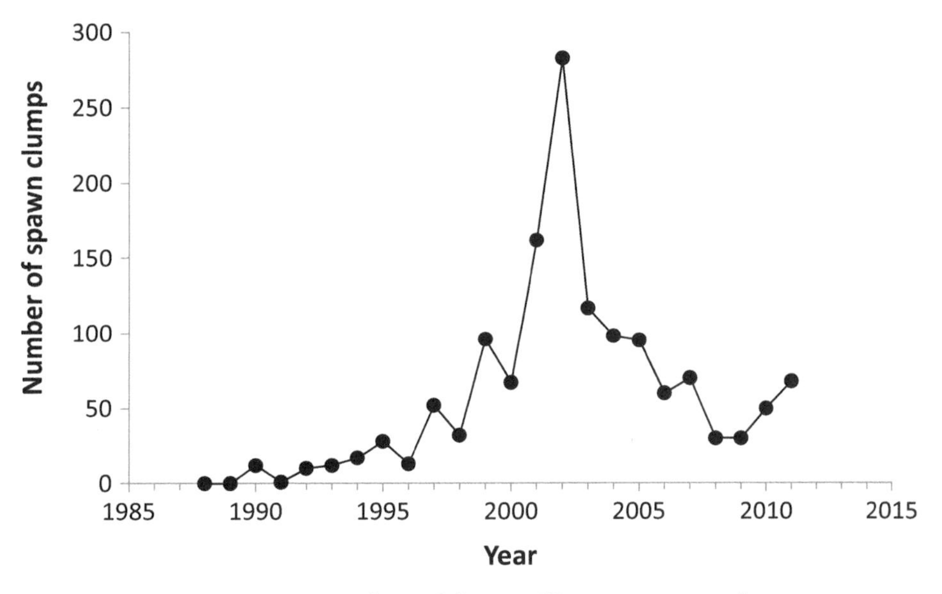

Figure 5.12 Numbers of clumps of frog spawn counted each spring in Bury toad pond, 1988–2011.

This, almost certainly, was a case of the frogs colonising from gardens in the area. It is possible that frogs from our second garden were involved in this process – the distance between the ponds was roughly 500 m (Figure 5.11). Toadlets from Bury Pond turned up in numbers in our second garden in the early 1980s, so frogs or froglets might have been capable of managing the reverse journey, if not directly to Bury Pond, then perhaps via garden ponds nearby. Breeding began in 1980 in that garden, and frogs first spawned in Bury Pond 10 years later. In addition, in 1993, I supplied frog spawn from the plastic pond in our third garden to a friend who lived next to Bury Pond, which probably helped to swell that frog population in later years.

Even during the three springs of 2001–2003, when the number of spawn clumps in Bury Pond exceeded 100, only two or three dead frogs were counted on the roads. In contrast, although toads had decreased hugely in numbers by that time and peak counts of live animals in the pond were only 1–6, road casualties still numbered 13 or 14 (Figure 6.10). Although, there are other explanations for the difference between the species, this does suggest that frog migrations were much shorter than those of toads. The estate around the pond was notable for the number of walls, fences and other barriers that had been constructed. The frogs nominally had access to about 8 ha of housing and gardens without crossing a through road, but they would have had to negotiate three cul-de-sacs. On that basis, density in 2002 at the peak of spawn production may have been about 70 per hectare, without taking frogs breeding in garden ponds into account. This is an example of opportunistic expansion of a frog population into a larger non-garden pond in a suburban situation.

Mixed pairs were never common but were seen in four years after 1990 (1998, 1999, 2003 and 2006). Typically, they involved a male frog on a female toad. As toads became rare, so such pairs will have had a proportionately greater effect on the output of the toad population. So could the frogs have been a factor in the toad decline? This is revisited in Section 6.3.6 but is not considered to have been important in the early part of the toad decline when they were numerous but frogs were rare.

5.4.2 *Frog colonisation of Horse Pond, Ramsey*

Horse Pond in Ramsey was a traditional breeding site for toads (Figure 6.6b). By the 1980s, toads still had a population numbering hundreds of adults. I was unaware of any frog spawn in the pond during that decade, although, while monitoring toads in 1985, I had recorded three male frogs in the water. During the early 1980s, a large number of semi-domesticated mallards were based at the pond, sustained mainly by being fed by people. I was asked for advice by the parish council, who were worried about the impact the ducks seemed to be having on the appearance of the pond. Not wishing to be known as someone who hated ducks, I restricted comments to what would happen if the ducks were or were not reduced in number, including pointing out the wildlife value that the local populace might enjoy from a balanced pond. The duck population was duly thinned out in 1984.

The first frog spawn in Horse Pond was recorded in 1992, which, from its appearance, was probably laid naturally and not introduced by people (see Section 5.2.1). Horse Pond was more than a kilometre from either of our homes in Ramsey (Figure 5.11), and frogs must have colonised from much nearer ponds, perhaps from those in gardens of new housing close by to the north-east. The only spawn which I recorded as 'dumped' in the pond was one clump in 1993 and nine in 2003. On the first occasion, the single clump was silted, damaged and floating free; the second time, a resident informed me that the spawn had been introduced. These clumps are omitted from the data in Figure 5.13. The number of clumps increased to 19 in 1994, but the duck flock was also recovering, peaking at 39 in 1997. I saw no frog

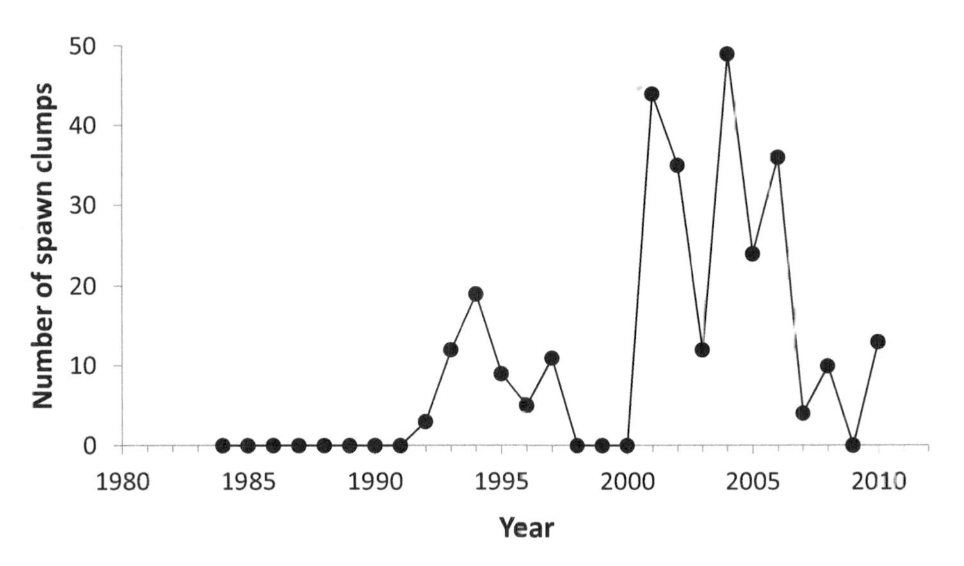

Figure 5.13 Numbers of clumps of frog spawn counted each spring in Horse Pond, Ramsey, 1984–2010.

spawn during 1998–2000, perhaps as a result of increased attention from ducks. Savage (1961) reported that a decrease in his local frog population in 1958 was associated with a large increase in breeding ducks, but was uncertain whether this could be cause and effect. The flock at Horse Pond numbered less than 20 from 2001 to 2010, and frog spawn peaked at 49 clumps in 2004, but then decreased. Dead frogs were more evident on the roads than previously, with a peak count of 11 in 2004. Fish had been seen occasionally in the earlier years, but from 2002 many goldfish lived in the pond, presumably introduced by the public; and in 2007 shoals of small ones gathered at the frog spawning site. As brown rats (*Rattus norvegicus*) were also seen occasionally, the frogs had motor vehicles and many other predators to contend with, and appeared to be losing the battle. The site was checked in 2021, but there was no sign of frog spawn in the pond or dead frogs on the road. Frogs may conceivably have played a small part in the decline of toads in the pond (Section 6.3.3), but mixed pairs were never seen.

5.5 Conclusions

Three of the principal factors controlling frogs breeding on St Neots Common during the 1970s and 1980s were collection of spawn by people, desiccation of breeding sites between hatching and froglet emergence, and predation of tadpoles. Frogs breeding in the plastic pond in our third garden appeared to be controlled to some extent by density-dependent mortality of ova and tadpoles – frogs could be too successful for their own good. In 2002, adult density in our garden and neighbours' gardens was thought to approach 100 per hectare, but a high proportion of the population tried to breed in the most confined of the five ponds, with an unfortunate outcome.

The loss of frogs at St Neots Common was both sad and difficult to comprehend. Had I been monitoring the population during the time it declined, I might have some evidence of the causes. As it stands, I can only speculate. In 2005, the ditches had evidently not been cleared for some time and the lagoon had been fenced to deny cattle access; perhaps as a consequence, reed had proliferated (Figure 6.4). If lack of management had resulted in an increase in aquatic invertebrate predators, this could have affected survival to the froglet stage so leading to population decline. Another factor that occurred to me was that an increased incidence of flooding on Islands Common might have carried away significant quantities of spawn. Flooding on that part of the common was fairly frequent during the study in the 1970s and early 1980s and was especially serious in 1979, but there was no evidence that such intermittent spawn loss was translated into a later decrease in population. During 1980–1982, 46–74% of all spawn was laid in the temporary pool on Lammas Meadow, which desiccated prior to froglet emergence. If such losses continued in a regular fashion beyond the end of the study, then this may possibly have led to population decline. It is, however, difficult to see that this factor alone could have caused extinction. The temporary pool existed during the breeding season in 2005 (as it did in 2021), but all of the spawn that year was laid on Islands Common in places where significant desiccation losses were unlikely. No loss to collection was suspected in 2005: indeed, the 12 clumps recorded in the southern pond on Islands Common were considered to have been deposited by people. In short, the apparently total loss of frogs from the common was perplexing.

There is another factor that might act, perhaps in combination with some of those listed above, to explain how frogs could thrive on the common in the 1970s but decline and disappear in more recent decades. Otters declined widely in this country, including in Cambridgeshire, from the late 1950s until the 1970s because of the use of organochlorine insecticides (Jefferies and Woodroffe 2008). Since then the species has re-established itself in the county (Arnold and Jefferies 1997, Bacon 2005, Hows *et al.* 2016). Otters have been recorded in the vicinity of St Neots Common (Arnold and Jefferies 2000, Walker 2008), a fact proudly announced on English Nature's notice board at the main entrance to the site. Otters will take frogs, sometimes in large numbers (e.g. Weber 1990, Slater 2002), and may conceivably be implicated in the demise of this frog population by focusing on breeding concentrations in the spring.

As will be seen in the next chapter, toads disappeared earlier than frogs at St Neots, probably primarily because of increases in road traffic and house building. In Ramsey and Bury, toads declined from abundant to rare for the same reasons, whereas frogs increased from rare to abundant. These two very different responses over time were due to factors such as the frog's ability to exploit garden ponds and its lower vulnerability to infrastructure changes and road traffic. Thus, at St Neots Common, no frogs were recorded dead on the roads, presumably because they did not need to cross. At Bury Pond, frog carcasses were found on roads close to the breeding site, but in much smaller numbers than toads. The main reason why frogs generally seem less vulnerable to declines caused by road traffic may be that they remain nearer to the breeding ponds. Nevertheless, death on the road could have contributed to population declines of frogs at Horse Pond, as the busy B1096 swings around two sides of the breeding site (Figure 6.6b). In my survey of 33 frog sites undertaken in 1975 across a wide area of southern England (Section 3.4.2), traffic casualties were recorded at six, including one site in Staffordshire where road vehicles appeared to kill at least 20% of the adult frogs during that breeding season.

When we came to live in this area in 1968, frogs were already very rare. Henry Berman, who was then recorder of amphibians and reptiles for the Huntingdonshire Fauna and Flora Society, reported no sightings at all in the annual reports for 1965 and 1966 (Berman and Thomas 1966, Berman *et al.* 1967). Henry was a well-known and enthusiastic biology teacher at the secondary school in St Ives and his information largely came from the catchment area of his pupils. Although this meant he seemed unaware of the large population at the more distant site of St Neots Common, it will probably astonish readers today that not one of his students appears to have seen a frog over a period of at least two years.

I tried to improve the fortunes of the county's frogs in gardens and in the wider countryside by introducing spawn or tadpoles or by supplying interested colleagues and others with tadpoles of mixed parentage from St Neots (i.e. the tadpoles originated from several spawn clumps). A number of introductions resulted in populations becoming established. On St Neots Common, people were helping themselves to spawn and probably taking tadpoles as well. This became less prevalent in the late

1970s and early 1980s – and by then surplus spawn from garden ponds was being deposited on the common.

By the early 1980s, the species was continuing to recover in the county by utilising gardens, but remained rare in the countryside, especially on fen farmland (Cooke 1982). My wife and I have had a succession of garden ponds with frogs, and individuals will have dispersed into the surrounding suburban areas. I have discussed above the possibility of frogs from our second garden moving 500 m to our third garden or to Bury Pond. Even if frogs from the second garden or their descendants were not involved in the initial colonisation of Bury Pond, it was very likely that my donation of spawn in 1993 to a garden next to that pond helped to increase reproduction there later in the decade. Our last house move was to Bury in 2007, and it soon became apparent that frogs were in our part of the village, at least in small numbers. Once I had installed another small pre-formed pond, a few frogs arrived to spawn without any extra help from me.

The process of establishment by people in suburban garden ponds must have been repeated in countless situations in Britain. Surveys have demonstrated that frogs are catholic in their choice of breeding site (Section 4.2.2; Beebee 1979, Swan and Oldham 1993). This means that, where there is a high density of garden ponds, their flexibility allows populations to survive and spread even if there is a turnover of suitable breeding sites. In contrast, toads prefer larger, deeper sites, so if a traditional breeding pond is lost they may need to walk further to find another, which could be much more difficult and, at best, might involve dealing with extra barriers and roads.

Because of the threat of ranavirus and the fact that it is unclear how the disease transmits, amphibian conservation organisations in this country have advised that frogs, tadpoles and spawn should not be transferred between sites. While I agree with this recommendation, I do not regret my part in distributing spawn and tadpoles around this area in the past. Looking at a couple of alternative scenarios:

- Had there been no vogue for creating garden ponds, it is likely that frogs would have continued to be very rare in many suburban areas. I suspect that conservation organisations would have (1) concentrated on creating new habitat for them in countryside or on the edge of towns, and (2) tried to establish populations by introducing spawn, much as was done at the Boardwalks reserve in Peterborough (Section 4.4.2).
- On the other hand, had garden ponds been created without spawn being introduced, then frogs in this area would have colonised much more slowly at best.

The possibility that moving spawn may also inadvertently encourage the spread of invasive non-native species of plants is another reason for caution. And there is also concern relating to the genetics of frog populations in gardens (Hitchings and Beebee 1997). These authors found substantial genetic differentiation and reduced genetic diversity in populations in gardens in Brighton when compared with rural populations. Tadpoles suffered higher mortality and a higher incidence of abnormalities indicating reduced fitness and genetic drift. This was blamed on the reduced ability for individuals in an urban environment to travel between populations. The structure of the urban environment presented much larger barriers to gene flow than those that occurred in the countryside. Observations from my survey of performance of frogs in various breeding sites in 1975 (Section 3.4.2) provided some support for the frequency of abnormalities being higher amongst tadpoles in garden and other ornamental ponds in towns. Samples taken from each of four such sites revealed abnormal tadpoles in three, whereas only three of 16 samples examined from 11 countryside sites contained abnormal animals.

While there is some concern about the long-term future of garden populations, at least children growing up in recent decades have known the frog as a real animal rather than as something that exists only in books or on screens. However, evidence of frogs returning to the countryside has been more limited, especially in areas such as the Fens and the former county of Huntingdonshire where the landscape has been much modified since the 1930s (Cooke 1999). It is important that, as landscape-scale initiatives gain further support and people consider the future of farmland after Brexit and other influences, frogs are not forgotten amidst the clamour to foster populations of more glamorous or popular wildlife species. In order for them to return to the countryside in a sustainable way, amphibians need sufficient areas of freshwater and terrestrial habitat with corridors of suitable and safe habitat, plus a commitment from individuals and organisations to monitor and manage the animals and the habitat appropriately. Recent landscape initiatives offer hope for recovery locally beyond built-up areas, as demonstrated, for example, by the Wicken Fen Vision and the Cambridge Nature Network (Wildlife Trust for Bedfordshire, Cambridgeshire and Northamptonshire 2021), and the Great Fen project (Section 6.4).

CHAPTER 6

Toads in a hole

6.1 Introduction

Here, events and changes are described for two of my local areas that had relatively large populations of common toads during the 1970s and 1980s. In both situations, populations seemed to suffer markedly: this was believed to be mainly because of the twin hazards of development and road traffic, despite the original habitats being very different. The account of toads on St Neots Common (Section 6.2) describes a time when I was more interested in the site's frog population (Section 5.2), and only recorded the toads because they were there and they were conspicuous. I am glad that I did, because I learnt something about them and the hazards they faced. The toads at Ramsey and Bury (Section 6.3) presented an opportunity to observe animals that had long been bound into the fabric of the suburban area where I lived. They were everywhere, and most of the human population was aware of them, even if some locals referred to them as 'frogs'. When I started observing these toads in 1974, and even 20 years later in the early 1990s, it did not enter my head that one day I would stop working on them because too few survived to study. What happened to toads in these two areas can be compared and contrasted with the fortunes of frogs at the same sites (Chapter 5).

Rosemarie and I moved to what was at that time the county of Huntingdon and Peterborough in 1968 when we bought a small bungalow in Sawtry, a village lying next to the A1 (Figure 1.5). As I started looking for frog spawn or tadpoles on which to work in 1969 (Chapter 3), I naively thought all I had to do was drive out in spring into the hilly rural area to the west and look in a few ponds. In fact, when going out in daytime, I found precisely nothing. As I broadened my search, I began to stumble across toad populations rather than frogs. Perhaps I should study toads instead? But then the whole point of working on frogs was because they were thought to have declined, so perhaps I had no right to expect them to be easy to find. When I eventually discovered the county's principal frog population on St Neots Common (Figure 1.5; Section 5.2.1), recording the toads at the same time was a bonus.

I soon learnt that frogs and toads in the countryside typically differed greatly in behaviour, and monitoring their population trends required different approaches. Breeding frogs often materialise in a field pond almost magically, and usually do not relish people trying to catch or even count them during daytime. More helpfully, they produce spawn in situations where it can be relatively easy to find and count. Then, by the time their tadpoles are part grown, they too can be difficult to catch in quantity because predators rapidly reduce their numbers and, not surprisingly, survivors disperse. Toads, on the other hand, are very obvious and amenable as adults (Figure 6.1a), but produce spawn that is often laid in places where it is difficult to access, and is, by its nature, equally difficult to quantify (Figure 6.1b) – although some researchers have managed to count the strings (e.g. Reading 1984). But once tadpoles hatch, they become easy to find and may shoal through a breeding site in hordes of thousands (Figure 6.1c).

Amphibians, particularly toads, are unfortunately prone to being killed by road traffic when on migration. Counting road casualties is a grisly occupation, but can be informative. The first person to count dead toads on roads in this country may have been Moore (1954). He began to notice dead toads on the A350 running through Iwerne Minster in Dorset in 1938 and counted them from 1950 until 1953 as part of a detailed study on their migration. By making a number of assumptions, he suggested that around 500 were killed in 1951 on the main road and side roads, out of a total of about 2,000.

Back in the 1970s, I regularly counted live (and dead) adult toads on St Neots Common, despite worrying that such counts might be unreliable and uninformative for monitoring purposes. Without doing more detailed studies, it was not possible to know what proportion of toads was being counted, how this changed through a season, and how detectability was affected by circumstances altering from year to year. Although I published an article about frogs on St Neots Common, largely based on counting spawn clumps (Cooke 1985a), I did not then have the time or the confidence to do the same with my data on toads. However, with the experience of monitoring live toads at the Boardwalks reserve in Peterborough (Section 4.4) and at sites in Ramsey and Bury (Section 6.3), I have now decided to pull together the information. It is also true that counting toads became an acceptable method of monitoring – as is discussed, for instance, by Richard Griffiths and Howard Inns (1998; but see Schmidt 2004).

The extent of my knowledge about toads in the early 1970s was more or less restricted to what appeared in the fourth edition of Malcolm Smith's New Naturalist volume (1969). He constructed a superb review from his own observations and the many fragments of other information available to him. But when compared with Trevor Beebee and Richard Griffiths' later account (2000), it is revealed to be descriptive natural history with little quantitative material. Smith died in 1958, having revised the second edition of his book in 1954. Up until then, subjects such as population ecology had simply not been considered, but Deryk Frazer undertook a detailed study of a toad colony near Maidstone in Kent from 1955 to 1961 (Frazer 1966). Frazer's pioneering paper reported on subjects such as population size, survival from year to year and migration. During the early 1970s, however, I may have been the only person in the country undertaking long-term surveillance of a toad population – and the first person to record for longer than 10 years.

Figure 6.1 (a) Spawning toads (photo by Mihai Leu); (b) a tangle of spawn, with some lifted clear of the water by emerging vegetation; (c) tadpoles shoaling.

6.2 Toads on St Neots Common

6.2.1 *Counts of adult toads*

As with the observations of frogs on St Neots Common (Section 5.2), I started counting live toads in 1971 and finished after the breeding season of 1984. Conditions on the common were recorded each spring as regards water depth and flooding, amount of plant growth and water turbidity, management of waterbodies, and degree of accessibility of toad breeding areas. Virtually all of the toad breeding activity occurred on Islands Common, the eastern half of the site (Figure 5.1) – and there they largely favoured the lagoon. Daytime visits were made roughly twice per week through the frog and toad breeding seasons, and then on a few occasions during each summer. Single toads, of which nearly all were males, and pairs were counted whenever seen. Peak counts of pairs and all toads are shown in Figure 6.2.

When surveillance began in 1971, the ditch on Islands Common had been dredged the previous winter, so making the toads easier to count. Amounts of vegetation in the ditch increased up to about 1973 and then remained fairly stable; few toads used the ditch in 1972. Part of the lagoon was dredged in 1973 but this was some distance from where the toads spawned and should not have affected counting. In 1976, they temporarily shifted their main breeding area to deeper, well-vegetated water in the lagoon, where they were more difficult to count; towards the end of the mating season, much spawn could be seen at the rear of the lagoon. There were no obvious reasons for relatively low counts on the common in 1977, 1978 or 1983; 1981 was, however, a long, drawn-out breeding season, which may have resulted in a lower peak number of toads. Counting problems with blooms of filamentous algae were recorded in 1982. After considering these issues, it seems reasonable to suggest that the population peaked in 1972 and 1984, but low counts in 1976, 1981 and 1982 may have been relatively more affected by conditions.

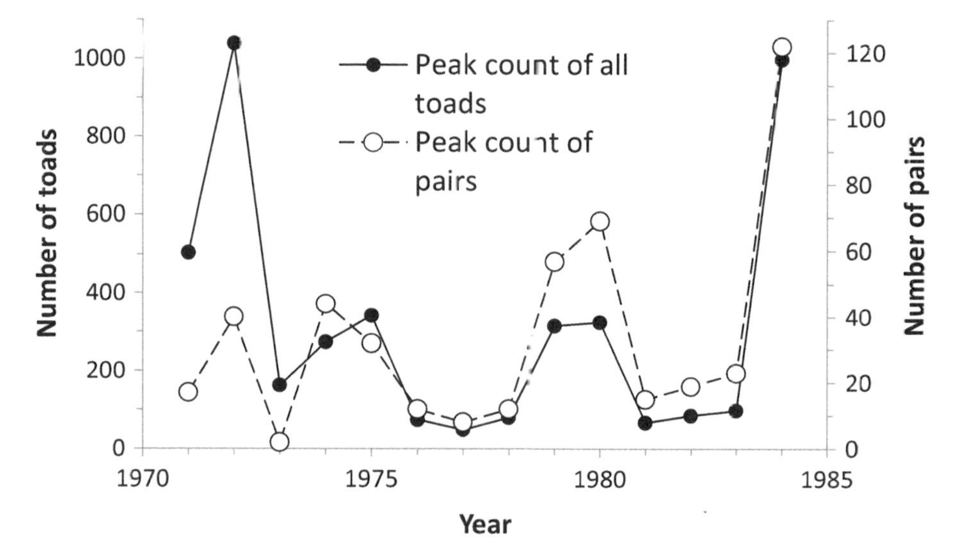

Figure 6.2 Peak numbers of all toads (black dots and solid line) and pairs (white dots and broken line) counted alive on St Neots Common during the breeding seasons of 1971–1984.

Table 6.1 Numbers of toads dead on roads beside or near St Neots Common, 1974–1984 (see text for methods).

Year	Dead toads on roads		
	Common Road	Road to north	Roads to east
1974	90	—	—
1975	65	—	3
1976	56	—	—
1977	95	—	8
1978	69	—	4
1979	70	32	—
1980	95	15	4
1981	99	17	—
1982	59	12	1
1983	150	45	13
1984	223	32	21

From 1974, fresh casualties were counted on Common Road (the B1041), usually twice per week, and the cumulative total was taken as an index of numbers killed (Table 6.1). In 1975, I realised that some toads were coming from further afield to the east and crossing another road about 500 m from Islands Common; casualties were counted there on single occasions during seven years between 1975 and 1984. In 1979, dead toads were noticed on the B1041 where it crossed the River Great Ouse to the north; casualty counts were made on single occasions each year thereafter.

6.2.2 *Subsidiary observations*

Basic information on numbers of adults counted in the breeding sites or dead on the roads does not go very far towards explaining why the population may have changed. Routine observations were not made on amounts of spawn produced, the degree to which it survived, what happened to the tadpoles or toadlet emergence. However, ad hoc records were kept, and thus there was information on the extent of spawn failure. Toads spawned in beds of reed sweet-grass (*Glyceria maxima*) in the lagoon and, as the vegetation grew, some spawn was lifted clear of the water and often became colonised by fungus (Figure 6.1b). Ova also became covered in fungus if spawn was frozen. Spawn with fungus did not hatch. Such problems were noted in 1974, 1975, 1977 and 1978, but the worst years were 1980 and 1981, when 20% of the spawn was estimated to have been affected. In several years, spawn appeared to have been broken up by people collecting it: the worst occasion was in April 1979, when Easter coincided with fine weather. Frog spawn that was broken up on the common had a low hatch rate (Section 5.2.2), and the same probably applied to toad spawn.

Samples of toad tadpoles were caught and checked for abnormalities in 1975 as additional observations in the survey of potentially polluted breeding sites for frogs

(Section 3.4.2). An unusually high incidence of distended bodies was found in a tadpole sample in June. A distended body usually stemmed from having a bubble of air in the gill chamber. Such tadpoles failed to recover when kept in captivity. The relevance of these observations to tadpoles on the common was not investigated, but no shoaling was noted that summer. Tadpoles were examined again in 1976 but nothing unusual was recorded.

The fact that toads concentrated their breeding activity in the deeper, permanent lagoon meant that their spawn and tadpoles were less likely to be affected by desiccation than was recorded for frogs, which often spawned in shallower, transient sites on the common (Section 5.2.2). Predation of toad tadpoles was not studied, although they will have been exposed to a high density of predators of a range of species. Some, such as fish and newts, find them distasteful (e.g. Section 4.3), but a number of invertebrate predators can significantly reduce tadpole populations. At Llysdinam pond in mid-Wales, for instance, adult and larval dytiscid beetles were largely responsible for heavy losses of toad tadpoles over just a few days (Griffiths *et al.* 1988, Beebee and Griffiths 2000). I have witnessed intense predation pressure from larval dytiscids on toad tadpoles around mid-season and learnt to my cost how efficient they could be if they managed to gain access to my tadpoles caged in field situations. In an ideal world, perhaps I would have decided to undertake summer netting of the lagoon to gain information on relative abundance of tadpoles and their predators (e.g. see Table 10.2). As it was, I recorded shoaling by toad tadpoles in the lagoon during the summers of 1973, 1976, 1980 and 1982. This observation was taken to imply that these were comparatively good years for reproduction, but I appreciate that invertebrate predation may have significantly reduced tadpole numbers later in the season. Good numbers of toadlets were recorded in 1972 and 1979. Years that appeared to be good for tadpoles or toadlets were not necessarily those years when high numbers of adults had been counted in spring.

To try to help interpret fluctuations in the count data for adult toads, three samples of males were measured, in 1971, 1975 and 1980. Measuring a sample may provide some information on relative amounts of recruitment from previous years. As was noted in the account of toads at the Boardwalks reserve in Peterborough (Section 4.4.3), toads continue to grow for a time as adults, so small toads tend to be younger than larger ones if conditions remain the same. Male toads can mature at two or three years of age, but may not mature until they are older, as was demonstrated by Paul Gittins and his colleagues (1982, 1985) in Wales, by Chris Reading (1988) in Dorset, and by Rob Oldham and myself (1995) at the Boardwalks reserve. There is also much intra- and inter-population variation in length displayed by toads of the same age, so we should not read too much into data on length distribution. Gittins *et al.* (1985) reported males up to six years old, and Rob Oldham and I found them up to seven years at the Boardwalks.

Data for males caught at St Neots were:

- 1971: median length = 54 mm, range = 48–68 mm, sample size = 63
- 1975: median length = 54 mm, range = 44–63 mm, sample size = 64
- 1980: median length = 59 mm, range = 47–68 mm, sample size = 95

Toads caught in the breeding season of 1980 tended to be larger than those in 1971 and 1975, but toads caught in 1971 and 1975 were not significantly different from one

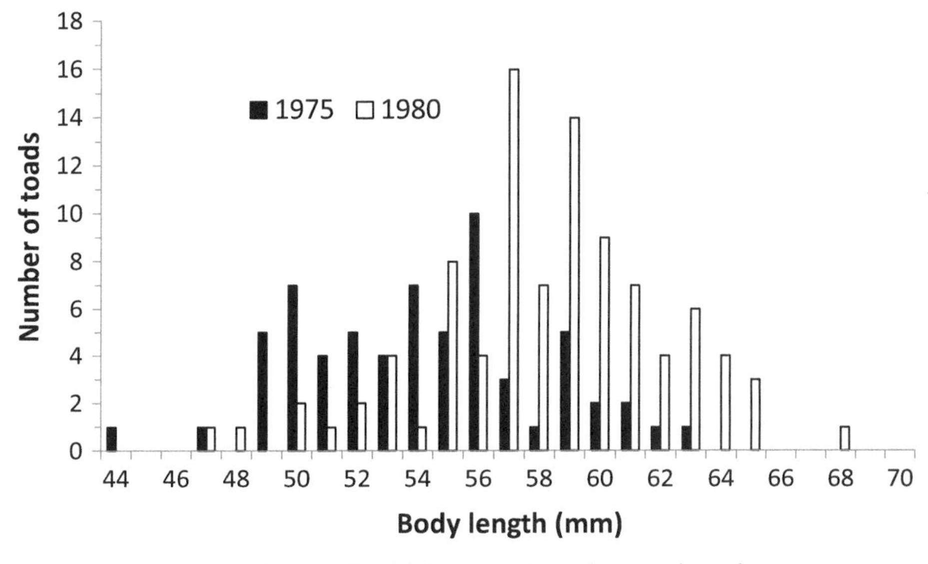

Figure 6.3 Length distributions of samples of male toads caught on St Neots Common in 1975 and 1980.

another. Length distributions for 1975 and 1980 are shown in Figure 6.3. The inference was that a preponderance of younger adults occurred in 1971 and 1975 when compared with the situation in 1980, although there was of course no information about the feeding conditions under which these samples of males developed. It is conceivable that the toads measured in 1980 were of similar age to those in 1975 but enjoyed much better food resources.

With hindsight, what would have helped would have been removing a toe as a year-mark from the male toads and saving the toes for age analysis. But the technique of estimating age had not been employed by that time in this country – and anyway the whole process would have been beyond my resources and experience. Later in the 1980s and early 1990s, I depended very heavily on Rob Oldham when 'we' carried out such a study at the Boardwalks reserve (Section 4.4).

In fact, the males caught in 1980 were toe-clipped, but this was done to estimate numbers present, and the toes were not kept. Again, with hindsight, I should have preserved the material for possible later investigation. On the same day, 31 pairs were also caught and each male was toe-clipped. All toads were released back into the site roughly where they were caught. After two hours for males and three hours for pairs, new samples were caught and checked for toe-clips. This gave estimates for total numbers of 230 for individual males and 160 for pairs, yielding a total estimate for that day at peak season of 550. The count that day prior to sampling of 325 represented 59% of the estimate. Up to that point in the season, 103 toads had been counted dead on all roads, which was 16% of the combined total of live and dead toads. Although not all live toads in the population will have been on the common that day, the number counted dead on the road was probably much lower than the real figure and was treated as an index, so 16% was likely to be an underestimate of the real percentage killed on the roads up to that time.

Records were kept of dead toads found on the common near or in the breeding sites. Small numbers were noted in most years, with higher totals of 11–14 in 1971, 1976 and 1977. In late March and early April 1972, the mutilated corpses of at least 80 toads were found. They were often in piles at the water's edge; the skin with its toxic glands was avoided and little flesh had been eaten. Brown rats were suspected as being the predators. In the same breeding season, one of my helpers encountered a man collecting toads on the common. The man admitted he had caught 160 adults, but did not reveal why.

6.2.3 *Conclusions*

Below is a speculative attempt to interpret some of the various items of different and occasionally conflicting information on how and why numbers changed. The numerous small males seen and caught on the common in 1971 suggested good recruitment from breeding in the late 1960s. There were even higher counts of single toads again in 1972, but peak number of pairs was not unusually high. Nevertheless, the population seemed buoyant at that time, with high numbers of toadlets seen in 1972. Counts of adults decreased considerably in 1973, and the 10-year period 1973–1982 showed relative stability, with ranges for peak counts being 49–342 and for casualties counted on Common Road 56–99. Sizes of males in 1975 suggested considerable recruitment to the breeding population that year, perhaps from the good reproduction in 1972 (and 1973). There were many larger males in the sample in 1980, possibly emanating from the successful breeding year of 1976. Good reproduction was indicated in the summers of both 1979 and 1980, which may have initiated the increase in numbers returning to breed in the 1980s. Although counts were modest in 1981 and 1982, I noted that individuals tended to be small. Huge numbers of tadpoles were seen shoaling in the lagoon in the summer of 1982. In 1983, large numbers of adults were killed on the roads, but numbers counted in the lagoon were not unusually high. But in 1984, high numbers of both sexes were counted in the lagoon and road mortality was even worse than in 1983. Again, males were recorded as mostly being of small size. If many of the females were from 1979 and 1980, they would then have been four or five years of age.

Like the frog population, the toads will have benefited from the general stability of the landscape of the common, apart from a certain amount of seasonal flooding. Spawn collection by people was not a problem, as it was for the frogs, but there was at least one example of large numbers of adults being taken and another incident of many being predated. The extent of deaths on the roads was surprising and distressing, particularly in 1983 and 1984. And the number of toads that were evidently trekking over the river bridge to the north was unexpected. Their chances of surviving that route cannot have been good, with the road being narrow, winding and busy with traffic.

Curiously, if counts in the study were ranked for the two species, a positive relationship was apparent between numbers of frog spawn clumps on Islands Common (Section 5.2) and peak counts of toads. When counts of clumps of frog spawn were comparatively high, then so were counts of adult toads. I did not expect such an outcome in view of their different lifestyles, the rather different threats that they appeared to face, and the fact that they often selected different places in which to spawn. It may simply have happened by chance, but could repay further investigation elsewhere.

My work on toads was fairly superficial and required little extra input on top of that needed to monitor breeding by frogs. Much more detailed studies were under way by the early 1980s at breeding sites at Llysdinam (Gittins *et al.* 1980) and Llandrindod Wells (Gittins 1983a) in Wales, as well as in Dorset (Reading and Clarke 1983). By then, annual meetings were being held between researchers and conservationists that enabled people to be better informed of new work and initiatives, and helped steer ideas and even national policy. Paul Gittins' career was tragically cut short, but Chris Reading continued with his studies on toads until 2020.

One feature of these intensive studies of toads was that animals were trapped and marked by toe-clipping, either with a year-mark or individually, which might require several toes to be clipped. Such a technique clearly had a lengthy history, as Deryk Frazer (1966) had used it in the 1950s and described it then as 'time-honoured'; he also sewed coloured beads to toads' backs in order to identify individuals. A Panjet inoculator was used by Paul Gittins (1983b) to 'tattoo' a blue dye at up to six sites on a toad, a method considered to be less damaging than removing several toes. I tried the Panjet technique in the early 1980s but found the marks were too indistinct; on the other hand, Rob Oldham used it in later years at the Boardwalks reserve (Section 4.4.1). Because toads lack the contrasting colour patterns of our other amphibians, recognising individual toads has always presented more of a problem. Over the years, Chris Reading used plastic tags, toe-clipping and passive integrated transponder (PIT) tags (e.g. Reading and Clarke 1983, Reading and Jofré 2021).

6.2.4 Postscript

In 2005, I visited the common on five occasions between 14 March and 7 April to see how the frogs and toads were faring after a gap of 21 years. Frogs were spawning in lower numbers than previously (Section 5.2.4), but I saw no evidence of toads either alive on the common or dead on the roads. It is conceivable, but unlikely, that toads arrived at the common to spawn after 7 April. Toad activity on the common was usually well synchronised with that in Ramsey (Section 6.3.1). In 2005, casualty counting at Ramsey just after peak inward migration was on 21 March; so if toads still bred on the common in reasonable numbers, I would have expected to have recorded them during several visits.

Some of the waterbodies were more vegetated in 2005 than previously (Figure 6.4) and lack of management and too much aquatic vegetation can increase invertebrate predation on toad tadpoles and reduce the level of a population (Beebee 2012). Nevertheless, areas of open water survived, including in deeper parts of the lagoon, so lack of management of the waterbodies was not considered capable at that time of eliminating the toad population. Therefore, I tended to assume that the lack of toads in 2005 was primarily due to land-based reasons such as high rates of mortality on the roads and changes in terrestrial habitat: by 2005, the toads had lost significant areas of their former terrestrial habitat to the east and north-east to new housing (Figures 5.1 and 6.5). Land converted to housing included a large field of allotments that was likely to have been well frequented by toads. If that land was

Figure 6.4 The lagoon at the north end of Islands Common, St Neots, fenced and dominated by reed in March 2021, as it had been in 2005. Contrast this with Figure 5.7, which shows a typical edge of the lagoon in the early 1980s.

Figure 6.5 A view of the northern end of Islands Common in spring 2005. This is a similar view to that in Figure 5.2, taken about 30 years before, but the trees in the distance have gone and new housing will have destroyed some terrestrial habitat, led to an increased risk of road death, and made access to the common more difficult from distant terrestrial habitat.

cleared at the wrong time of year, significant mortality of all terrestrial life stages could have resulted. The new housing probably also hindered access to the common from an easterly direction.

In 2020, several local naturalists who visit the common told me that they had not seen a toad there for many years. In 2021, I visited on 17 March and 7 April, primarily to check for frogs and their spawn, but no evidence of toads was seen. Extensive housing had been constructed during the intervening years immediately to the north-west of the river (marked as 'Factory' in Figure 5.1), further reducing terrestrial habitat formerly used by toads. Thus, both aquatic and terrestrial habitats were far less suitable in 2021 than they were in the early 1980s, and presumably road traffic increased with the extra housing. It is hard to understand, however, why a small residual population did not persist, especially if those toads remained confined to the common. There was a reference in the annual report of the Huntingdonshire Fauna and Flora Society to a toad or toads being present on the common in May 2014 (Orbell and Orbell 2015), indicating a presence, if not necessarily a significant or continuous one. In Section 5.5, I have suggested that predation by otters could have been implicated in the loss of frogs on the common, and as this predator will also occasionally kill toads in large numbers (Slater 2002) the same could apply to the loss of toads. In the case of the toads, however, there was ample direct evidence of road traffic killing significant numbers, whereas there was no direct evidence of otters doing so. Otters were not considered as potential culprits of the toad mortality incident noted in 1972 because of their extreme rarity in this region at that time.

6.3 Toads in Ramsey and Bury

6.3.1 Toads at Field Road

By 1973, we were still living in the village of Sawtry but looking for a larger home for our growing family. We were especially interested in the Ramsey area (Figure 1.5) as its schools had a good reputation and houses were less expensive than in some other towns and villages within range of Monks Wood, where my work was based. Ramsey was also an attractive proposition for me because I knew it had at least one good toad population which I hoped to study. Our attempt to buy at auction a property that backed onto the main breeding site was thwarted by the owner's reserve price being too high at a time when house prices were rising quickly. We eventually bought a newly built house in a small estate on the edge of the town (Figure 5.11), and I immediately started trying to get to grips with the toads. My initial intention was to undertake long-term surveillance in more depth than I was able to do at St Neots. Being close to home, I could do much of it in my own time. This, however, turned out to be rather more difficult than I had anticipated.

I knew of two breeding sites very close together. The more important one was a large ballast pit created beside Field Road in 1864 when one of Ramsey's two railway stations was created. This station was Ramsey North, which connected up with what is now the East Coast Main Line. The station and track were dismantled just before we moved to Ramsey and the industrial dereliction that remained provided the toads with some good terrestrial habitat. For many years, the pond had served as a source of water for Ramsey's fire brigade and in the 1970s was known to most local people as the 'Fire Pond', but I have always referred to it as Field Road Pond. This road was and still is one of the busier suburban roads in the town. Next to the pond was a mill, and on land owned by the mill was a second, smaller pond of more recent origin. A few toads could be seen at times in the water. I was interested in this pond primarily from the point of view of pollution (Section 3.4.1), but it was filled in during the winter of 1979/80.

Close to these ponds, Field Road joins the B1040. Judging by the positions of amphibian casualties on the roads, most of the toads (and the newts) preferred to breed in Field Road Pond. Unfortunately for me, as can be seen in Figure 6.6a, this pond was at the time surrounded by a thin but virtually impenetrable belt of trees, shrubs and bramble. Initially, I attempted to fight my way through and net the pond from the bank, as the water tended to be turbid. However, the boys from the town seemed much more adept at negotiating the banks and catching toads than I was. My attempts stopped abruptly when a low overhanging branch that I used as a natural jetty suddenly broke and I ended up in the water. At that point, I decided it was better to rein in my ambitions and count road casualties rather than toads in the pond.

In the spring of 1974, I counted freshly dead toads on the nearby roads almost every day until after the adults had left the ponds. Counting showed lengths of road which were the worst hotspots and indicated where toads were spending the rest of the year. It told me something about weather conditions that encouraged toads to migrate, but it was not a programme to which I could commit each spring. I decided in future to keep an eye on movement and then do a single count of all dead toads just after the main inward migration had occurred in order to determine what this might reveal about changes in both migration routes and the size of the toad population. In a small number of years, I was too hasty in counting and casualties increased, so I had to repeat the process and use the second count. For the first five years, not much changed, with counts of dead toads remaining below 100, but this was followed by a period when counts were generally higher, sometimes around 200. At the time, of course, I did not know that I would keep on counting annually until 2010 (Figure 6.7).

Figure 6.6 Toad ponds in Ramsey: (a) Field Road Pond fringed by scrub and bramble in spring 1974; (b) historic Horse Pond in 2005 showing the B1096, ducks and, on this particular day, a fountain.

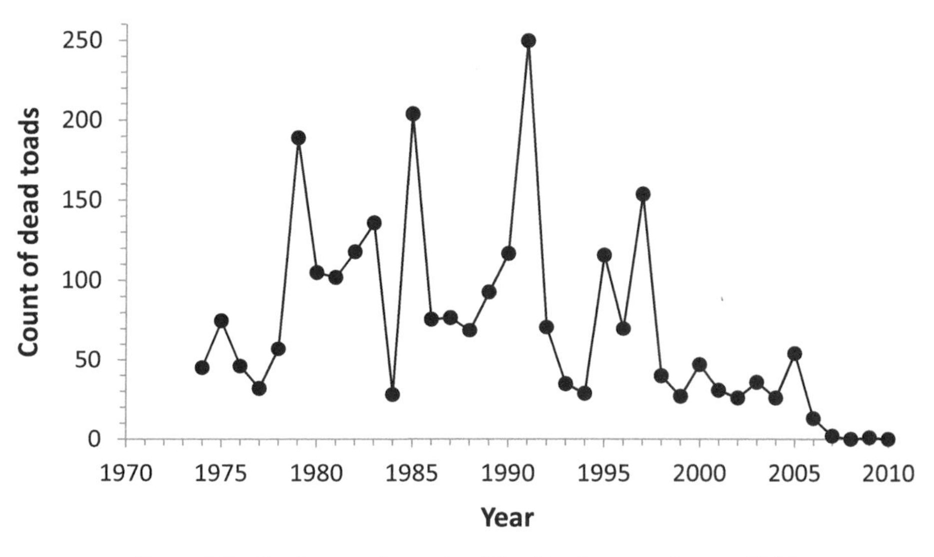

Figure 6.7 Single annual counts of dead toads on roads near Field Road Pond, Ramsey, just after peak migration, 1974–2010.

6.3.2 *Toads at Horse Pond*

In the early 1980s, I began counts in and around another toad pond in Ramsey, which I knew as Horse Pond (Figures 5.11 and 6.6b). This is an ancient site, about 900 m from Field Road Pond, which has a sloping edge and bottom so that horses could be easily led in to drink. It is 25 × 45 m, but often has turbid water, probably because of its resident population of ducks. Nevertheless, in 1982, 1984 and 1985, I undertook night counts of toads in the pond and counts of casualties on the B1096 next to the pond. Counts ranged 80–175 and 14–121 respectively, indicating a medium-sized breeding population. In view of the issue of intermittently turbid water, I decided to concentrate on monitoring numbers killed on the B1096 each year. I knew from counts at Field Road that relatively little input was required. The large increase in road casualties near Horse Pond and Field Road Pond from 1984 to 1985 was both alarming and intriguing, and I wondered whether counts might be synchronised over a longer period. Counting casualties on the B1096 continued up until 2010 (Figure 6.8), as at Field Road.

6.3.3 *Toads at Bury Pond*

During the very early 1980s, Rosemarie and I sometimes found toadlets in our garden in late summer. Toadlets often walk a kilometre on leaving their natal pond (Oldham 1985), but there was more than 1 km of housing between our home and Field Road Pond – and Horse Pond was even further away (Figure 5.11). I suspected there must be a significant breeding site closer to home. On making enquiries, I discovered that toads bred in washing lagoons behind a vegetable packing plant in the attached village of Bury, about 500 m from our house (Figure 6.9a). I first visited the site in April 1982 and found large amounts of toad spawn, and in subsequent years the site changed little and the toads continued to breed. In 1986, I began counting toad casualties on the main road, the B1040 (Figure 6.9b), which ran past

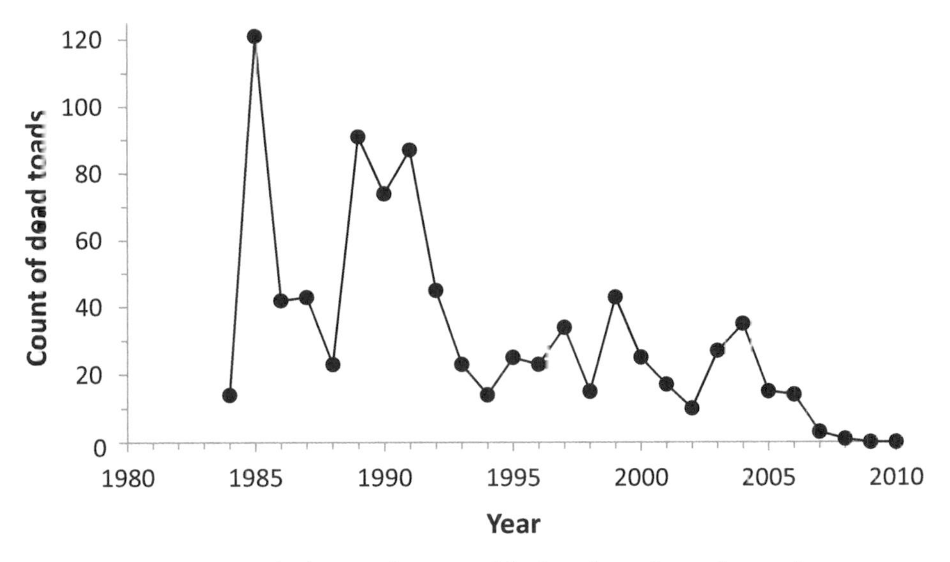

Figure 6.8 Single annual counts of dead toads on the road around Horse Pond, Ramsey, just after peak migration, 1984–2010.

Figure 6.9 Bury Pond: (a) originally created as a vegetable washing lagoon, it is here seen in 1991 after being retained within a new housing development; (b) the busy B1040 runs past the pond, which is situated behind houses on the left.

the packing plant. The following year, a friend told me that the land occupied by the plant had been granted planning permission to be used for housing. I passed on this news to both the NCC and the local Wildlife Trust, and discussions resulted in part of one of the lagoons being set aside by the builders to form a small reserve, which was eventually managed by the Wildlife Trust. Since then I have referred to the site as Bury Pond (see also Section 5.4.1). The packing plant was pulled down and replaced by housing during 1988 and 1989, and toad monitoring began while houses were being built.

Events in Bury Pond occupied me through the 1990s. I was aware that it was an opportunity, almost on my doorstep, to determine how the population coped with the new estate which was built around the pond. House building was nothing new for the area, with recent bursts in the 1970s and 1980s (Figure 5.11). The colony itself was quite new, as the washing lagoons did not exist until the 1950s. Bury Pond is roughly 1.3 km from both Field Road Pond and Horse Pond, but the toads probably colonised naturally from a nearer, smaller population, such as one on Ramsey golf course about 800 m away. Bury Pond is roughly 25 × 15 m, but only has a fringe of terrestrial habitat 5–10 m in width, so the site cannot support large numbers of toads outside the breeding season (Figure 6.9a).

Water in the pond was usually clear, so another aim was to test how well trends in casualty counts on a surrounding ring consisting of the B1040 and more minor roads reflected what was going on in the pond, and whether there was a good relationship between peak counts of live toads in the pond during daytime and at night. Monitoring results are summarised in Figure 6.10. There were significant positive relationships between pairs of the three measures. Each variable was highest in 1991 and then decreased substantially. Night counts were stopped when it became clear that they were adding little extra information about the toads. The spawn string index (Section 4.4.3) decreased to the year 2000 and was zero thereafter. Casualty counts decreased to a very low level in 1996, but did not become zero until 2009. During the late 1990s, toads still bred in the pond, but the scarcity of road casualties suggested that most of these animals were living close to the pond within the ring of roads. By the year 2000, when I wrote an article on the effect of the new housing on toads breeding in this retained pond, it was clear that the initiative had failed to protect a viable colony (Cooke 2000). There were very few sightings of live toads in the pond from 2001. Low numbers of toads were recorded dead on roads passing through the estate to the west up until 2008; these may have been heading for a garden pond or ponds close to Bury Pond.

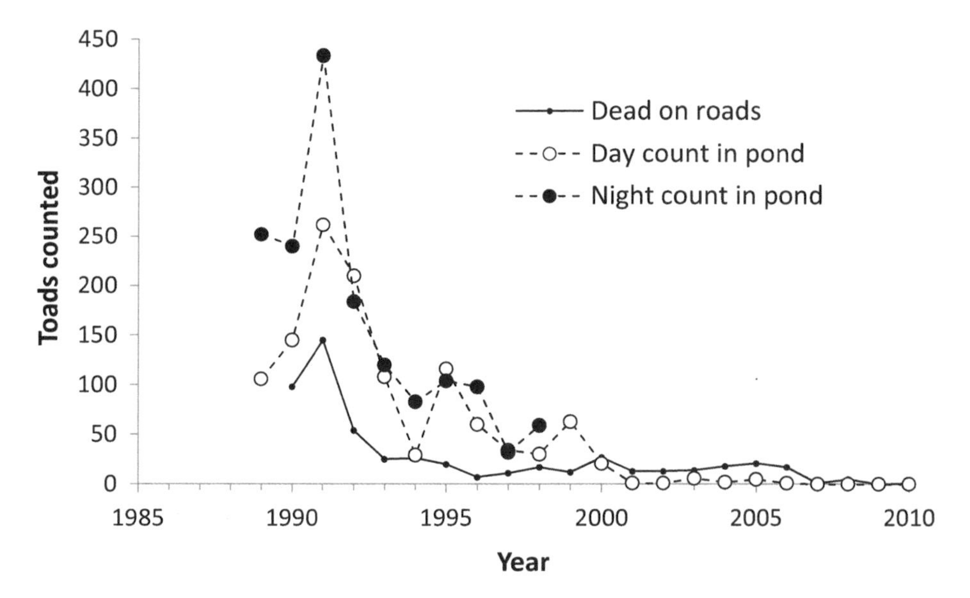

Figure 6.10 Peak-season counts of live toads in the water and dead toads on the roads at Bury Pond, 1989–2010. Night counts stopped after 1998.

6.3.4 *Toad rescue*

At this point in the story, I need to say something about toads being helped across roads by local people. During the 1970s and 1980s, conservationists in Britain became aware of the carnage that was occurring on roads near some toad breeding sites. I can remember being alarmed by the sudden increase in 1979 in casualties on roads near the pond at Field Road (Figure 6.7). Although it was not considered to be of conservation importance at the time, it was certainly an animal welfare problem, and many people reacted to it by helping toads across roads (Gittins 1983c, Langton 1989, Beebee and Griffiths 2000, Petrovan and Schmidt 2016). Other protective techniques introduced over the years included erecting warning signs, installing tunnels under roads (Figure 12.7), using barriers to steer migrating toads away from danger, and even moving colonies to safer locations. Simply catching and helping toads across a road was the easiest and most popular activity, and was first operational in Ramsey in 1987. I liaised closely with the local Wildlife Trust coordinator about when and where 'toad lifting' might be most effective; and more than 1,300 toads were moved across our local roads in 1988 and 1989, which turned out to be the busiest years. I well remember agreeing to supervise the local Cub pack and show them what to do beside Field Road Pond. Unfortunately, the pack leader was unavailable on the best evenings for toad migration, and by the time the event took place the toads were happily in the pond. I collected a bucket of bemused males from the shallows and went around behind Cubs putting the wet animals on the road surface and drawing attention to them so the Cubs could return them to the water. I hope the toads were not too inconvenienced – the Cubs enjoyed it and so did I.

In 1988, I wrote an article describing the casualty counting and toad lifting at Field Road Pond, and speculated about how effective the rescue operation might have been in the 1980s. By making a number of assumptions, I estimated that one toad in eight might be killed crossing Field Road: so, for every eight toads moved across the road, one was being saved from death (Cooke 1988). In 1993, permanent toad warning signs were erected close to Field Road Pond and Horse Pond. Soon after, I wrote another article based on the first 21 years of data from Field Road, arguing that whereas observing dead toads could confirm whether they still migrated to a pond in spring, casualty counts should be used with caution to indicate population trends because of complicating factors such as toad lifting and an increase in road traffic (Cooke 1995b). At that time, there was no significant long-term trend in numbers killed. However, I had not appreciated that most of the toad lifters had been deterred from continuing to volunteer by a relative dearth of toads during the period 1992–1994 (Figure 6.7), and few toads have been helped since 1995.

6.3.5 *A pause for thought*

Returning the narrative to the early years of this century, I had, by that time, long-term information from the three breeding sites in Ramsey and Bury, and my views had changed markedly since the mid-1990s. Casualty counts at all three sites had declined significantly from 1991. The decrease in casualties around Bury Pond had been demonstrated to be associated with a decline in counts of breeding toads and eventual loss of the colony. A little count information was available for toads in Horse Pond: the minimum number counted during spring 1982–1985 had been 80, whereas in two night counts in 2004 none was seen. During the last two counts, the water was clear enough to count 34 and 55 adult frogs. So there seemed little doubt that toads declined markedly at Horse Pond too, and it appeared reasonable to assume that similar decreases in casualty count around Field Road Pond meant that the breeding population had suffered the same fate – despite the lack of counting of toads in the water.

By 2003, questions concerning processes needed to be answered. First, I mulled over again what influenced whether a toad might be killed crossing a road and subsequently recorded by me. In 2001, Danish researchers Tove Hels and Erik Buchwald had published a paper that listed the factors affecting the chances of a toad becoming a road victim: the rate of passage of motor vehicles, the width of their tyres, the speed of the toad and its angle of crossing. Whether a dead toad was recorded (i.e. the detection rate) depended on amount of traffic, how soon a count was made after the toad was killed, and the persistence of evidence on the road, which in turn was influenced by the weather and by the activity of scavengers. Hels and Buchwald recorded a detection rate of 53% on a road with a traffic flow rate within the range of rates on roads in Ramsey and Bury. Locally, I will have missed an unknown proportion of dead toads, but as my monitoring was undertaken annually and as systematically as possible, trends in casualty counts should reflect population changes if all other factors remained constant (see also Beebee 2013a). However, traffic rates can change from year to year: for instance, County Council statistics revealed an 11% increase between 1992 and 2002 on the B1040 through Ramsey and Bury. And this led me to wonder what would happen to trends in casualty numbers if road deaths were high enough to affect population size.

When traffic flow rates increase, more toads in a stable population are likely to be killed until losses become unsustainable; then the population begins to decrease and numbers killed on the road first stabilise and then decrease. Working with Tim Sparks, a statistician at the Centre for Ecology and Hydrology (CEH), I realised that when trends in overall casualty counts around Field Road Pond were 'smoothed out' they indicated an increase to the early 1980s followed by a gradual decline – in other words, trends that were consistent with steadily increasing levels of traffic detrimentally affecting the population (Cooke and Sparks 2004). However, just because numbers of road casualties decrease, it does not necessarily follow that (1) road traffic is responsible (it may be some other factor) or even that (2) the toad population has declined (behaviour may have changed). What can be learnt from road-kill data has recently been reviewed for wildlife in general by Schwartz *et al.* (2020).

Another phenomenon had been flagged up by 2003: numbers of road casualties around Ramsey and Bury seemed to decline faster over the years on the busiest roads. This observation precipitated a survey to determine traffic flow rates on roads where casualty counts were made, as is discussed below in Section 6.3.6.

Tim Sparks and I concluded that for the technique of casualty counting to be useful, an observer should not wait for statistical proof before considering further conservation action (Cooke and Sparks 2004). If declines in casualty counts (or in counts of live toads) occurred for three consecutive years, this should trigger real concern. Around Field Road, for instance, this point was reached in 1994 and, had we known then what we know now, there should have been increased efforts to help toads across the roads rather than progressively diminishing interest because of the reduced numbers of toads to help. Silviu Petrovan has told me that this was seen to occur with many other toad patrols operating under the national scheme.

6.3.6 *Unravelling the causes of the population declines*

The fact that there were synchronous declines at all three sites suggested there may have been a common cause or causes. Tim Sparks and I considered causes of decline one by one; these are the factors that might have contributed:

- The area of the breeding site had been reduced for Bury Pond.
- Depth of water was affected at Field Road after the pond was drained in 1992 by the police when they were looking for a missing person.
- Amount of aquatic vegetation had been reduced by ducks in Horse Pond.
- Loss of terrestrial habitat occurred at all sites, but particularly around Bury Pond.
- Migration became more impeded at all sites, but particularly around Bury Pond.
- Traffic had increased on roads around all sites, leading to an increased likelihood of death.
- New roads had been created close to Bury Pond.
- Creation of garden ponds may have led to movement away from traditional sites.
- The Field Road population suffered periodically from collection of adults and spawn by children.
- No toadlet emergence was recorded at Bury Pond during 1993–1998.
- Frog populations increased at Horse Pond and Bury Pond, with mixed pairs seen at the latter site.
- Climate change may have increased the likelihood of mixed pairs occurring by closing the gap between breeding dates of the two species.

There was no reason to suspect pollution was involved; neither was there evidence of disease or high levels of predation on adults (except by motor traffic). The only action to counteract the effects of the damaging factors was helping adults across roads in the late 1980s and early 1990s. Volunteers were especially active on roads near Field Road Pond.

Overall, it was concluded that road traffic played an important part in population declines, with loss and modification of habitat also having significant effects (Cooke and Sparks 2004). These were common factors at all three sites. As regards habitat loss, approximately 40% of land within 250 m of Bury Pond had been converted to relatively high-density housing between 1975 and 2003; the figure for the other two ponds was about 10% (Figure 5.11). Although gardens offered some suitable terrestrial, and maybe aquatic, habitat, much of the land was covered by buildings and roads. And, perhaps even more importantly, many barriers to free movement were constructed, such as walls, fences and kerbs. Gully pots introduced a new mortality factor around Bury Pond. When a toad encounters the barrier of a roadside kerb, it is inclined to walk along below it and is at risk of falling into a gully pot. I spent an unexpected amount of time rescuing toads from these death traps in the early years of the new estate. This is now generally recognised as a threat to migrating amphibians, and there are ways of avoiding the animals becoming trapped by careful design of gully pots and kerbs (Baker *et al.* 2011).

As mentioned above, road traffic appeared to have a disproportionate effect on toads crossing the busiest roads, especially the B1040 close to Bury Pond (Figure 6.9b). Toads counted dead on this road decreased from 48 in 1989 to 9 in 1993 and never exceeded this number again. Using the flow rate recorded by Cambridgeshire County Council in 2002 for this road in the model of Hels and Buchwald (2001) suggested a toad had a one in three chance of crossing the road alive (Cooke and Sparks 2004); if it was lucky enough to reach the breeding pond, it still had to make a return journey. Unfortunately, toads following that migration route seemed to suffer unsustainable losses, and that was the route to the closest areas of open terrestrial habitat. At a breeding site in the Netherlands, the level of mortality of female toads in a single season was estimated at 29% on a road carrying less than 5% of the traffic of the B1040 at Bury (van Gelder 1973).

The enhanced effect on toads crossing the busiest roads probably arose because they have been demonstrated by other researchers (e.g. Oldham 1999b, Beebee 2012) to tend to migrate to and from a breeding site along the same route. Then, if some of the routes cross roads, differential mortality could lead to less use of more lethal migration routes in the longer term. Thus, Rob Oldham and Mary Swan (1991) described a situation at Coleorton in Leicestershire where a huge population of toads bred in a pond of 1 ha. Toads were trapped during 1982–1989 as they migrated into the pond from various directions and as they dispersed again after breeding. Those toads that were based to the south of the pond had to negotiate a busy road, the A512, on which traffic increased annually during the study. Toads either attempted to cross the road surface or migrated under a bridge. Over time, the proportion of toads moving under the bridge increased, with progressively fewer toads successfully negotiating the road. As road traffic increased, so did the proportion of incoming toads that were killed on the southern half of the carriageway – in other words, they failed to get halfway across. Young and Beebee (2004) mentioned a stable toad population in which migration across a nearby road had decreased, apparently because greater numbers of toads stayed nearer the breeding site.

In March 2004, following up the information on changing patterns of toad migration, I selected 12 roads in Ramsey and Bury where toad casualties had been monitored since 1990, and counted traffic flow rates on each night of the week (Cooke 2011). One observation to come from these counts was that Friday was the busiest night and Sunday the least busy, the difference being more than two-fold. So which day of the week peak migration occurred would affect the extent of toad mortality. When proportional annual change in casualty counts was calculated for each of the 12 roads, the rate of decrease in casualties was greater on the roads with more traffic. Around Field Road Pond, for example, reductions in casualties occurred more quickly on Field Road itself than on two quiet cul-de-sacs close to the pond, implying that toads were safer and survived better if they utilised gardens in houses beside the cul-de-sacs – but numbers decreased over time even in those situations. Data for all 12 roads were not completely independent, as what happened on one road may have affected events on another, and some migration routes will have crossed more than one road. Nevertheless, this was further evidence of road traffic having an appreciable impact on local toads. It added to the other evidence against road traffic, such as the long-term trend in overall road mortality around Field Road Pond and the predicted mortality rate of toads trying in the past to cross the B1040.

This is an unremarkable, relatively quiet area, and if road traffic can affect our toads, then the same could happen to other populations with roads nearby. Observations from near Ottawa in Canada appeared to provide another example of the process of amphibian populations being affected by roads: locations with high rates of road traffic had smaller populations of frogs and toads and fewer casualties per kilometre of road (Fahrig *et al.* 1995). The impact of road deaths on amphibian populations more generally was reviewed by Beebee (2013a), who concluded that the long-term effect on population dynamics could be severe. Trevor Beebee had at the time just completed a 15-year study of toads breeding in Offham Marshes (Beebee 2012). The population declined from the late 1980s into the 1990s and remained low despite conservation efforts. Trevor estimated that, in the 1990s, an adult toad's chances of crossing an adjacent main road, the A275, were in the region of 50% per crossing. A volunteer patrol moved toads across the highway in the early 1990s, but numbers of toads decreased year by year and the patrol disbanded, just as happened at Ramsey. Surviving toads were effectively cut off from suitable terrestrial habitat beyond the road.

I greatly regret losing our local toad populations. The fact that the three main ponds have survived is an ever-present reminder of what has been lost. Results indicated that the initiative at Bury Pond had failed by the year 2000, and by about 2002 it became clearer that all three colonies were struggling. There were few toads left crossing the busiest roads by that stage, so stopping, let alone reversing, the declines would have been difficult. I wish the outcome had been different, but my counting did at least seem to provide an example of the destructive potential of road traffic, which has, I believe, been useful in the debate to protect toads more generally. Had there been further conservation effort beyond 1994, my observations might instead have demonstrated a longer, slower decline into oblivion. With hindsight, the populations were in any case declining and in trouble by 1994, despite eight years of variable numbers of toads being helped.

6.4 Conclusions

The toad colony on St Neots Common and the three colonies combined in the Ramsey/Bury area will have exceeded 1,000 adults in the more distant past, yet these populations have declined to (virtual) extinction in recent decades. It is interesting, but sad, to utilise casualty counts on single roads in the two areas as simple comparative indices of change: comparing number seen in 2005 with the annual average for 1974–1984, the decline was 100% on Common Road, St Neots, and 88% on Field Road, Ramsey. The main factors causing losses in these two areas were considered to be increases in road traffic and destruction and modification of terrestrial habitat, but other factors will be important in other situations: for example, neglect of aquatic and terrestrial habitats at the Boardwalks reserve in Peterborough apparently nearly led to failure of that colony (Section 4.4).

Traffic flow rates are also worth comparing in the two areas. County Council statistics for the five years 2013–2017 revealed that total number of motor vehicles was 50% higher on the B1041 at St Neots than on the B1040 in Bury. At the latter location, a toad was calculated to have a one in three chance of making a successful crossing, thereby emphasising the likely impact of road traffic at St Neots. An additional risk factor at Neots was that toads moving over the river from the north would have been walking down the length of a road bridge with side walls.

These are examples of a wider phenomenon of toad declines across much of England, which was highlighted by Jo-Anne Carrier and Trevor Beebee (2003) and by Susan Young and Trevor Beebee (2004). Because of concerns about population declines, the toad was listed in 2007 as a Priority Species under the UK's BAP. More recently, Silviu Petrovan and Benedikt Schmidt (2016) used data on numbers of toads moved across roads by volunteers to demonstrate that widespread population declines had occurred between 1985 and 2014, despite huge numbers being helped; without the toad patrols, declines would have been even worse. Petrovan and Schmidt suggested that roads next to exceptionally large and important toad populations should have underpasses where necessary rather than conservation action having to rely on volunteers. Froglife's Wildlife Tunnel Campaign in 2020 attempted to ensure that new developments, where roads impinge on wildlife sites, have tunnels that are monitored and maintained by the developers. The UK BAP was shelved by the government in 2016, but was superseded by strategies for the four constituent countries which could result in special protection for toads and for biodiversity more generally in road-building schemes (Froglife 2021). This certainly does not, however, mean there is no place for volunteers, who have done a brilliant job at countless locations across Britain and in other countries. In Cambridgeshire, we have a well-known and particularly long-running rescue operation at Madingley that has been organised by William Searle since 1988.

Road traffic is a broad-spectrum killer of all terrestrial life stages of the common toad, including toadlets and juveniles, and increases in road traffic will add to any impact on population size. In 2019, another review by Petrovan and Schmidt highlighted the importance of survival at the toadlet stage in affecting population growth. So far, the emphasis of toad patrols has been on protecting adults rather than ensuring the persistence of populations. These authors argued that we need to know much more about the ecology of young toads and how their survival might be enhanced when roads need to be negotiated. Toadlets could be particularly at risk on roads, because they often disperse in daytime and will probably move more slowly than adults across a road surface.

So how might we be able to conserve toads better in the future, specifically in Cambridgeshire? An adult common toad is likely to spend more than 90% of its time on land, and (open) woodland comprises ideal terrestrial habitat (Oldham 1999a, Salazar *et al.* 2016; Figure 6.11). Several local toad colonies have existed in comparatively large nature reserves for many years without changing very much and without necessarily receiving any attention. In such situations, they live in relatively stable habitat and are almost entirely safe from motor traffic – and from new housing. Thus, Monks Wood NNR is one of the largest woods in Cambridgeshire; it has little suitable freshwater in which toads can breed, but the population has ticked over. I once accidentally stood on a toad in the middle of a woodland block in Monks Wood while I was recording vegetation growth along a transect. But apart from freak bad luck like that, toads in such situations should have much longer lives than those trying to share a suburban landscape with us – and even that toad survived the experience. Roger and Sarah Orbell have compiled an annual amphibian and reptile report for the Huntingdonshire Fauna and Flora Society since 2009.

Figure 6.11 A relatively safe common toad in woodland. Photo by Marc Baldwin.

These very useful accounts reflect the stability of toad populations in the big, essentially terrestrial reserves, including Brampton Wood, which is a Wildlife Trust site of more than 100 ha.

At Woodwalton Fen NNR, toads have always bred in the ditches in reasonable numbers since I began visiting the site roughly 50 years ago. This reserve is an integral part of the Great Fen, a long-term restoration project being led by the Wildlife Trust that will transform a landscape of 3,700 ha back to wetland and low-intensity farmland (Figure 1.5). In 2015, I searched for breeding toads inside and immediately outside Woodwalton Fen (see Orbell and Orbell 2016): toads were widespread in their usual haunts in the reserve but did not appear to have migrated outside its boundaries into ditches between newly sown grass fields on the adjacent farmland. In the spring of 2016, however, toads were heard calling for the first time on reclaimed arable land further from the reserve (Bowley 2020). In the north-east corner of the Great Fen, new pools, excavated in part with assistance from Froglife under the Million Ponds Project (Section 12.7; Figure 12.7d), have also been colonised naturally by toads (and smooth newts), which I suspect were breeding in the large fen river nearby. A thorough survey of non-brackish main rivers and district drains on the Fens may show the toad to be more abundant and widespread than people imagine (Cooke and Cooke 2008; Figure 2.6). While their locations are unknown, they cannot be protected.

The original concept of the Great Fen was inspired by principles of rewilding (Bowley 2013). Rewilding means different things to different people but can be viewed as being at one end of a spectrum of conservation practices with 'interventionism' at the other (Fuller and Gilroy 2021). These authors acknowledged that rewilding was a benefit by increasing the amount of wildlife-friendly habitat, often by removing farmland from production, but questioned whether it necessarily helped species in trouble. It has recently been pointed out that rewilding could be seen by the UK government as a cheap alternative to supporting traditional management in marginal and upland farming areas, and lead to loss of investment where there was a greater need for conservation, such as in the urban–rural fringe (Williamson 2022). Rob Fuller and James Gilroy concluded that rewilding alone 'will not deliver the conditions that all species require', so a combination of rewilding and intervention is likely to be needed to meet biodiversity aims. Such an approach should help toads and other amphibians (re)colonise the Great Fen in the longer term. One of the areas in the north of the Great Fen is known as Froghall, possibly reflecting past abundance of amphibians.

Like the ponds created on the Great Fen, new farm reservoirs may also be successfully colonised by toads – an example of this occurred little more than 1 km outside Ramsey (Cooke and Sparks 2004). Toads can colonise new ponds on farmland relatively quickly (Baker and Halliday 1999). Farm reservoirs are often well away from public roads, although toads living there may be exposed to other hazards associated with cultivation. Roger and Sarah Orbell's local annual reports also demonstrate that toads still occur in gravel pits along the River Great Ouse, where Peter Ferguson and I recorded them nearly 50 years ago (Cooke and Ferguson 1974). And toads may still hang on in small numbers within the suburban environment in Ramsey and Bury, although new building continues: the area once occupied by Ramsey North Railway Station close to Field Road Pond was finally cleared for housing in 2020, after lying derelict for about 50 years. Elsewhere in the county, surveys of amphibian sites by Steve Allain and other members of the Cambridgeshire and Peterborough Amphibian and Reptile Group in and around the city of Cambridge reported toads from 8 out of 10 sites surveyed during the period 2014–2017 (see Allain and Goodman 2019).

Nevertheless, massive declines in toad populations have been reported from parts of Cambridgeshire since the 1940s (Sections 2.3 and 2.4). The populations that existed at St Neots Common and in Ramsey and Bury in the 1970s were among the area's largest colonies. Hopefully we can learn from these accounts of the incidental impacts of human activities. The fortunes of the frog have been turned around by the creation of large numbers of small artificial ponds, but unfortunately garden ponds have not provided the same level of help for toads. There is, however, something to be said for greater emphasis on actively promoting the conservation of existing toad populations in nature reserves and elsewhere more generally. If our farmland becomes friendlier to wildlife in the future via various initiatives, incentives and schemes, it may become a much better place for toads than has been the case over the last seven decades. In 2021 the local Wildlife Trust announced a new Nature Network within 10 km of Cambridge which will cover 9,000 ha and will include 13 SSSIs and many Local Wildlife Sites 'linked together by nature-friendly farmland and wildlife-rich towns and villages'. And, more widely, the toad is already beginning to benefit from recent large-scale strategies and projects to create ponds and terrestrial habitat such as via District Licensing Schemes for crested newts (Section 9.7).

CHAPTER 7

The natterjack years

7.1 Introduction

I started working for NCC in 1978 and was eager to get to grips with natterjack toads, which I regarded as attractive and interesting animals that were badly in need of conservation. This was only a few years after papers detailing the studies of Trevor Beebee and Keith Corbett had described the extent of their population decline in the British Isles during the twentieth century (Prestt *et al.* 1974, Corbett and Beebee 1975, Beebee 1976). By then the natterjack was virtually extinct on heathland and had also decreased significantly on sand-dune and marsh habitat, but it had received protection via the Conservation of Wild Creatures and Wild Plants Act 1975.

An important piece of casework cropped up almost immediately on the Sefton Coast, Merseyside. A golf course was proposed for an area south of Formby known as Cabin Hill, which was home to what was thought to be the largest population of natterjacks in Britain. I was aware that I knew little about the species at that time, and NCC's response in this case relied heavily on the knowledge of people outside the organisation, particularly Phil Smith of Liverpool Polytechnic (Figure 7.1a). This experience turned my thoughts to how everyone could be better informed. I asked NCC's relevant regional and national staff plus people in the BHS and County Naturalists Trusts if it would be helpful to have an annual report that included fairly detailed accounts of monitoring, conservation issues, threats and research from the various natterjack sites around Britain. My reasons behind volunteering to compile such a report were not entirely altruistic, as I realised that visiting the sites, communicating with people who had an interest in the species and writing a report would help me to do the job better. The response was universally favourable and the 1979 report was the first of five, the final two being written with the help of Brian Banks and Tom Langton. These annual reports were not formally published, but simply copied and circulated to anyone who was involved with the species by March of the following year.

After completing the report for 1983, I felt that they had probably served the purpose for which they were intended. By then, there was more dialogue between practitioners, and contributors and readers would have realised (1) how each population

Figure 7.1 Natterjack breeding sites on the Sefton Coast, Merseyside: (a) a newly sculpted site near Formby, September 1981, with Phil Smith on the left; (b) a slack at Ainsdale NNR, May 1985.

fitted into the national picture, (2) what was needed to monitor and conserve them, and (3) where to go for advice. Also, Trevor Beebee had just published an excellent monograph on the species (Beebee 1983), to which people could refer. He had been collecting and collating site information back to 1970 to produce a Site Register, so there was a solid base on which to build. I was able to secure funding from NCC for him to produce copies of the first two editions of the Site Register. The second edition appeared in 1989 and covered the period 1970 to 1989 inclusive.

In the early 1990s, Brian Banks, Trevor and I wrote a review on the conservation of the natterjack toad in Britain over the period 1970–1990 (Banks *et al.* 1994). By the time of publication, NCC had been reorganised and I had relinquished responsibility for herpetological advice, so the review marked the end of my formal involvement with the species. But natterjack monitoring and conservation has of course continued apace since that time.

Much credit for the conservation of the natterjack must also go to John Buckley (Figure 12.5), whom I first met in the 1970s as a member of the BHS Conservation Committee. From 1995 until his retirement in 2017, he worked for the Herpetological Conservation Trust (HCT) and ARC and had an infectious enthusiasm for natterjacks. Trevor and John co-authored updates of the national Site Register (Beebee and Buckley 2001, 2014a); the latter covered the period 1970–2009. The quality of the monitoring information is probably unique in terms of its thoroughness, and its spatial and temporal extent. It enabled John and Trevor to collaborate with Benedikt Schmidt to analyse data on population size (based on counts of spawn strings) and breeding success (based on estimations of toadlet abundance) for all natterjack sites in Britain during 1990–2009 (Buckley *et al.* 2014). And the conservation message from that modelling was reasonably reassuring too: compared with the declines earlier in the twentieth century, there had been little if any change in the national population between 1990 and 2009.

In view of my modest and relatively brief contribution to natterjack conservation, the approach I have taken in this chapter is rather different from that of the rest of the book. Here, I will ramble through, on a very personal level, what it was like in the 1980s to be out monitoring natterjacks, and mention those aspects that especially interested me.

7.2 Looking for and after natterjacks, 1978–1991

7.2.1 *Surveying and surviving*

During this period, natterjacks still occurred in a few sites in southern and eastern England, from Saltfleetby–Theddlethorpe on the Lincolnshire coast down to Woolmer in Hampshire and across to Winterton on the Norfolk coast. From home in Cambridgeshire, I could visit any of these sites comfortably in a day unless I needed to monitor in the evening, in which case I might stay overnight. The bulk of the British sites, however, were on coastal dunes or marshes stretching from Redrocks on the Wirral peninsula northwards as far as Southerness on the north Solway coast. I made at least one trip to visit these north-western sites each year from 1979 until 1989, and had my favourite hostelries as bases for visiting different parts of the species' range.

Much time was spent on the Sefton Coast because I was also interested in studying the northern population of the sand lizard. When I joined NCC in 1978, members of BHS's Conservation Committee were very well focused on sites for the sand lizard and smooth snake in southern England, but the lizard's northern stronghold in and around Ainsdale and Birkdale tended to be less well covered. Conversations with Keith Corbett convinced me that much of the sand lizard's dune habitat was eroding and blowing away inland, so I decided to make that my reptile niche and tried to visit the area in the spring and late summer each year to monitor both the lizard population and the structure and composition of habitat and vegetation where the lizards lived (Cooke 1991, 1993a). I eventually concluded that the main issue was stabilisation of the dunes (termed 'fixing') rather than erosion. When I visited for this type of work, I sometimes monitored natterjacks or great crested newts in the evenings on the Sefton Coast (Figure 7.1). And providing they had a vacancy, I stayed for bed and breakfast at the Franklyn Hotel on the Promenade at Southport, and lived on fish and chips – you can always get good fish and chips at the seaside in the summer. Another feature of the area was the lone men or pairs who frequented the quiet areas of dunes, enjoying the sunshine as much as the reptiles did. When the men saw me with my green wellingtons, clipboard, surveying equipment and the rest, their interest was aroused, especially if I happened to be taking fixed-point photographs as part of the project to determine how the habitat changed over the years. Sometimes they imparted potentially interesting information on how they perceived the dunes to be changing or where they might have seen natterjacks or lizards.

For visits further north, I tended to stay at the Old Ship Inn at Ravenglass for the south Cumbrian sites, the appropriately named Four Winds Guest House at Allonby for the northern ones, and the Nith Hotel in Dumfries for the Scottish sites on the north Solway coast. A real treat was to stay in the Wildfowl Trust's accommodation at Caerlaverock.

On several occasions, I travelled with Tony Bell from Monks Wood. One of his duties was recording sparrowhawks (*Accipiter nisus*) nesting in several well-spread areas of the country. The sparrowhawk had disappeared from several regions of England because of organochlorine pesticide poisoning decades earlier, and by the 1980s Tony was expecting recoveries to begin. One of the study areas was in Cumbria, so by travelling together, I could help him with the sparrowhawk survey and check any nests that we found, and he could help me monitor natterjacks. The fact that he still kept a house on the Wirral and monitored the natterjacks at Redrocks meant he was not a stranger to the species. When we travelled together, we often drove out after our evening meal looking for natterjacks crossing roads. I can remember coming across the species in several places on the B5300 south of Silloth in Cumbria; subsequent investigations helped to demonstrate the widespread nature of natterjacks on the Mawbray to Silloth coastal strip.

The weirdest hotel I stayed in during my time in NCC was a place somewhere in the wilds of Surrey. I was on a reptile trip and a room had been reserved for me by regional staff, but the hotel gave the impression it had not accommodated any guests for quite some time. I was directed along corridors and across a ballroom in which long-neglected plants had died and were now stretched plaintively down the side of their pots and across the floor. The carpet in the bedroom initially seemed to be grey with dark spots, but on closer inspection the spots turned out to be dead flies. The hotel was miles from anywhere and, when I opened a window and looked out, the only sounds were the cries of unidentifiable animals. I decided to sit up all night writing notes.

In the 1980s, I looked hard for natterjack breeding sites that were not in the Site Register. Most of these were probably long-standing sites that had been overlooked. I studied maps of the Cumbrian coast in particular, searching for promising areas without records that were close to known breeding sites. I trawled through the Bowness peninsula several times; on each occasion, I expected to find evidence of them, but never did, although no visits were made at night. I had much more luck around the shores of the Duddon estuary. For instance, I believe I was the first person to draw attention to them on and around the golf course at Dunnerholme. This was a particularly productive site in the 1980s and was one of the colonies studied by Jonty Denton during 1988–1990. Jonty was a postgraduate student working under Trevor Beebee at the University of Sussex, and I will return to his research below.

A site on the other side of the Duddon estuary provided me with the worst experience I had while out looking for natterjacks. The site in question was at Lady Hall and was approached down a quiet, narrow lane off the A595; I always parked close to the verge just past a farmstead, and walked down a track towards the marsh. Hills rose steeply beyond the trunk road, with Black Coombe attaining a height of 600 m. The distance from the car to the area where natterjacks then occurred in small numbers was roughly 1 km. On the afternoon of 12 May 1985, I parked the car as usual, walked down the track and checked for natterjacks: I found and measured the lengths of three adults and saw tadpoles – then I happened to look back up towards the hills. Smoke was billowing from a fire which, as far as I could tell, was very close to the car – or might even be the car. The vehicle was one of NCC's Vauxhall Chevettes, but I was more worried about my possessions in it, including all of my precious fixed-point photographs of lizard habitat on the frontal dunes at Birkdale (I always took the collection of slides into the field to help me find previous views and see how they had changed). In those days, I was still fairly fit, and I ran hard up the track in my wellingtons with my pond net and survey gear. It was only when I reached the road that I had a clear view of the situation: the fire had spread down the verge about 30 m and was within paint-peeling distance of the car. I dropped my gear in the road, opened the car door, got in, and it started first time. The car roared in first gear down the road for 50 m or so and stopped. I had not closed the driver's door, and feeling an urge to lie down, I collapsed on my back in the road with my feet still in the car. After I had recovered to something like normality, I was aware of a presence next to me. It was a boy, about 10 years old, and all he said was, 'Did you start the fire, Mister?' I could not croak even a short reply and he trudged back to the farm. I have often wondered what his version of the story was like. I never went back to that site again.

Another natterjack site that I never relished visiting was beside the nuclear plant at Sellafield on the Cumbrian coast (Figure 7.2). I recall monitoring the natterjacks soon after there had been an incident involving an 'abnormal' discharge of radioactive material into the Irish Sea and the beach was closed to the public. I walked along the shoreline regardless and remember thinking it probably was not a good idea as an onshore wind whipped spume off the sea. On another occasion, I paddled

Figure 7.2 Pools created for natterjacks at Sellafield on the Cumbrian coast, 1987.

around netting in a stream quite close to Sellafield; it flowed from an area enclosed by fencing and trees, which I later discovered was where low-level radioactive waste was stored. And I had a field trip to Cumbria in 1986 soon after radioactivity from the explosion at Chernobyl had drifted over and been deposited on the wetter parts of Wales and north-west England.

As a chemist by training, I have never been fazed by toxic chemicals, but I am more nervous about environmental radioactivity. In the 1980s, Dave Horrill was based at ITE Merlewood in Cumbria and was an expert on levels and effects of environmental radioactivity. He had been a friend from when we had both worked at Monks Wood, and I asked him how dangerous the levels of radioactivity might be in some places on the Cumbrian coast. He replied, probably with tongue slightly in cheek, 'If I dropped my lunchtime sandwich in the marsh, I wouldn't pick it up'. The consensus view has always been that environmental radioactivity in Britain is insufficient to impact wildlife. But in the early 1980s I began to worry about natterjacks, because the adults live for a number of years and those in certain colonies may be in contact with contaminated substrates. My concerns extended to both point-source contamination emanating from Sellafield and diffuse pollution from Chernobyl.

On behalf of NCC, I commissioned Dave Horrill and his colleague, Francis Livens, to determine levels of selected gamma- and alpha-emitting radionuclides in small numbers of natterjacks found dead at Ainsdale, Ravenglass and Sellafield. At the same time, I collated natterjack breeding success at sites on the north-west coast in relation to (1) distance from Sellafield and (2) the time when the contamination from Chernobyl settled in the north-west. The effects of Sellafield releases and the Chernobyl explosion were apparent as isotope fingerprints in the natterjacks. Thus concentrations of caesium, ruthenium and plutonium isotopes were higher in 'pre-Chernobyl' natterjacks at Ravenglass (8 km from Sellafield) than at Ainsdale (103 km from Sellafield). In the Sellafield sample collected a few days after contamination from Chernobyl arrived in Britain, amounts of caesium and ruthenium isotopes were roughly 30 times higher than the sample taken at Ravenglass a few weeks before; this was indicative of significant deposition from the explosion. Plutonium was emitted at Chernobyl in particulate form and will not have reached Britain, but its isotopes were present in higher amounts in the Sellafield natterjacks than at the other sites, again suggesting point-source contamination from the plant. Analytical details were included in the review published with Brian Banks and Trevor Beebee (Banks *et al.* 1994). Although elevated amounts were noted, Val Kennedy, Dave Horrill and Francis Livens concluded in a general review on environmental radioactivity for NCC that all isotope levels were much lower than those likely to have physiological impacts (Kennedy *et al.* 1990). This was borne out by my analysis of natterjack breeding success, which demonstrated no effects either close to Sellafield or following the explosion at Chernobyl (Banks *et al.* 1994). There was never any suggestion that natterjacks in the samples had died because of exposure to radionuclides. Also, when I checked out the Sellafield site in 1987, I found tens of thousands of metamorphosing tadpoles and toadlets, the latter often forming balls of scores, if not hundreds, of animals. And, I almost forgot – I was diagnosed with chronic myeloid leukaemia in 1999.

A striking difference between then and now is how much consideration is given to health and safety. In those days, I had never heard of personal risk assessments or mobile phones – let alone tracking using a phone app. People at headquarters will have known I was out somewhere in the north-west of England, and my wife will have been aware that, for instance, I should have left Southport and was heading in the direction of Ravenglass. I was certainly more conscious of where public phone boxes were located in those days. English Nature first supplied me with a mobile phone in 1993 when I was lone working in woods for a year; but that phone was not very mobile, being only slightly smaller and lighter than a house brick.

7.2.2 *A few small contributions*

Maurice Massey (Figure 7.5) was Chief Warden for NCC's East Midlands Region, and my trip with him to Saltfleetby–Theddlethorpe in Lincolnshire was always one of the most enjoyable days of the year. Visits seemed to coincide with good weather, the reserve warden Tim Clifford was invariably hospitable, and the site was interesting and important. It was the most northerly outpost for the natterjack on the east coast – but the population was small and struggling to survive. This meant that a few toads were recognisable as individuals and each string of spawn was precious. A recurring issue at the site was tadpoles loss, especially to invertebrate predators. On some visits, it was possible to sit and watch life-and-death struggles happening in a pool. Many years later, it was shown that tadpoles of the inbred population at Saltfleetby had reduced growth and survival when compared to those taken from Ainsdale (Rowe and Beebee 2003). The problem of tadpole predation led, during the time of my involvement and since, to rearing tadpoles in predator-free conditions either on site or elsewhere. I was pleased to be able to help during 1981–1983 by rearing samples of tadpoles through to the toadlet stage at my home in Ramsey. In 1981, 40 tadpoles were taken in a bucket and put in one of my tadpole cages in the garden pond. They were returned to Saltfleetby when the first ones started changing into toadlets. Survival and time to metamorphosis were very similar to those reared by the warden at Saltfleetby.

In 1982, I contributed by attempting to rear 700 tadpoles from the small hind-leg stage to metamorphosis. This time I constructed a shallow pond from a liner on the surface of our lawn, and the dimensions were 300 × 200 cm with an average depth of 6.5 cm. Tadpoles were fed on pelleted rabbit food and sausage meat, and they also grazed microscopic life on the liner surface and scavenged any dead invertebrates in the water. While I had captive tadpoles, it seemed a good idea to see how accurately I could count living, moving tadpoles under optimal conditions. At the same time, I undertook a capture–mark–recapture trial using neutral red to dye the initial sample; this was done to test its potential use in less clear water in field situations. Both counting free tadpoles and estimating by using neutral red as a marker indicated totals close to the real number (Cooke 1983). The red colour was soon lost by the dyed tadpoles. Tadpoles were returned to Saltfleetby once some started to acquire front legs, overall survival being 99%.

Another sample was taken to Ramsey for rearing in 1983. This time there was higher mortality, mainly from blackbird predation, and overall survival was 69%. The temporary pond that year measured 300 × 170 cm with an average depth of 12 cm (Figure 7.3a). Some heat stress was noted in early June, so shade shelters were provided and were used by the tadpoles. On 17 June, a rather cloudy day, the shelters were removed and tadpole behaviour and temperature were assessed at hourly intervals from 08.00 hours until 23.00 hours. Up until 15.00 hours, roughly 80% of the tadpoles were within 20 cm of the pond's edge where the water temperature was higher than in the centre by about 0.5 °C. During this time, water temperature had increased generally, peaking at nearly 27 °C at the edge. By 16.00 hours, however, water at the edge had cooled and the deeper water was marginally warmer for the rest of the day; tadpoles responded by occurring in increasing numbers in the deeper part of the pond until, by 22.00 hours, only about 20% remained at the edge. This demonstrated how the tadpoles were sensitive to water temperature and moved around so that most were found in the warmer areas. Natterjack tadpoles can survive temperatures of 29–33 °C without harm (Beebee 1983).

Invertebrate predation was a chronic problem at a number of sites, not just at Saltfleetby. During 1982–1984, Brian Banks worked on this issue as a research student at Drigg dunes in Cumbria and at Woolmer in Hampshire. NCC provided some funding for the study. Brian undertook predator trials with natterjack tadpoles as prey, and found that important invertebrate predators included dytiscid beetle adults and larvae, notonectids (greater water-boatmen or backswimmers), and dragonfly larvae (Banks and Beebee 1988). Adult crested newts also took significant numbers. In waterbodies which did not dry out, survival of natterjack tadpoles was inversely related to the density of invertebrate predators, which was highest in the most permanent sites. Tadpoles that have managed to escape from predators may show signs of conflict such as damaged tails. I recorded affected tadpoles at several other sites, such as North Walney in May 1981.

At Woolmer during the 1980s, I was able to help the natterjack conservation effort by organising financial support from NCC, for example for the introduction of golden orfe (*Leuciscus idus*) to reduce numbers of predatory invertebrates, and by

Figure 7.3 Examples of species and habitat management: (a) the author's son, Steven, beside a temporary pool used for rearing natterjack tadpoles from Saltfeetby at home in Ramsey, 1983; (b) spraying the herbicide Krenite to control birch regeneration on heathland at Woolmer in Hampshire, September 1981, with Rob Marrs spraying and Trevor Beebee watching.

arranging for the spraying of herbicide to control young birch (*Betula* spp.) which was invading terrestrial habitat (Beebee *et al.* 1990, Banks *et al.* 1993). This herbicide was called Krenite, and it acted by suppressing bud development the following spring, thereby killing the sprayed trees without having side effects (Figure 7.3b).

When poking around looking for tadpoles in quiet places, I sometimes came across odd sights, and one of the strangest was common toad tadpoles apparently grazing the surface of what remained of a child's doll upside down in a pool on the north Solway merse (Figure 7.4a). If you happen to be surveying during the tadpole season, it is most important to be able to differentiate between natterjack and common toad tadpoles. And this was a problem I encountered at several Cumbrian sites, including Allonby. Deciding between the two species is easy enough if the natterjacks have pale 'chins', but most do not. I often found the 'jizz' of the animal to be of use in this respect: to my eye, natterjack tadpoles seemed more streamlined, and between frog tadpoles and the dumpier common toad tadpoles in general shape, when viewed from the side. However, I had problems with identification based on the gap in the second upper labial row of 'teeth'. For instance, Frazer (1983) repeated the drawings of tadpole mouthparts from Smith (1969) and stated that the number and shape of labial ridges differed between species. The gap was depicted as about 32% of the width of the second upper labial row for the natterjack and 5% for the common toad. Going back to the beautiful and original illustrations in Boulenger (1897), the gaps were 29% and 4% respectively. Tadpoles that I suspected were common toads appeared in the field with a hand lens to have a wider gap in the second labial row than was indicated in these standard reference books. I brought samples of tadpoles from five sites back to the laboratory in 1983 for examination under a microscope. The tadpoles that I believed to be natterjacks had a gap equal to 25–30% of the width of the labial row, which agreed reasonably well with the old published drawings. On the other hand, those tadpoles that I thought were common toads had a gap of 12–20% rather than 4–5%. I did not bring back live tadpoles to rear beyond metamorphosis, so this issue was not fully resolved. Later, drawings of tadpole mouthparts in the identification chapter in the *Herpetofauna Workers' Manual* (Buckley and Inns 1998) showed a labial gap of about 12% for common toad and 26% for natterjack toad, supporting my suspicion that some common toad tadpoles may have wider gaps than earlier drawings suggested. In conclusion, labial gap width

Figure 7.4 Examples of odd things seen while surveying: (a) common toad tadpoles on an upside-down doll floating in a marshy pool on Preston Merse, north Solway coast, in May 1982; (b) stripeless natterjack caught near Mawbray in Cumbria, also in May 1982.

may be of more limited help in identification in the field than we have been led to believe.

Brian Banks, like Jonty Denton, was a student being supervised by Trevor Beebee at the University of Sussex. Both Brian and Jonty were involved in the 1980s with studying the relationship between age and size in adult natterjacks. This was a topic on which I collected as much information as possible from my field trips during the period 1979–1983, when my interest extended to growth from the toadlet stage to maturity as well. I attempted to work out what size animals might be throughout their seasonal period of activity and not just during the breeding season. This was done by using fragments of my and other people's information to construct an overall view. Here are two examples of useful observations:

- There was exceptional reproduction of natterjacks in the reserve at Caerlaverock in 1981, with the warden, Malcolm Wright, estimating that up to 20,000 toadlets left the ponds. Emergence in 1980 had been limited to a few hundred toadlets, so the bulk of juveniles in 1982 should have been one-year-olds. These proved easy to find, and Malcolm and I measured several samples of 14–32 juveniles in the summer: length ranged 20–30 mm on 19 May, 23–38 mm on 7 June, and 27–38 mm on 30 July.
- At Redrocks, Tony Bell reported there was good emergence in 1977 but none in 1975, 1976 or 1978 and very limited success in 1979, so when 12 males were found to have lengths between 47 and 54 mm in 1980, it was reasonable to assume these were probably all three years old. When the next measurements were done in 1982, 22 males ranged 56–73 mm and 15 females were 56–67 mm. This appeared to show (1) the lengths of five-year-old males from 1977, with no three-year-olds from the limited recruitment in 1979, and (2) five-year-old females of similar sizes.

By 1983, I had pieced together the available information to produce an indication of lengths that might be expected for the first six age classes: 10–25 mm during the year of emergence, 15–40 mm at one year of age, 25–58 mm at two years, 45–65 mm at three years, 50–70 mm at four years, and 55–75 mm at five years. Males less than 45 mm did not appear to have thumb pads. Brian Banks and Jonty Denton aged toads directly by counting growth rings in digits (Banks and Beebee 1986, Denton and Beebee 1993), and it seemed there were differences between the various populations, with, for example, much larger toads being found in the south at Woolmer. And within a colony there could also be differences in growth rate between years. Nevertheless, my approach seemed to fit most colonies reasonably well, and it could always be fine-tuned to adjust to local circumstances. It was used in 1983 to assist the warden, Keith Payne, when trying to interpret breeding data for the Ainsdale reserve over the period 1971–1983 (Figure 7.1b).

Ainsdale was the reserve where a Cambridge postgraduate, Anthony Arak, studied mating behaviour in the natterjack in the early 1980s. I spent an interesting evening with him as he described what he was doing, and we visited several slacks. In one, he pointed out an individual that he knew well: he called it 'Bottom Bend' because the dorsal stripe had a kink at its end. We left that slack and went some distance to the next one. On shining my torch into the water, the first natterjack I noticed had exactly the same kink in the base of the stripe. Bottom Bend could not have got there before we did! It is unwise to assume the appearance of an animal is unique, especially on the basis of one distinguishing character, if there are many animals in the population. In 1983, I co-authored an article on stripeless natterjacks in England, having found one at Mawbray in Cumbria in May 1982 (Cooke *et al.* 1983; Figure 7.4b). It was definitely a natterjack as (1) it had typical colouration elsewhere on its body, (2) it ran like a natterjack, (3) it gave the release call when handled, and (4) it had parallel paratoid glands. Although very unusual, it was not unique, as several others have been found in this country. I also caught a female with a yellow stripe only 5 mm in length in the same sample at Mawbray.

Anthony Arak worked at night because he was interested in calling and other breeding activity (Arak 1988), and natterjacks are mainly active at night. Most of my monitoring was, out of necessity, undertaken during the daytime, when natterjacks are much quieter and more difficult to find. In order to try to induce males to call and give away their presence, I often carried around a tape recorder and played a recording of a chorus of croaking. This was particularly useful at places such as Saltfleetby where they lived at low density in a well-vegetated environment (Figure 7.5). Trevor Beebee commented in his monograph that they can also be stimulated to call by noise from passing cars or aeroplanes (Beebee 1983). Once, at Sandscale in Cumbria, I was mobbed by scores of black-headed gulls (now *Chroicocephalus ridibundus*) which were nesting nearby, and their cries appeared to set off a chorus of male natterjacks.

7.3 Conclusions

So what did this monitoring and conservation activity by everyone achieve? The review I co-authored with Brian Banks and Trevor Beebee (Banks *et al.* 1994) assessed the period 1970–1990. The number of documented sites increased because more were discovered (having probably existed for many years) than were lost, and in addition there were a few translocations to new sites, such as to heathland at Sandy in Bedfordshire (Figure 7.6). In reality, overall numbers of natterjacks probably changed little during this period. Site protection improved, with the percentage of

Figure 7.5 Maurice Massey playing a tape recording of natterjack calling to try to induce a response from males in the marsh at Saltfleetby–Theddlethorpe on the Lincolnshire coast.

Figure 7.6 (a) Newly created natterjack pool on heathland at Sandy, Bedfordshire, May 1980: the site of the first successful translocation of the species in Britain; (b) a natterjack at the New Forest Reptile Centre, demonstrating good camouflage in a damp heathland environment (photo by Marc Baldwin).

sites that were SSSIs rising from 60% to 83%. Problems still existed, however, over some landowners being unaware of the presence of natterjacks and with habitat degradation both within and outside SSSIs.

The amount of knowledge that had accrued by the 1990s enabled Trevor Beebee and Jonty Denton (1996) to prepare a *Conservation Handbook* for the species. A Species Action Plan for the natterjack was formulated in the 1990s (Denton *et al.* 1997) and an updated plan is expected in the early 2020s (ARC 2021a). Thanks to further introductions, habitat and species monitoring, and a considerable conservation effort to protect and enhance sites, our national population of natterjacks remained more or less stable up to 2009 (Buckley *et al.* 2014).

Nevertheless, as pointed out by John Buckley and Trevor Beebee (2004), stability should be regarded as a modest achievement in view of the massive declines suffered by the species prior to 1970. Reg Wagstaffe studied the flora and fauna of the Sefton Coast from the 1920s until the 1960s; he was Ian Prestt's father-in-law and told Ian that 'until the early 1930s the natterjack population could be numbered in hundreds of thousands' (Prestt *et al.* 1974). This compares badly with a recent census estimate of roughly 1,000 breeding adults (Beebee and Buckley 2014a). Sadly, a new review of natterjacks on the Sefton Coast from 1987 until 2017 revealed significant declines in spawning and toadlet production over the last 10 years or so (Smith and Skelcher 2019). This was largely the result of a reduction in spring rainfall, meaning dune slacks (pools) were less likely to hold water until toadlets could emerge; and stabilisation of the dune system led to growth of vegetation on the natterjacks' open terrestrial habitat and loss of the opportunity for new slacks to develop (see Section 11.3). In addition, Trevor Beebee has told me that natterjacks have also declined during the last decade at several Cumbrian and Scottish sites.

It seems that the battle to conserve natterjacks is a continuous fight of monitoring, reviewing, acting to improve the habitat, and monitoring again. At least it is a species about which much is known of its habitat requirements – and if everything is in order, then the species responds. The problem seems to be that currently everything does not stay that way for very long.

CHAPTER 8

Newts of fen and forest

8.1 Introduction

This is the first of three chapters on newts, and it begins with studies to investigate how pond characteristics might influence the distribution of newts in Britain. Information was gathered from desk studies and from field surveys in two contrasting areas of the country: the Fens in Cambridgeshire and the New Forest in Hampshire. Following the groundwork done in the 1970s, studies in these two areas were extended in order to determine how and, if possible, why newt populations might change in the longer term.

This introduction to the first chapter on newts is probably a good place to delve into a couple of points of terminology, starting with the name 'great crested newt'. For many years after I began working on it, I referred to the species by its older name, 'warty newt', which was favoured by Malcolm Smith in the first New Naturalist volume on the amphibians and reptiles in the 1950s. By the 1980s, things were moving on and Deryk Frazer referred to the 'great crested newt' in the second New Naturalist book. In the third New Naturalist in 2000, it was introduced by Trevor Beebee and Richard Griffiths as the great crested newt but was then usually referred to in the text simply as the 'crested newt'. Nowadays, it may be referred to as the northern crested newt to distinguish it from closely related species (Jehle *et al.* 2011). I have tended to favour the name crested newt in more recent articles, but here I call it the great crested newt initially in each chapter before shortening its name to crested newt (except in Chapter 1). As regards what to call the aquatic young of newts: following on from Smith (1969), Frazer (1983) and from the title of this book, they are referred to here as 'tadpoles' rather than 'larvae', which is how they have usually been addressed in recent years. They then turn into 'efts' when they leave the water (Figure 10.10b).

8.2 Is the palmate newt a montane species?

When I started taking a serious interest in amphibians at the end of the 1960s, I bought several books on the subject. Three of them published during that decade were Walter Hellmich's *Reptiles and Amphibians of Europe*, J. W. Steward's *The*

Tailed Amphibians of Europe and the fourth edition of Malcolm Smith's *The British Amphibians and Reptiles.* I was not familiar with the palmate newt (Figure 8.1), so I read everything I could find about the species, but quickly became perplexed about its distribution. The first two books said it favoured hilly or mountainous country. Malcolm Smith began with the statement 'The palmate newt is a montane species'; but then qualified it by saying it lived at sea level in some parts of England, often being found with the other two species of newts. He also went on to state that in parts of northern England it lived in the hills and was replaced at lower levels by the smooth newt. These statements were intriguing, as I could not understand why topography by itself would be important, and so in 1973 Peter Ferguson I extracted altitude data from record cards for newts held by the national BRC at Monks Wood (Cooke and Ferguson 1975). This was a time before records were digitised on computer.

Altitude data were often included on the record cards; and for cards without this information but with a six-figure grid reference, we determined altitude from large-scale maps. All altitude data were recorded in feet (another anachronism), but are converted approximately into metres in Table 8.1. There was no significant difference between records for smooth and crested newts, but the palmate newt had fewer records up to 100 feet and more records above 500 feet. Three of the five records above 1,000 feet were for the palmate newt. So these differences were consistent with the view that palmate newts tended to be found more often at greater altitudes in Britain. But was there evidence of them living at higher altitudes where all three species were found together? Peter and I checked regions in various parts of

Figure 8.1 A female palmate newt, showing its unspotted pink throat – a female smooth newt has a spotted throat. Photo by Marc Baldwin.

Table 8.1 Altitude records for the three newt species in Britain, extracted from BRC records, 1960–1973 (adapted from Cooke and Ferguson 1975).

Range of altitude		Percentage of records		
Feet	Metres	Palmate newt	Smooth newt	Crested newt
Up to 100	Up to 30	18	30	29
100–500	30–150	66	60	65
More than 500	More than 150	16	10	6
Number of records		97	226	136

Britain where the three newts lived together and found no difference between their altitude preferences, but there were insufficient records from northern England to test Malcolm Smith's claim that the smooth newt replaced the palmate at low levels. Very recently, however, I have been told by Pete Carty of the National Trust that such a distribution occurs on the Long Mynd in Shropshire, where the altitude reaches about 1,700 feet (500 m).

The differences between species seen in Table 8.1 were largely due to a low proportion of records for palmate newts coming from the relatively flat areas of eastern England. We concluded that climate or properties of the soil substrate might be the underlying reason for the differences in distribution rather than altitude *per se*. Comparing the BRC's distribution maps with pedological and geological maps for Britain suggested that the palmate newt was very rarely found where there were saturated soils on chalk or clay, or calcareous soils on chalk or limestone.

8.3 Characteristics of breeding sites

In 1960, Kate Creed, who was then a biology student at Cambridge University, visited the New Forest with a party of other students to study the fauna, flora and physical properties of a sample of ponds. Her personal study focused on the newts: she looked at their abundance and egg laying in April in 10 ponds and returned to check for tadpoles in all of them plus an additional site in June (Creed 1964). She was advised and guided in her work by Deryk Frazer of the Nature Conservancy. Early in 1974, Deryk invited me to accompany him to repeat her fieldwork in April that year. I cannot remember our dialogue in any detail, but I probably said I would be pleased to do so, but why not expand the study to as many ponds as possible so as to try to understand breeding preferences of smooth and palmate newts – I would not have expected to find many ponds with crested newts, based on Kate Creed's data. However, I realised crested newts could be compared with smooth newts later in the spring in a group of experimental ponds at Woodwalton Fen in Cambridgeshire (Figure 1.5).

So in April 1974 we duly met at Lyndhurst in the New Forest. Deryk Frazer arrived with a large-scale map with 88 waterbodies marked on it, having consulted Colin Tubbs, who was the Conservancy's man on the ground and a mine of information. We had permission to drive around the Forestry Commission's tracks and were armed with a very precious key which unlocked all the many barriers. Deryk's battered campervan was used for transport, its only advantage over my Conservancy vehicle being that it had a Calor gas stove so we could brew tea at regular intervals. For my part, I came with a programme of work and with equipment. We would visit every pond and measure pH and oxygen levels in the water; and the pond would be 'scored' for size, depth, amount of aquatic vegetation and extent of open water. Each pond would be netted for newts for 15 'man-minutes' during daytime, focusing on those areas most likely to shelter them, such as beds of submerged vegetation. Sight records of newts would not be included because of difficulties in distinguishing between female smooth and palmate newts (Figure 8.1). Finally, water samples would be taken back to Monks Wood for analysis of concentrations of sodium, potassium, magnesium, calcium and iron.

The spring of 1974 was exceptionally dry, and 36 sites were either dry or too overgrown for netting, reducing the sample size to 52 sites: 50 were ponds and two were sections of streams. Ten of these sites were isolated, and as no newts were caught, water samples were not taken. Of the 42 remaining sites, newts were found in 29: palmate newts in 22, smooth newts in 14 and crested newts in three. Oxygen levels were at least adequate in virtually all of the sites. No newts were caught in the eight sites with pH of less than 3.9 (but see Section 8.5). Smooth newts occurred more frequently when pH was greater than 6.0. There was no difference between the two small species in the physical characteristics of the ponds they selected. Both tended to avoid small ponds with an area of less than 100 m^2 and selected sites with much aquatic vegetation. Depth of water and extent of open water did not seem to affect where they bred. Where they did differ, however, was that palmate newts tended to be found in sites with low metal concentrations, whereas smooth newts favoured metal-rich sites. We calculated an 'ionic score' for each site, which varied from zero for sites that had low levels of each of the five metals that were analysed up to five

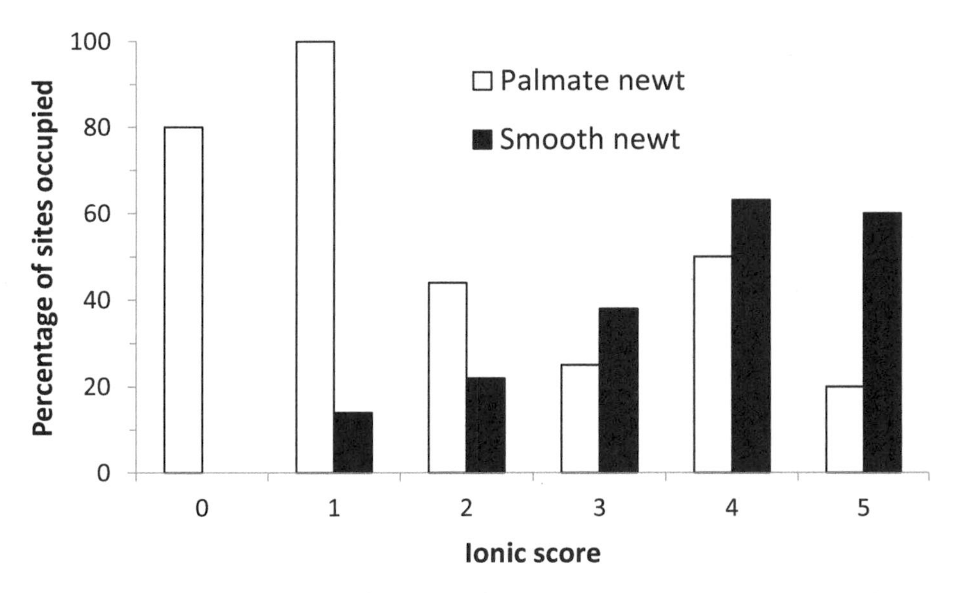

Figure 8.2 Palmate and smooth newts in the New Forest: percentage occupancy of 42 sites with varying ionic scores, 1974 (constructed from data in Cooke and Frazer 1976).

for sites with high levels of each metal. Smooth newts were rare in ponds with low scores while palmate newts were less frequent in those with high scores (Figure 8.2).

The metals that seemed to be most associated with the presence or absence of the two newt species were calcium and potassium, which made us wonder if newt distribution could be related to water hardness. To test this suggestion, we calculated the frequency with which each species was recorded in the 100 km squares that had the best overall cover in the most up-to-date national distribution maps and compared those data with a measure of water hardness derived from a map in *The Atlas of Britain and Northern Ireland* (Bickmore and Shaw 1963). Smooth newts were shown to be less commonly recorded in soft-water areas, while palmate newts were less common where there was hard water. Where water was of intermediate hardness, both species were reasonably common. Their different distributions on the Long Mynd in Shropshire are probably the result of water chemistry (Pete Carty personal communication; Section 8.2).

Crested newts mirrored the distribution of smooth newts. Deryk and I noted that, as a general rule, crested newts were rarely found in the absence of smooth newts. Between us at that time, we knew of 118 crested newt sites – and smooth newts were present in at least 94 (80%). Later in the spring of 1974, I netted the 20 experimental ponds at Woodwalton Fen for 15 minutes each to try to determine whether smooth and crested newts selected ponds with different characteristics. The fen is in a region with hard water, and the ponds typically have a pH of more than 7; the nearest palmate newts known to me are about 20 km away (Figure 2.8). The ponds at Woodwalton Fen had been dug in 1961 on a five-by-four grid with about 25 m between them. They were all 5 m in diameter, so variation in size was not a consideration, but depth and amount of aquatic vegetation and open water varied and were recorded. Such small ponds might have been ignored by newts in the New Forest. Here, however, they had been colonised naturally by smooth newts from 1963 and by crested newts from 1970, according to Norman Moore (1991), who had arranged for the ponds to be created for his research on the ecology of dragonflies.

Sight records were included with newts caught in the fen ponds because crested newts and smooth newts can be distinguished when seen clearly in small ponds. During the first netting operation in May 1974, I recorded both species in 9 ponds, so in June I repeated the exercise for the other 11 ponds, where one species or no newts had been recorded. The combined operation resulted in smooth newts being found in 18 ponds and crested newts in 10 (Cooke and Frazer 1976). The latter species tended to be found in ponds with more open water, and there was an indication that they also preferred the deeper ponds. A preference for ponds with deeper water had been previously reported (Steward 1969, Beebee 1973). Later, in their major national report, Mary Swan and Rob Oldham (1993) concluded that crested newts preferred sites in the size range 25–750 m^2 whereas smooth and palmate newts bred in a wider range of sizes of waterbodies including small garden ponds, but palmates tended to be found in nutrient-poor situations. All species required aquatic plant growth. The fact that crested newts show a preference for larger, deeper ponds without fish was referred to by Will Atkins (1998) as a 'catch 22', as the probability of fish being in a waterbody tends to increase with size. Gradually our knowledge of the requirements of the three species accumulated from the 1960s to the 1990s.

Deryk Frazer and I speculated in our paper why palmate and smooth newts were often found in waterbodies of different chemical composition. For instance, was it that palmate newt tadpoles could survive at greater levels of acidity, or were the two attracted to the odours of different algae that flourished in water with different chemistries? I did not follow up these avenues of research because their results might not have been critical for conservation of the species. There was always something more pressing and apparently more relevant to do, no matter how attractive such studies seemed. Looking back on this period, I am quite pleased that I did not spend more time trying to tease out the fine detail of why the three species had these distributions, as Trevor Beebee, in his review of amphibian ecology and conservation 20 years later (1996), admitted that complete explanations remained elusive.

8.4 Further studies in the ponds at Woodwalton Fen

8.4.1 Introduction

The 20 experimental ponds proved a useful resource for studying smooth and crested newts. The main focus was on monitoring colonisation of the ponds and trying to maintain good populations, while attempting to understand more about the species and how to conserve them. Later, after the crested newt had become protected under the Wildlife and Countryside Act 1981, regular monitoring of the species at the site was required by the NCC and its successor bodies.

This is a unique situation with a total water area of about 400 m^2 and a combined edge of roughly 300 m in a grassy compartment of a couple of hectares. I have never come across newts of any species in the nearby farmland ditches, but crested newts breed in ditches and ponds further north in the reserve. During the time I was involved with the populations of the two species in the 20 ponds, they probably experienced relatively little immigration, despite presumably having been initially colonised by reserve residents. Emigration is more likely to have occurred, especially when the population was high or if the ponds were less suitable for breeding. The following account of observations is in chronological order.

8.4.2 Crested newts become more widespread

By 1974, a body of data on newts in the ponds had started to build up from the recording done that year (Cooke and Frazer 1976) and also from observations made in relation to experimental treatment in 1972 of four ponds with the herbicides dichlobenil and diquat (Cooke 1977; Section 3.2.3). Smooth newts were found in 18 ponds in 1974, including all four treated with herbicide; crested newts were in 10 ponds, including three treated ponds. There was therefore no evidence of long-lasting effects of herbicide use on newt presence. Two ponds held large numbers of ten-spined sticklebacks in 1974: one of these ponds was devoid of newts, while the other had just smooth newts. That was the first time that I had surveyed all 20 ponds, but I was involved in such surveys eight more times up to 2005, 31 years later (Table 8.2).

The early netting programmes provided some insights into differences in behaviour displayed by the two species because I was recording separately what was netted in the zone of aquatic vegetation around the edges of the ponds and what was seen

Table 8.2 Survey results for newts in the 20 experimental ponds at Woodwalton Fen: numbers of ponds where presence was confirmed, 1974–2005.

Year	Month	Method	Number of ponds with newts	
			Crested newts	Smooth newts
1974	May	Netting	10	18
1979	May	Netting	17	20
1982	June	Torch counting	18	4
1985	May	Torch counting	19	8
1987	May	Torch counting	20	17
1992	July–September	Netting tadpoles	18	19
2001	May	Torch counting	14	5
2003	May	Torch counting	16	10
2005	May	Torch counting	6	1

during the day in the centres of the ponds, which could not be reached with the net (Cooke and Frazer 1976). Relative to smooth newts, adult and immature crested newts tended to be seen in the centres rather than netted around the edges. Data suggested that small crested newt tadpoles tended not to be caught because they stayed away from the edges, while larger tadpoles more often came to the edges of the ponds. Thus, no tadpole of less than 2 cm long was ever caught, and more than half of those netted were 4 or 5 cm in length. I never discovered where the smaller tadpoles were, but I assumed they tended to be in the weed in deeper water out of the range of the pond net. They would in any case have needed to keep out of the way of cannibalistic adults. This behaviour tended to be different from that at night-time, when larger tadpoles could be seen behaving in a pelagic fashion (out in open water) (e.g. see below and Section 10.4).

The ponds dried out during the hot, dry summer of 1976, but then refilled in the wet autumn and winter. In 1979, I joined my colleagues Al Scorgie and Christine Brown to repeat the netting and observation exercise of 1974 (Cooke *et al.* 1980; Figure 8.3). This time, fish had evidently not recolonised following elimination in the drought of 1976, and smooth newts were in all ponds. Crested newts had continued to colonise and were found in 17 ponds, although the bulk of those netted were immature specimens. This was the first time that smooth newts were recorded in all ponds – 18 years after the ponds were created.

Most of the later emphasis was on crested newts. Keith Corbett first recommended that I should use counting by torchlight rather than netting as a means of assessing or monitoring status of the species, and I was certainly heavily into night work by June 1982 when the experimental ponds were tackled again. Crested newt adults and immatures were counted in 10 ponds. Numerous crested newt tadpoles were seen in the centres of 15 ponds, meaning this species was in 18 ponds. In contrast, smooth newt adults were only seen in four ponds, but the count was in June so it was possible that most smooth newts had left the ponds by that time. Another

Figure 8.3 Christine Brown and Al Scorgie netting one of the experimental ponds at Woodwalton Fen in 1979. The barbed-wire fence was to prevent grazing cattle, maintained for conservation reasons, from accessing the ponds. The bund behind was designed to keep floodwater out of the pond compartment during periods when much of the reserve was used as a flood storage area. For another view of the pond area, see Figure 3.4.

possibility was that using a torch was not as efficient for smooth newts as it was for crested newts and that netting was a better method for confirming presence. Searching for eggs as a means of detecting breeding by newts was not yet in use at that time.

Torch counting was next undertaken in May 1985, when crested newt adults and immatures were in 19 ponds, while single adult smooth newts were seen in only eight ponds. This was the first of five May-time counts up to 2005, which allowed cautious comparison over the years (Table 8.3). The pond where crested newts were not recorded had the most turbid water, so newts were quite likely to have been missed. The fairly low number of smooth newts and low pond occupancy suggested either that a real decrease had occurred during the 1980s or that the apparent decrease was due to a change of method. The count in 1985 was a family affair, as both my wife Rosemarie and son Steven helped.

Counting by torchlight was again used in May 1987, when it was more or less a repeat performance but with the warden, Ron Harold, and my fellow night-counter Roy Bradley (Figure 12.4b) taking part. That night, crested newts were counted in every pond, with a total of 179 adults and immatures (Table 8.3). Smooth newts were found in 17 ponds, with a total of 49. This outcome did not completely resolve the issue of the suitability of the method for smooth newts, but it did suggest that the smooth newt population dipped in the early 1980s, before recovering much of the losses by 1987.

Table 8.3 Total torchlight counts of newts in the 20 experimental ponds during May in five years, 1985–2005.

Year	Crested newts		Smooth newt adults
	Adults	Immatures	
1985	57	7	10
1987	113	66	49
2001	56	1	6
2003	53	5	13
2005	19	2	1

8.4.3 *Newts and shade*

The ponds eventually began to scrub over (Moore 2001). All had small bushes around their edges by 1980, and management was undertaken on an ad hoc basis to prevent scrub from impeding Norman Moore's observations on the dragonflies. I was consulted about management in 1987 and requested that five ponds were left unmanaged for a while to see what effect this had on the newts. Steven went to Nottingham Polytechnic in 1990 to take a combined degree course in biology and chemistry, and we both saw an opportunity for an interesting college project at the end of his second year in the summer of 1992. His college supervisor was Rob Beattie, whom I already knew as a member of the amphibian researchers' discussion group that had formed in the 1980s (Section 12.4). By that time, Steven had become an experienced fieldworker. One of the main aims of his project was to determine the effects of scrub cover on newt breeding success in the 20 ponds. Each pond was netted for tadpoles for five minutes on six occasions from mid-July until late September. One pond was dry by late July, but the others held water throughout the survey period. He limited netting time so as to minimise disturbance to the aquatic vegetation and the dragonflies, which were still being studied by Norman Moore. Steven found (Cooke 1993):

- Ponds with high amounts of scrub cover had lower densities of newt tadpoles, this being more marked for crested newts.
- Excessive scrub cover was associated with delayed hatching of ova for both species.
- Lengths of crested newt tadpoles were not significantly affected by scrub cover.
- Lengths of smooth newt tadpoles decreased in ponds as scrub cover increased, but tadpoles were larger than expected in the most scrubbed-over ponds, perhaps because of reduced interspecific and intraspecific competition.

Steven and I later enlisted the help of Tim Sparks, who undertook more advanced statistical analysis of the relationship between crested newt productivity and scrub cover, and demonstrated that numbers of crested newt tadpoles were highest if there was about 5% cover, but larval numbers were low when cover exceeded 30% (Cooke *et al.* 1994). These conclusions fitted qualitatively with those of Swan and Oldham (1993) in their national survey of breeding sites: the species preferred ponds with a little shade (up to 25%) to those with no shade or more than 75% shade.

The technique of netting tadpoles on six occasions for five minutes each time revealed crested newts in 18 ponds and smooth newts in 19 (Table 8.2). The frequency with which single netting sessions detected tadpoles did not decrease until September; during July and August, the two species were found in 13–15 ponds and 15–17 ponds respectively during each netting session.

8.4.4 Newt counts decline

The last three counts with which I was involved were in 2001, 2003 and 2005. The site manager Alan Bowley took part, as did Roy Bradley, and Steven was able to come in 2001, but then moved away from the area. During the winter of 2000/2001, there had been a programme of scrub reduction. Pond occupancy was disappointing in 2001, with crested newts in 14 ponds and smooth newts in only five. These totals could be partially explained by fish being seen in three ponds, while two of the three ponds that were completely scrubbed over by 2000 had no newts in 2001 despite scrub clearance. It is probably worth saying something about fish colonisation at this point. The reserve acts as a flood storage area during times of high precipitation, and I have known temporary floodwater to be more than a metre deep in parts of the reserve. Although this can enable fish to colonise ponds and ditches over most of the reserve from the adjacent river, the compartment with the 20 ponds is surrounded by a bund and should be protected from flooding (Figure 8.3). But fish have still found a way to colonise.

Further scrub clearance occurred later in 2001, leaving only four ponds with more than 10% scrub cover by the spring of 2003. Torchlight counting that year revealed fish to be very abundant in two ponds where no newts were seen. Fish were also recorded in two other ponds in smaller numbers. Nevertheless, pond occupancy by newts had recovered to a total of 16 ponds with crested newts and 10 with smooth newts.

Counting was difficult in 2005. The night of 10 May was cold and clear. The temperature at the reserve office was 7 °C when we left for the ponds (I only went out with a torch if air temperature was above 5 °C). However, there was ice on the all-terrain vehicle by the time we returned, so weather conditions that evening were not good. Neither were pond conditions ideal, as vegetation cover at the surface was at least 80% on 14 ponds. It was not a complete surprise that crested newts were seen in only six ponds and just a single smooth newt was recorded. No fish were seen. This was also the lowest total night count of crested newts, with just 19 adults and two immatures; the highest total had been in 1987 (Table 8.3). The count in 2003 had, however, been very similar to the night counts in 1985 and 2001, so I was not unduly concerned by the low count in 2005.

For various reasons, that was the last time I surveyed those ponds. However, I checked water levels occasionally during winter when I visited the reserve to monitor deer. Norman Moore reported in 2001 that, apart from the drought year of 1976, no ponds desiccated up to 1989; but then six dried in 1990, two did so in 1991 and 1995, and three were dry in 2000. And there may have been some desiccation in 1997 and 1998. My own records indicate that, up to 2015, some ponds probably dried out in 2003, 2006, 2009, 2010 and 2011. This will have meant that at least predatory fish populations were unlikely to survive for long.

One of the problems with these small ponds is that they have an exceptionally high ratio of shoreline to water area. Once the scrub is established it can be difficult to eradicate, even with herbicide treatment of the cut stumps. Anyway, eradication was not then my general aim, as some scrub on the north side of a pond can be beneficial by ensuring clear areas persist and the pond is not choked by submerged vegetation (e.g. Figures 9.3d and 9.9b). In November 2014, there was another programme of scrub removal, but several very overgrown ponds were left untouched. In discussion with the Alan Bowley in December 2014, I suggested creating a single large pond in an adjacent field. The higher area-to-edge ratio of a large pond would reduce the frequency of the need to control fringing and overhanging scrub.

Terry and Helen Moore took over torch counting from 2014 to 2017, but had practical problems, particularly with accessibility to the edge of ponds: up to six ponds were not safely accessible each year, mainly because of encroachment of scrub and brambles. Relatively low numbers of newts were recorded with, for instance, crested newts being seen in 3–8 ponds each year. Newt populations in the 20 ponds seemed to be ticking over at a low level.

In 2020, the reserve manager, Katy Smith, told me that scrub had been gradually reduced over the previous three years. Woodwalton Fen is within the Great Fen (Section 6.4) and Henry Stanier, monitoring officer for the Great Fen, confirmed to me that crested newts were still breeding in the small ponds in the early 2020s, meaning they had survived there for more than 50 years. They are also known to breed in three ponds further north in the reserve (Figure 8.4). The newts' future

Figure 8.4 The author at one of the larger ponds at Woodwalton Fen that contained breeding populations of crested and smooth newts in 2020. This is more than 1 km from the group of 20 small experimental ponds. In some compartments, the terrestrial habitat is dominated by reed and sallow carr. Photo by Rosemarie Cooke.

in the reserve looks secure, and population sizes of the two species are probably considerably greater than before the experimental ponds and the other ponds were created.

8.5 Further studies in the New Forest ponds

8.5.1 Aims and surveys

Section 8.3 describes surveys of the newts in the New Forest undertaken by Kate Creed in 1960 and Deryk Frazer and myself in 1974. Deryk and I subsequently made further separate visits to the Forest: we were both interested in changes in numbers over time, but whereas he focused on attempting to understand distribution more completely, I concentrated on changes in water acidity.

In 1974, Deryk and I had found that smooth and crested newts occurred more rarely in water that was more acidic than pH 6, although it was recognised that it might be some related factor rather than acidity *per se* that was responsible (Cooke and Frazer 1976). Deryk returned to the Forest in 1975 to repeat observations at some of the ponds (Frazer 1978). One of the points made in his paper was that at least three ponds studied by Creed appeared to have acidified by more than one pH unit since 1960.

By the early 1980s, there was worldwide concern about acid rain and, in particular, its effects on freshwater habitats (Section 3.6.3). There was no evidence to suggest that acid conditions in the Forest were associated with pollution, although there was a power station close by at Fawley, and it was also known by that time that conifer plantations could cause acidification of local freshwaters. Thus it seemed worthwhile checking whether pond acidity had changed. Accordingly, I visited the Forest in May 1982, and measured pH in 37 of the ponds recorded in detail in 1974. I concentrated on those in clusters as they required less time to check than isolated ponds; they comprised groups of waterbodies at varying distances of 6–27 km from the power station. An additional cluster of five ponds was added to the sample, these being 6–7 km from Fawley. In order to improve accuracy of measurement, three water samples were taken from each site, pH was recorded with two meters, and the average of the six values was calculated.

Derek Frazer revisited the Forest in 1983 and 1984 to net for newts (I had not undertaken netting in 1982). He was told about additional sites by Forestry Commission staff and covered 43 of our original sample in 1974, plus 66 additional ponds that were suitable for netting (Deryk Frazer personal communication). He also listed scores of others which he regarded as being unsuitable for newts or for netting; many of these were dry. Newts were found in 59 ponds, with palmate newts in 49 and smooth newts in 21, again indicating the greater flexibility of the palmate newt in choice of breeding site across the Forest. By this time, it was clear there was no definitive list of New Forest ponds, and whether many existed in any particular year depended on weather conditions. In all, Deryk found eight ponds with crested newts.

By 1989, it seemed to be time to have another look at the ponds in the New Forest, so Sarah Woodin and I revisited 34 ponds that had pH and newt data for both 1974 and the early 1980s; pH measurement and netting of newts were repeated.

8.5.2 *Changes in pond acidity*

As far as possible, I have compared pH data from Kate Creed in 1960 with what Deryk Frazer and I found in 1974. In 1960, pH papers were presumably used to measure acidity, and some of the results were ranges rather than precise values (Creed 1964). She surveyed 11 ponds in 1960, nine of which we recorded in 1974 using a pH meter (Cooke and Frazer 1976). Although one pond had a drop in recorded pH of greater than two units and there was a decrease in pH overall, it was not significant in the statistical sense. I was aware that even with meters the accuracy of recorded pH levels had a degree of uncertainty, and I later learnt that desiccation could affect pond pH over a period of years (van Dam 1988).

Comparisons of pH values measured during the three surveys in 1974, 1982 and 1989 could, however, be undertaken with rather more confidence. Of the 37 ponds recorded in 1974 and 1982, for example, average pH changed minimally from 5.6 to 5.5; pH decreased in 20 ponds, increased in 16 and remained the same in one. There was, however, one element of concern. A subset of nine ponds had a pH value of greater than 5 in 1974, but based on the chemical composition of their water, they were likely to be poorly buffered against acidification. On average, these ponds became more acidic in 1982 by 0.8 of a unit. Although such a change might have implications for life in freshwater, its relevance for the sustainability of newt populations was unclear.

The power station at Fawley is to the east of all the ponds, so the prevailing south-westerly wind should carry pollution away from the Forest, but of course the wind does not always blow from that direction. In 1982, the average pH of pond clusters, of 9–12 sites each, increased in a westerly direction away from the power station: 6–7 km, pH 4.4; 10–11 km, pH 5.1; and 12–17 km pH 6.8. But then the most distant group at 24–27 km had an average pH of 4.8, so the trend did not extend beyond 20 km. Consulting a large-scale geological map did not shed much more light on the issue; in some cases, ponds on brick-earth substrate had a relatively high pH, but heathland ponds on plateau gravels had variable pH values.

Most of the concerns about acidification and its effects on newts in the Forest disappeared after the survey in 1989. Average pH for the whole sample was 5.8 (having been 5.6 in 1974 and 5.5 in 1982). Much of this slight change was due to an increase in pH of the subset of nine poorly buffered ponds: pH in five of these ponds increased by more than one unit, demonstrating the extent to which they could fluctuate. One of these ponds was at Janesmore (Figure 8.5). The pattern of how acidity in the pond clusters varied with distance from the power station remained the same as in 1982.

8.5.3 *Changes in newt abundance*

As regards the newts, data are assembled in Table 8.4 to show the number of ponds where newts were netted and the number of newts caught in 1974, 1983–1984 and 1989. There were no significant changes over time. Crested newts were relatively rarely caught, but Deryk Frazer had already demonstrated that netting alone was not a very effective method of either detecting crested newts or providing comparative information on numbers in the Forest's ponds. We continued with the method for convenience and so as to survey in the same way throughout.

Figure 8.5 Janesmore Pond is one of the larger ponds in the New Forest. It was in the poorly buffered subset, and its pH changed from 6.1 in 1974 to 6.2 in 1982, and to 7.5 in 1989. The pond had little aquatic vegetation, and perhaps as a consequence only low numbers of smooth and palmate newts were caught.

Table 8.4 Details of newts netted in the same 34 ponds in the New Forest in three time periods. Ponds were netted for 15 minutes.

		Number of ponds with newts (number of newts caught)		
Year	Source of data	Crested newts	Smooth newts	Palmate newts
1974	Cooke and Frazer (1976)	3 (11)	10 (52)	17 (100)
1983–1984	Frazer (unpublished)	4 (6)	9 (42)	14 (70)
1989	Cooke and Woodin (unpublished)	4 (4)	10 (33)	26 (96)

Each of the ponds with smooth and/or crested newts in 1989 had a pH of greater than 6.0. Palmate newts were found across the whole range of pH values from 3.3 up to 8.3; the former figure was the lowest pH at which I have ever found newts to be present. In the nine poorly buffered ponds, palmate newts were the dominant species, being found in seven ponds in both 1974 and 1989; smooth newts were in two in 1974 and three in 1989.

It is difficult to make direct comparisons with numbers of newts caught by Kate Creed in 1960 as she did not record time spent netting. Of the 10 sites she netted, two could not be relocated with certainty in 1974 and three others were netted in 1974 but not subsequently. In the five remaining sites, she caught more than 100 newts in two and more than 10 in the other three, whereas we netted more than 10 newts in four sites. This discrepancy in numbers caught may be due more to duration of netting than to a reduction in numbers. She found all three species in three sites. We also found all the species in three sites, but not the same three sites.

8.5.4 *Postscript*

That was not quite the end of the story. In July of 1998, a month after leaving English Nature, I bumped into an old NCC colleague, Jonathan Spencer, at a gathering in London. By that time I had just left English Nature, while Jonathan was working for the Forestry Commission and had some responsibility for funding research and survey work. I explained about the surveys of the newts in the New Forest and he said that if I wanted to repeat the exercise in 1999, then he might be able to provide expenses. The status of the newts was of interest in part because of the global declines in amphibian species that had been reported over the previous 10 or so years. However, plans were thwarted by illness six months later and that survey never happened.

8.6 Conclusions

Initial studies with Deryk Frazer revealed that the palmate newt was only a 'montane species' because of its preference for areas with soft water, which occurred more frequently in upland areas, whereas the smooth newt and crested newt tended to be hard-water species (see Beebee and Griffiths 2000 for more recent discussion). Surveys underpinning our conclusions were undertaken in the New Forest and at Woodwalton Fen, and in both areas surveying was continued to determine how populations changed in the longer term.

At Woodwalton Fen, the 20 ponds had been created in 1961 for the purpose of research into the ecology of dragonflies. Smooth newts quickly began to colonise naturally in 1963, with crested newts following in 1970. Smooth newts were recorded in all 20 ponds by 1979, and crested newts managed the same feat in 1987, so the species were found in all ponds 16–17 years after being initially reported. For the rest of the twentieth century, pond occupancy by crested newts remained at a high level, but the number of ponds where smooth newts were found fluctuated considerably. From 2005, numbers of ponds with newts were low for a variety of reasons: one count was done under poor weather conditions; fish inhibited newt presence in some ponds; while scrub and bramble developed around and over the ponds, causing problems by reducing accessibility for recording, lowering water temperature, and shading out aquatic vegetation. Few or no tadpoles were produced in ponds with much scrub. Management of scrub was needed at frequent intervals, and cleared ponds seemed to take time to become attractive again to newts. Conserving newts in small ponds at a location where scrub is established is likely to require more management input than would a single large pond, but then these particular ponds were not created with newt conservation in mind. Management of shading scrub was identified as a 'key element' of pond management by Jehle *et al.* (2011). Although Steven's study indicated that the optimal level of scrub cover for tadpole production was 5% rather than zero (Cooke *et al.* 1994), it can be difficult in practice to maintain cover at such a level.

Another very important issue relating to survival of amphibian populations is the proximity of other breeding sites of the species. Griffiths and Williams (2000) modelled crested newt populations and found, not surprisingly, that isolated colonies have a greater risk of extinction. It is advantageous if there are several sub-populations breeding in sites sufficiently close for some exchange to occur – in other words, a metapopulation. Later, Griffiths *et al.* (2010) described a metapopulation in Kent that survived throughout a 12-year study despite only one sub-population regularly breeding successfully. The crested newts through the whole 206 ha site at Woodwalton Fen can be regarded as a metapopulation, with those in the area of the 20 experimental ponds being, at least in past decades, the most important sub-population. It is not easy to envisage how crested newts could become extinct in the reserve. For instance, in the unlikely event of persistent long-term flooding, predatory fish populations might become widely established in other waterbodies in the reserve. The experimental ponds, however, are better protected against flooding; and although scrub encroachment has been a problem which has affected the size of the sub-population, there is much variation between ponds and some show little sign of becoming covered with woody growth. In addition, reserve managers are well aware of the need to conserve the sub-population, and of the habitat requirements of the species.

The situation in the New Forest is very different. Large numbers of ponds of varying sizes and water chemistries occur in the National Park. In the studies reported, there were several areas where a number of ponds occurred within a few hundred metres of one another and metapopulations of newts may have resulted. The terrestrial environment seemed to change little over time, while amounts of water in the ponds varied depending on season and levels of precipitation.

In the New Forest, my main consideration was how pond acidity affected newt distribution. Palmate newts were found down to pH 3.3 whereas smooth newts were only rarely found below pH 6.0. For a time, I was concerned that acid deposition from industry might decrease the pH of ponds and affect newt distribution and status. However, although acidity fluctuated in the group of nine poorly buffered pools, these were primarily used by palmate newts, which were unaffected. Over a period of 15 years, acidity did not increase overall in the whole range of ponds examined. Crested newts were and are probably more widespread and abundant than was indicated by the netting programmes, and newts generally should continue to be an important and interesting element of the New Forest's freshwater wildlife.

CHAPTER 9

Here be dragons: development and the great crested newt

9.1 Introduction

'Here be dragons' was a label often applied by early cartographers to unexplored regions of the globe as a general warning to the unwary. The map-makers may or may not have believed in dragons, but such an area was very likely to have dangers of various kinds. The phrase seemed appropriate as a title for this chapter because since the early 1980s the possible presence of that tiny dragon, the great crested newt (Figure 9.1), has caused apoplexy amongst developers and journalists – and created work for countless environmental consultants. The annual cost of such work has been estimated to be between £20 million and £43 million (Lewis *et al.* 2017).

In the first Covid summer of 2020, Boris Johnson delivered his 'Build, build, build' speech to encourage the UK to recover from the huge economic problems caused by the virus. In it, the Prime Minister attacked the 'newt-counting delays' to the planning process that had been a 'massive drag on the prosperity of our country'. This was despite one of the central tenets of his programme being to build 'greener'. Jim Foster, Conservation Director of ARC (Figure 11.1), immediately responded that the speech seemed 'at odds with the government's stated ambition' in the Environment Bill (ARC 2020a). He continued that, 'Conserving newts brings a range of benefits to other wildlife and importantly to people, and doesn't sterilise large areas of land'.

This chapter deals with broader issues than just development, but it is not my intention to produce a treatise on the protection and conservation of our crested newt populations, particularly as Jim Foster has achieved that in his review with Robert Jehle and Burkhard Thiesmeier (Jehle *et al.* 2011). Instead, it begins with a brief description of my involvement with the Wildlife and Countryside Act of 1981 and how that affected what I did on newts for more than 30 years. The following three sections discuss specific case studies. Chapter 10 deals with my main and longest study on crested newts at Shillow Hill in Cambridgeshire.

Figure 9.1 Even unpromising-looking ponds can harbour crested newts. Photo by Mihai Leu.

9.2 Protection for the crested newt – and its consequences

The first person to raise concern about the crested newt in Britain was Trevor Beebee in 1975: his enquiry indicated considerable losses overall during the 1960s and early 1970s. As NCC's herpetological adviser, I was consulted in 1980 about whether this species should receive the same protection under the impending Wildlife and Countryside Act as the natterjack, sand lizard and smooth snake, which were already protected under the Conservation of Wild Creatures and Wild Plants Act 1975. At that time, the crested newt was known to be quite widely distributed in Britain (Arnold 1973), whereas only 30–40 natterjack sites were documented in the whole country (Banks *et al.* 1994). So the crested newt was certainly not in the same category of rarity as the already-protected natterjack. I could appreciate that such protection would help conserve both the newt and pond life in general, but I could not envisage how it might work in practice. Knowing there were probably many thousands of breeding sites was not the same as knowing their locations. However, the UK had just become a signatory to the Bern Convention, a binding international agreement to conserve wildlife and habitats over much of western Europe; and the crested newt was specifically included on Annex II of the Convention, due to its decline in several countries. Therefore its legal protection was necessary in the UK, which was viewed as a potential European stronghold; and it duly appeared on Schedule 5 of the 1981 Act as a species that received the same level of protection as the natterjack toad. The other widespread species of amphibians received much more limited protection in relation to taking for trade and sale.

My main NCC role with the regard to the crested newt was facilitating the accumulation and dissemination of knowledge on where it occurred, what its terrestrial and aquatic requirements were, how and why populations were changing, and how the species and its habitat could be managed and conserved most effectively. This was an enormous task, and time and resources were limited. Gradually, information and responses to cases and queries were improved and refined. Much of the fact-finding was commissioned with Rob Oldham and his colleagues, Mark Nicholson, Richard Griffiths and Mary Swan, at Leicester Polytechnic, which resulted in a series of important contract reports in 1986, 1989 and 1993. But other people such Clive Cummins, Tom Langton, John Baker, Robert Bray, Tony Gent and Jim Foster were also closely involved with crested newts through the 1980s and into the 1990s. Keith Corbett (Figure 12.3) was the person who really drove the early interest in surveying for crested newts, especially at night. Tom Langton told me he remembers the huge amount of help and guidance he received from Keith when, as a teenager, he started surveying newt ponds in London in the late 1970s.

A different type of task in the mid- to late 1980s was to formulate a method for identifying sites worthy of designating as SSSIs on the basis of their crested newts. I was considerably aided in this by being able to consult with researchers and conservationists at annual meetings on amphibian research and recording that had started around that time. In the SSSI guidelines published in 1989, the eventual recommendation for choosing a site was that night counts should exceed 100 individuals over each of three breeding seasons (NCC 1989). If no site qualified in one of NCC's 'areas of search', then the site with the best population could be considered for notification providing a thorough survey had been done. This approach did not satisfy everyone, as it was clear that SSSIs would only protect a small percentage of the nation's total number of newts (unlike with the natterjack; Section 7.3), but that was never the intention of designation for this species. At the same time, I devised a scheme for assessing and designating the very best sites for assemblages of the widespread amphibians, including the crested newt. Jim Foster reported that in 2010 there were 51 SSSIs for which the crested newt was the prime reason for notification; and 32 of these sites were also Special Areas of Conservation (Jehle *et al.* 2011). Notification as an SAC through the EU's Habitats Directive conferred a higher level of protection: the sites formed part of the European Natura 2000 network (Beebee and Griffiths 2000). The reserve at Peterborough brick pits became both an SSSI and an SAC (Section 9.5.3).

I contributed to a certain amount of casework by advising and visiting sites in England and Wales with problems or exceptional populations of crested newts. But I lived in a good area of the country for crested newts and enjoyed honing my skills and improving the database by surveying and studying my local populations. By 1983, I knew of 46 sites or areas where the species occurred in the former county of Huntingdon and Peterborough (Cooke 1984). Although I describe the 1980s as the 'natterjack years' in Chapter 7, that decade was, in truth, more dominated by crested newts. However, as newt work mainly consisted of surveying in the evenings, it was largely done outside normal office hours. Rosemarie took to referring to herself as a 'newt widow'. Our son, Steven, often accompanied me.

Local studies could be divided into three main themes:

- First, I tried to obtain an overall indication of how and why the crested newt 'resource' might be changing in my area.
- Second, by monitoring certain populations in the longer term, I hoped (1) to understand whether introductions and remedial management were successful, and (2) to determine the stability of populations that did not appear to be significantly affected by human activity.
- And finally I resolved to take opportunities to gather information of relevance to the ecology of the species and its conservation in the broadest sense.

Out of the 46 local sites known in 1983, I was sufficiently familiar with 22 to be able to say how the populations had fared since 1970: three had increased in size, eight had not changed, seven had decreased, and four had been totally lost (Cooke 1984). In the last four cases, breeding sites were filled in because of housing development, expansion of playing fields and 'improvement' of farmland and a brownfield site.

In 1989, I decided to check 16 of the original sites to see whether the breeding site still existed and whether terrestrial land use had changed (Cooke 1990). In one case, the pond had been filled in, and in a second case it had been partially filled in. At two other sites, land use had changed, but not so that it would detrimentally affect the newts. In addition, 23 previously unrecorded sites in Huntingdon and Peterborough were documented between 1984 and 1989; at one, land use changed benignly during this period. Site loss per annum was estimated to be 4.6–4.9% during 1970–1980, 0.6% during 1980–1989, and 3.3–3.6% overall during 1970–1989. There was therefore a decrease in the rate of site loss after the Wildlife and Countryside Act came into being in 1981. Although not perfect, this legislation did benefit the species by giving it some legal protection and by drawing it to the attention of many people for the first time in their lives. Even the less serious and more excitable media coverage will have helped. For instance, I received a mention in a cartoon in the local newspaper: two men were drawn running away from an erect Godzilla-sized newt with one saying to the other, 'Quick, what is Arnold Cooke's phone number?'

I started torch counting in 1982, and from 1986 this method was used to monitor a small number of colonies annually (Table 9.1). Predation by sticklebacks was probably largely responsible for crested newts dying out in Site 3 – the site persisted but the newt population did not. Successional and weather changes led to Site 5 drying out. Numbers in Site 7 declined because of inappropriate or lack of management. Nationally, huge numbers of sites have been lost as a result of neglect. The local losses were partially offset by the successful introduction to Site 8 in 1985 (Section 9.4.1). Turbid conditions hindered counting in some years in Sites 1 and 2; newts were recorded in both ponds when they were checked again during the breeding season of 2005. If recording and breeding conditions remain reasonably stable then counts should broadly reflect changes in the size of breeding populations. When conditions change, however, some inter-year fluctuations can be considerable. For example, counts at the two ponds at Shillow Hill (Sites 4 and 5) illustrated the impact of weather: in the two years when Site 5 was dry in spring, counts at nearby Site 4 were unusually low because of reduced water levels. This is discussed at greater length in Chapter 10. Overall loss of populations for the small sample of eight sites was 1.1% per annum, based on those years of data and a prior knowledge of some of the colonies; if the introduction at Site 8 is excluded, loss was 2.6% per annum.

Table 9.1 Torch counts of adult crested newts at eight sites in the former county of Huntingdon and Peterborough, 1986–1998. Any immature newts seen are also included. Values are single counts, or averages where the number of counts is given in brackets (updating Cooke 1994).

	Night counts of crested newts							
Year	Site 1	Site 2	Site 3	Site 4	Site 5	Site 6	Site 7	Site 8
1986	3	12	9	150 (9)	6	102	17	3
1987	12	7	23	187 (9)	8	29 (2)	16	2
1988	7	0	8 (2)	140 (5)	3	16 (3)	11 (2)	3
1989	0	2	2 (3)	40 (5)	2	67 (4)	31	0 (2)
1990	4	3	0 (3)	35 (5)	0	55 (3)	30	0
1991	12	9	0 (2)	19 (5)	0	106 (3)	39	0
1992	13	12	1 (2)	3 (5)	Dry	74 (3)	29	3
1993	11	11	2	43 (5)	0	123 (3)	19	5
1994	3	6	0	28 (5)	0	75 (3)	0	10
1995	13	6	0	36 (5)	Drying	55 (3)	4	10
1996	5	7	0	71 (5)	0	41 (3)	18	19
1997	0	1	0	9 (5)	Dry	56 (3)	4	19
1998	—	—	—	102 (5)	0	35 (3)	11	27

Site 1, Monks Wood, pond 75; Site 2, Monks Wood, eastern hillside; Site 3, Ramsey, Field Road; Site 4, Shillow Hill, Top Pond (see Chapter 10); Site 5, Shillow Hill, Wood Pond (see Chapter 10); Site 6, Stanground, Buntings Lane, East and West Ponds (see Section 9.3); Site 7, Stanground, Peterborough Road, roadside pond (see Section 9.3); Site 8, Ramsey, Worlick Farm fish ponds (see Section 9.4.1).

Information on breeding-site loss from various parts of this country was reviewed by Jehle *et al.* (2011): overall losses were reported in all six local surveys mentioned, with a national loss estimated to be 1.4–2.0% per annum by the mid-1990s. An annual loss of 2% may not sound very serious, but it represents a loss of one-third of all sites over a period of 20 years. Tom Langton estimated that such an overall loss of crested newt breeding sites occurred between 1981 and 2009 (Langton 2009).

My formal responsibility for advising NCC on amphibian conservation ended around 1991, but this did not prevent involvement with the species for more than another 20 years on a personal basis. National interest in crested newts continued to accelerate in the 1990s and beyond. Some of the more notable publications and events relating to conservation of the species in the last three decades were as follows:

- A symposium on conservation and management was organised at Kew Gardens in 1994 by Tony Gent and Robert Bray and published by English Nature that year.
- The UK's BAP was launched in 1994, and included the crested newt as a priority species.

- A symposium on ecology and management of the species was organised in London in 1998 by the Research Committee of BHS and published as an issue of *The Herpetological Journal* in 2000 (Section 12.6.2).
- Two important guides that appeared in 2001 were a *Conservation Handbook* (Langton *et al.* 2001) and *Mitigation Guidelines* (English Nature 2001). The handbook was intended to supplement the SAP and its associated work programme. The guidelines were aimed at projects where there was potential for conflict between newts and the purpose of the development.
- A report was submitted to the European Commission highlighting the unsatisfactory state of surveillance, monitoring and general conservation of the crested newt in the UK (Langton 2009). There had been heated debate on how many breeding sites existed for the species (e.g. Fair 2007).
- *The Crested Newt* was published by Jehle *et al.* (2011). This book was a translation and an update of an earlier German version, with important UK material added by Jim Foster on subjects such as trends in status, threats, conservation measures, legislation, policy and management.
- Following the publication of development work on the analysis of environmental DNA (eDNA) in pond water (Biggs *et al.* 2014), the method was sanctioned by Natural England for the detection of crested newts.
- In 2016, Natural England introduced a new scheme for District Level Licensing as an alternative to the traditional method for dealing with developments. This was designed to save developers time and money and also provide an 'unprecedented level' of funding for creating and enhancing newt habitat. Among the initiatives stemming from this development was the formation of a Newt Conservation Partnership (Section 9.7).

Personally, collecting pond water for eDNA analysis does not sound as much fun as going out at night with a torch, and in some respects it is less informative. I still remember the thrill of discovering the underwater world of newts at night. The following three sections describe local studies in which torch counting was used to survey and monitor newt populations directly or indirectly in relation to building developments.

9.3 Stanground newt ponds

9.3.1 Introduction

Following up on some local information in 1982, I netted a pond in horse pasture on the edge of Stanground, a satellite village now absorbed into Peterborough (Figures 1.5 and 9.2) – and caught three crested newts in 15 minutes. The pond was then largely forgotten until April 1986, when I was informed that Persimmon Homes planned to build houses on the site. I immediately revisited the pond and confirmed by netting that crested newts still occurred. After discussing the situation with Tim Barfield of NCC's East Midlands Region, I carried out a night count of the site on 1 May 1986. An exceptional 93 crested newts were counted in the pond; an additional nine were seen in a long wet depression to the west in the same field, and another 17 were counted in a roadside pond a short distance to the north-east of the area proposed for development (Sites 6 and 7 in Table 9.1). In view of the importance of the population, NCC asked for the pond, the wet depression and about 1 ha

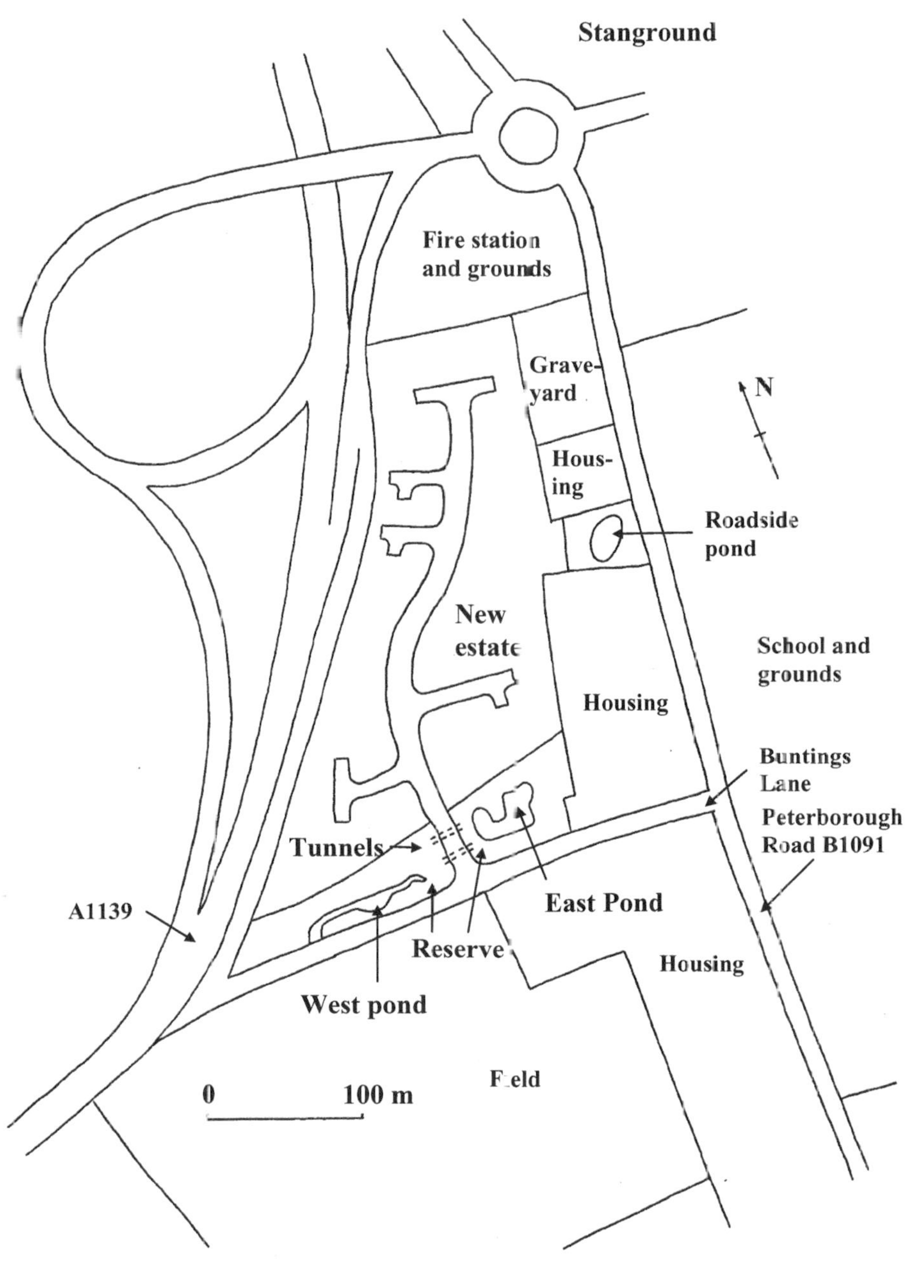

Figure 9.2 A map of Stanground newt ponds and the surrounding area as it was in the early 1990s.

of land to be safeguarded as a reserve within the development. Crested newts had ready access to suitable terrestrial habitat in a southerly direction, and we requested that walls and fences should not prevent the newts from reaching gardens to the north in the new estate (Figure 9.2). A water tap was also requested at the site because of the uncertainty of the effects of the development on local hydrology (Figure 9.3a).

Finally, it was agreed that two tunnels would be provided under the access road to the estate which cut the reserve into two parts. These were constructed from 30 cm diameter concrete pipe and were 14–15 m in length; low brick walls were built to guide newts towards the tunnels (Figure 9.6a).

9.3.2 *Monitoring and management*

Two counts were made in the breeding season of 1987: averages were 28 crested newts in the pond (East Pond) and one in the depression (West Pond), so counts were highly variable in 1986 and 1987 prior to building commencing (Figure 9.4). House construction began in the second half of 1987 and was finished by 1989. That year the reserve was declared open and handed over to the Wildlife Trust for management. The Trust appointed a local resident, George Young, as warden. This turned out to be an inspired appointment, and George was a constant presence at the site until he retired in 1995. I continued to undertake the monitoring and advise the Wildlife Trust on management until 1998.

The aim of the project was to maintain and, if possible, enhance the newt population (Cooke 1997b). Torch counting was the measure by which success or failure was judged, as pond conditions were usually favourable for this method. However, I decided that torch counting was insufficient on its own to indicate whether problems needed rectifying, so I set up monitoring of tadpole production and also kept detailed records of water levels. This approach was immediately vindicated when the estate's main drain was installed in April 1988 and the water table promptly went down by more than half a metre. The local Wildlife Trust arranged for both

Figure 9.3 Stanground East Pond: (a) re-excavation, February 1989, also showing the encased tap in the lower right-hand corner; (b) pond full, May 1989; (c) soon after the second drastic re-excavation, December 2005; (d) the view from under the hedge, July 2020.

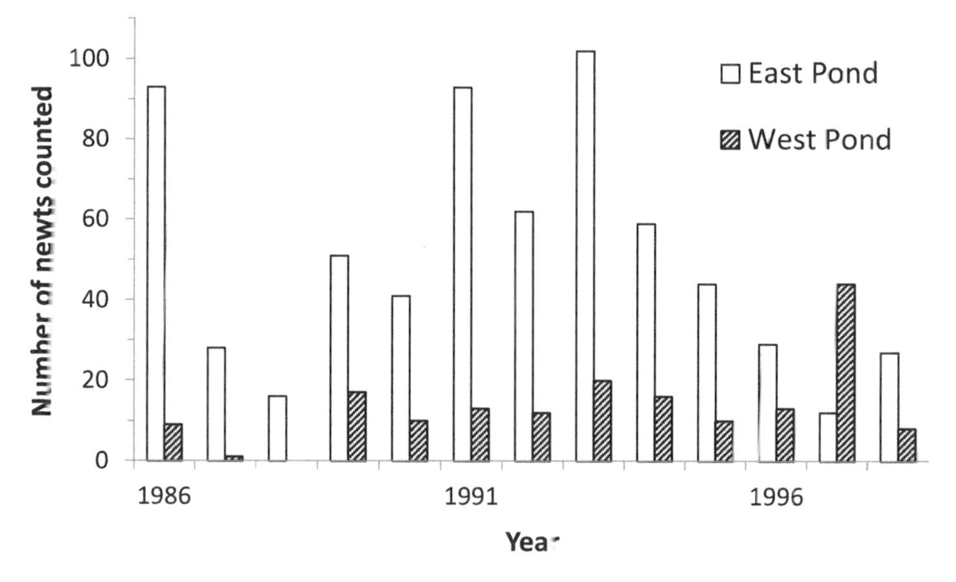

Figure 9.4 Average number of crested newts counted at night at Stanground: East Pond (white bars) and West Pond (hatched bars), 1986–1998 (updating and adapting Cooke 1997b). A single count was done in 1986, two in 1987, and three or four in the following years. The housing estate was built between late 1987 and 1989.

waterbodies to be re-excavated in 1989 (Figure 9.3a and b), which turned the western depression into a proper pond, but there was no successful breeding on the reserve in 1988 because of the reduction in the water level.

Average night counts in the two ponds, 1986–1998, and tadpole counts 1988–1998 are depicted in Figures 9.4 and 9.5 respectively. Roy Bradley, Howard Hillier and my son Steven helped with counting over the years. Combined counts of newts in the two ponds did not change significantly over the monitoring period, indicating that the population had been maintained. Numbers of tadpoles were highly variable, with little or no recruitment in three years in addition to the failure in 1988 mentioned above. In 1990, complete failure was due to a combination of natural desiccation and predation by three-spined sticklebacks. George Young pumped out East Pond in the following winter and separated fish and invertebrates by hand; the fish were taken elsewhere and the invertebrates were returned to the pond when it was refilled. While doing so, George became completely stuck in the mud – fortunately I was on hand to tie a rope around his body and haul him free. Desiccation again caused problems in 1995, while in 1993 invertebrate predators were suspected of being implicated in the near failure of recruitment. Perhaps George should not have put them back? I investigated this form of predation during 1996–1998, but found no evidence to blame dragonfly larvae, greater water-boatmen (*Notonecta glauca*) or beetles for significant losses of newt tadpoles (Cooke 2001b; see also Table 10.2).

The article I published on the results up to 1995 (Cooke 1997b) was the first to report extensive monitoring of the conservation of crested newts within a building development. I was careful not to rush into print before I had been monitoring for 10 years and had enough data to be confident about how the population had fared. Crested newts may be capable of living for 10 years, so some adults from 1986 might have survived

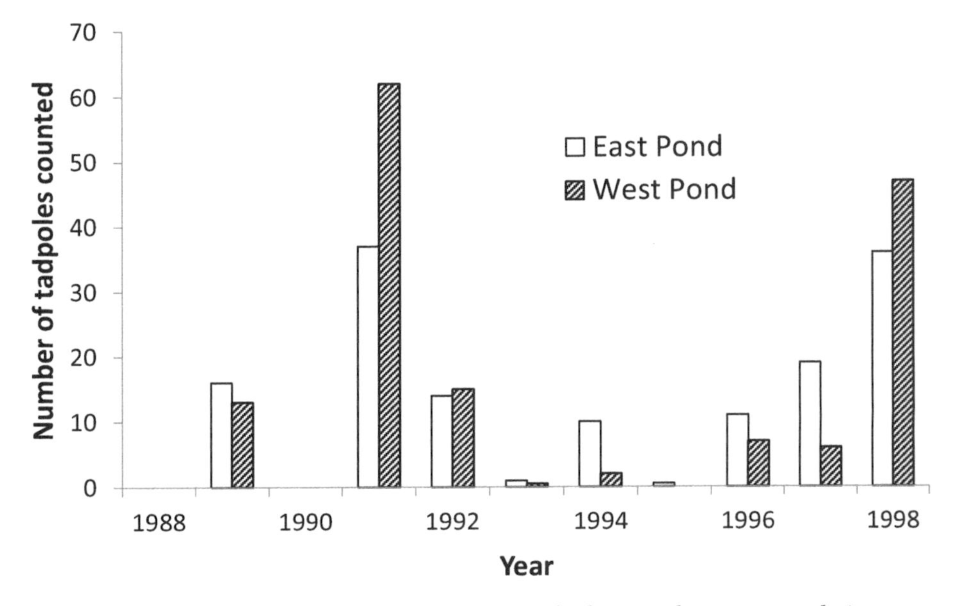

Figure 9.5 Average numbers of crested newt tadpoles netted at Stanground: East Pond (white bars) and West Pond (hatched bars), 1988–1998 (updating and adapting Cooke 1997b). Ponds were netted on four occasions in summer each year: the sump and western end of each pond were netted for a total of 10 minutes on each visit.

until well into the 1990s. But monitoring was consistent with considerable recruitment of new adults in 1991 and 1993, perhaps from the larval cohorts of 1989 and 1991 respectively. There seemed little doubt that the population was self-sustaining.

My recording at Stanground stopped in 1998, but monitoring by night counting was taken over first by Tony Gent and then primarily by Jim Foster, my successors at English Nature and Natural England. Monitoring results up to 2010 demonstrated that the population in the reserve continued to flourish to beyond 20 years post-development (Jehle *et al.* 2011). Brett Lewis, Richard Griffiths and Yolanda Barrios (2007) reviewed the success of this and a number of other crested newt mitigation projects. Sites were visited in spring 2005 and concern was expressed that 'little or no management' on the Stanground ponds meant the wildlife interest would 'continue to decline'. However, the ponds were re-excavated by the Wildlife Trust later in 2005 (Figure 9.3c), and the authors' concern was not reflected by the night count data presented by Jehle *et al.* (2011). It is true that crested newts in the roadside pond declined because of lack of sympathetic management, but that site was outside the reserve area and not managed by the Wildlife Trust – multiple ownership of habitat over which a population roamed could be a problem. For this type of remedial management to succeed, adequate monitoring and management must be in place and operational for as long and as often as necessary. Lewis *et al.* (2017) extended the original study on mitigation projects to 18 sites that had received remedial treatment either in the 1990s or in 2004 (so this sample did not include the Stanground site). These sites were all intermittently monitored over a period of at least six years post-treatment: there was an overall decline, with newts considered extinct at four sites. Population densities were lower than those in 'unmitigated' control ponds. These authors highlighted the following problems: pre-mitigation populations being non-viable, inadequate interventions and/or management, cumulative impacts of further developments, and exposure to new threats.

There had initially been much interest in the Stanground tunnels (Figure 9.6a), which were the first to have been built for newts in this country. I first saw newts in them on the night of 12 April 1989. During the breeding seasons of 1989 and 1990, I inspected the tunnels on 15–20 nights and recorded a total of seven newts in the tunnels on three occasions. By 1990, I had recorded only five dead crested newts and one live newt on the estate road, despite having made about 100 visits. At that time, I thought it likely that newts were using the tunnels to move between the two parts of the reserve. By 1993, however, with only one more sighting of a newt in a tunnel, I had come to the conclusion that the tunnels were simply too long, narrow, cold and dry to be used regularly, if at all, by the newts to transfer from one side of the reserve to the other. Later I discovered that tunnels should have a diameter of 100 cm rather than 30 cm (Baker *et al.* 2011).

Because of the tunnels, we had to endure early newspaper headlines such as 'Newts to get a tunnel of love'. On the few occasions that a local television crew turned up, they always wanted information on the tunnels. My last involvement with the television industry at the site was a chastening experience. I arrived at the reserve just before the agreed time and waited. After about an hour, no one had turned up so I drove back to the English Nature office in the centre of Peterborough and phoned the person who had contacted me about the interview. They explained that the crew would have come as arranged, but were diverted to a much more important story that cropped up that day. No effort had been made to contact me, which would have been difficult anyway as I did not have a mobile phone. But it hurt that they clearly had no intention of letting me know.

9.3.3 *Postscript*

After a gap of many years, I visited the ponds again in July 2020, finding East Pond had much emergent vegetation and surrounding scrub, but retained plenty of open water (Figure 9.3d). West Pond was overgrown but not entirely dry. Matt Hamilton of the Wildlife Trust explained to me that, while there had been routine clearance

Figure 9.6 Newt tunnels of contrasting sizes at Peterborough: (a) a tunnel at Stanground, December 1992; (b) one under the road at Hampton Nature Reserve between the two reserve areas, April 2014 – the grill allows mammals the size of small deer and badgers to use this tunnel but keeps people out.

of marginal scrub and dense vegetation, the ponds had not been re-excavated since 2005 because they usually held water through the summer. Some newt monitoring had taken place and the Wildlife Trust had created a management plan for Peterborough City Council to enhance and manage not only these ponds but also nine others within the city where crested newts bred. Matt told me that the last count for the ponds was 15 crested newts in early May 2019, when egg laying was observed.

Recently, there has been further house building to the south of the reserve, but good newt habitat persists to the south of West Pond that is doubtless exploited by newts breeding in the reserve ponds. With sympathetic and creative management, this southern area has potential to extend the range over which the newts breed and so help safeguard the population further.

9.4 Newts around Ramsey: from losses to a more strategic approach

9.4.1 Worlick Farm fish ponds

This section is an account of how attitudes and approaches to crested newt populations changed, one might say evolved, over a period of about 50 years in and around part of the town of Ramsey in Cambridgeshire (Cooke 2001a, 2001c). I moved to the town with my family in 1973 and my wife and I still live in the attached village of Bury (Figure 1.5). I learnt about the history of crested newt ponds in the 1950s and 1960s by interviewing long-term residents. It was clear from their descriptions that the species had been fairly abundant in the northern and western edges of the town and out onto the fenland beyond. As the town expanded and infilled, however, building work destroyed much damp habitat, while ponds on agricultural land were unused and neglected. Prior to the introduction of the Wildlife and Countryside Act in 1981, little or no thought was given to pond destruction, as typified by the in-filling of the pond behind the mill in Field Road during the winter of 1979/80 (Section 3.4.1). It may have been polluted and had few newts, but its loss was symptomatic of the lack of interest shown towards ponds by most people apart from schoolboys. As a result of the Act, I spent much time in the 1980s searching for and documenting local ponds with crested newts. Then, if threats became apparent, it was a case of trying to find a solution. An example of this was the protection of the Stanground newt ponds summarised in the previous section.

In the 1980s, my work with crested newts focused on how to react to situations as they arose, but this implied a hint of strategic planning too. In 1983, I was asked by the local landowner, John Fellowes, to advise how a site on one of his farms could be enhanced for wildlife. The site, which was on Worlick Farm just outside Ramsey, comprised six ancient ponds used to rear fish for the table by the monks of the long-lost abbey in the town (Figure 9.7). As the ponds were surrounded by about 5 ha of woodland, scrub and grassland and were listed as a Historic Monument, I realised they had potential as a secure site to receive rescued newts. The need for receptor sites in England was escalating as people learnt more about where newts occurred in places that already had planning permission. As a result of my advice, rubbish and accumulated silt were excavated from the ponds, and scrub and rank vegetation

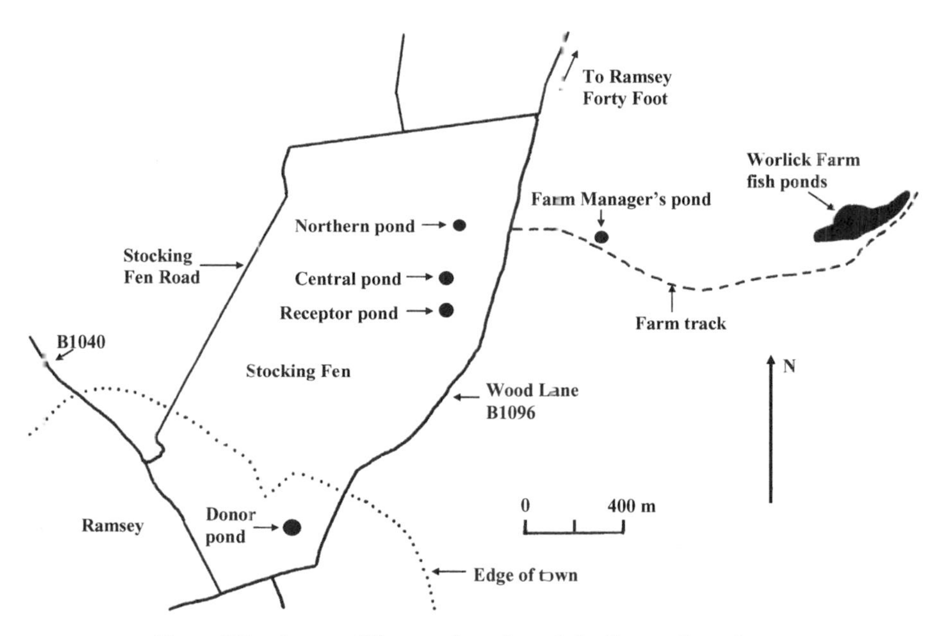

Figure 9.7 A map of the ponds and roads in the north-east part of Ramsey in Cambridgeshire and the fenland beyond.

removed from their southern edges in October 1984 (Figure 9.8). There was no evidence of fish or crested newts in the ponds.

Having prepared the site, I made it known that I had a suitable receptor site, if any of my herpetological contacts was aware of rescued newts without a home. I did not have to wait long. In 1985, BHS members were involved with a site at Swanscombe in Kent, where hundreds of crested newts were rescued before a large quarry was infilled. Most of these newts were released at sites in Kent, but I received a licence from NCC to take 38 adult newts to the fish ponds at Worlick Farm. I would imagine that Natural England would not look favourably at such a long-distance translocation today, but things were different then.

The prospect of monitoring the fate of no more than 38 newts on a 5 ha site with ponds extending to roughly 2,500 m² of water surface seems even more ridiculous now than it did then, but I decided it was best attempted by torch counting. Much of the land area was still impenetrable despite the clearance, so I ruled out searching for the newts on land, as well as netting the ponds for tadpoles. A farm cottage was on the edge of the site and its occupants informed me that they had found smooth newts in their garden but had never seen crested newts. Most of the sites discussed in this chapter had smooth newts, but I only rarely refer to counts of that species so as not to complicate the story unnecessarily.

Prior to releasing the crested newts into the ponds at Worlick Farm, I undertook one night count in 1985 and saw none at all. The timescale for monitoring was open-ended: I would continue until I considered it had succeeded or failed. At least one monitoring visit would be made each breeding season. In 1986, on my first post-release visit, I was pleased to count three adult newts; then two were counted in 1987 and three in 1988. During 1989–1991, counting became more difficult mainly

Figure 9.8 One of the six historical fish ponds at Worlick Farm, Ramsey, in 1984, prior to management of scrub and rank vegetation and removal of sediment.

because of growth of vegetation in the ponds and on their banks. I saw no crested newts during this period despite doubling up on visits in 1989. By 1992, I was ready to admit that the translocation appeared to have failed, but as it was one of eight local sites that I had been monitoring for at least seven years (Table 9.1), I kept going and was delighted to count three crested newts that year. One of those newts was a small male, the first indication of successful breeding. Counts then increased fairly steadily, culminating in 27 in 1998 (see Site 8 in Table 9.1), and I stopped counting after 14 years of monitoring. My conclusion was that the introduction had succeeded. In a later article about the introduction, I considered whether (1) some crested newts might have been there prior to the release in 1985, (2) they colonised naturally from elsewhere during the early 1990s, or (3) someone else introduced newts in the 1990s – but I believed these alternative scenarios varied from unlikely to not feasible (Cooke 2001a). The increase during the 1990s was even more impressive than was suggested by the counts, because the average percentage of pond edge accessed during monitoring decreased gradually from 58% in 1986–1988 to 34% in 1995–1998.

9.4.2 *Stocking Fen*

John Fellowes took a close interest in events at the fish ponds and invited me to advise him about improving two old field ponds on another part of the farmland. This work was done in the winter of 1989/90. The fish ponds are slightly more than 1 km to the east of the B1096 between Ramsey and Ramsey Forty Foot (Figure 9.7). To the west of this road is a 130 ha parcel of land on Stocking Fen on which six old stock ponds were marked on my 1:25,000 scale Ordnance Survey map from the 1970s. Virtually the whole area had been converted in the past to arable, and the old ponds were barely discernible in the landscape prior to the two being restored (the receptor pond and northern pond in Figure 9.7). These two were about 400 m apart and were linked by the remnants of an old ditch and hedgerow. The receptor pond seemed more suitable for crested newts as it had an open southerly aspect (Figure 9.9a).

In September 1988, John Fellowes had warned me that he was applying to renew planning permission to build houses in a paddock in the town where another old pond was situated (the donor pond in Figure 9.7). He had been told by a neighbour that crested newts occurred in the pond and asked me to confirm their presence. This I was able to do in the breeding season of 1989 (Table 9.2). John Fellowes undertook to safeguard the newts within the development, but in November 1990 he informed me that the plan agreed with District Planning Officers required the pond to be moved by up to 20 m. In its time, this pond probably held a flourishing population of crested newts, but by 1990 it had become increasingly unsuitable, being totally surrounded by large trees, which heavily shaded the pond and made access difficult. I expressed disappointment, because moving the pond increased the risk of failure. However, a solution that occurred to me was to move the newts to the receptor pond on the fenland about 1 km away; then, if the new pond in the development was suitable, some newts could be returned to the site in the town.

Therefore, under licence, newts were collected by net and bottle trap during 1991–1993 and eggs were transferred on strips of plastic. In total, 23 adult newts were transferred; average length for the seven females was 136 mm, and for 16 males 128 mm. No adult crested newts of less than 120 mm were caught and no immatures or tadpoles were encountered. These data on size indicated a senescent population with zero recruitment for several years (see Section 10.6.2). By the mid-1990s, the water in the pond was too inaccessible for further monitoring or collection. In spring 1998, development at last began and scrub was removed, enabling limited night counting in 1998 and 1999, but no newts were seen. The pond was eventually drained in July 1999 and the water used to create a new pond. I was present during this process but saw no newts or other amphibians on the site. The only item of interest in the drained pond was a large soggy mass, roughly cubic in shape with edges of about 50 cm. This was puzzling until I realised that it was a bundle of local newspapers, presumably dumped in the pond by an errant delivery person.

Figure 9.9 The receptor pond on Stocking Fen, Ramsey: (a) in 1990, just after renovation; (b) torch counting in 2001 – the area of the pond under an old willow was clear of aquatic vegetation and was attractive to courting newts (photo by Rosemarie Cooke).

Table 9.2 Average night counts of crested newts in the ponds where newts were recorded during the translocation from the donor pond in Ramsey to the receptor pond on Stocking Fen, 1989–2007 (updating Cooke 2001c).

	Average night count of adult crested newts (number of counts)		
Year	Donor pond	Receptor pond	Central pond
1989	5 (2)	Renovated	
1990	—	—	
1991	2 (10) Newts moved	0 (2)	
1992	1 (8) Newts moved	4 (4)	
1993	1 (8) Newts moved	5 (3)	
1994	—	2 (3)	
1995	—	7 (3)	
1996	—	15 (3)	
1997	—	18 (3)	
1998	0 (1)	14 (3)	
1999	0 (1) Destroyed	—	
2000		—	
2001		28 (4)	Created
2002		15 (4)	1 (4)
2003		9 (4)	3 (4)
2004		10 (3)	2 (3)
2005		5 (3)	2 (3)
2006		0 (3)*	1 (2)
2007		1 (3)	2 (2)

* Tadpoles netted in summer

Initially, the newts translocated to the fenland receptor pond settled down well and established a breeding population (Table 9.2). I was unable to monitor during 1999 and 2000; but when I returned to count the newts in 2001, it was as if I had never been away (Figure 9.9b). A sample of nine males caught one night in 2001 measured 117 mm on average; they were significantly smaller than those translocated in the early 1990s, which suggested a better-balanced breeding population. However, although I routinely checked the northern pond on the fenland, I never saw any evidence that newts had dispersed to it from the receptor pond. Research articles in the literature suggested that 400 m might be too far for colonisation to occur, so I asked if a new pond could be dug between the receptor and northern ponds (Figure 9.7). This was done, and a small breeding population of newts was recorded in this central pond from 2002 (Table 9.2). Unfortunately, at the same time, numbers counted in the receptor pond were declining for reasons which were not readily apparent. Counting conditions were not deteriorating markedly, no fish were recorded at that time, water levels were satisfactory, and there were no obviously harmful successional changes. I did wonder whether creation of the new central pond in 2001 had enticed a few newts away from the receptor pond. Newts remained unrecorded in the northern pond up to 2007, the last year in which all three ponds on Stocking Fen were monitored. Ten-spined sticklebacks and high turbidity were recorded in the receptor pond during 2008–2010 and I then, rather despondently, drew a line under the monitoring.

By the end of the monitoring programme, there was evidence of only small numbers of crested newts in the ponds on Stocking Fen. This was disappointing, but at least the newts were not dependent on the new pond that had been created in 1999 beside the old field pond in the town: the new pond was located in a garden and was highly unsuitable, with much of it covered by decking. Rob Oldham and his colleagues (2000) had published an immensely useful paper on deriving a crested newt Habitat Suitability Index (HSI) for a waterbody and its surroundings (see also Section 10.7.2). Ponds with night counts of more than 10 newts were usually associated with indices in the range 0.7–1.0, whereas ponds with an index of less than 0.4 did not contain newts. I estimated the indices of all four of the ponds in 2001: receptor pond 0.64; central pond 0.71; northern pond 0.67; pond created in 1999 beside the original pond 0.34. On that basis, the fenland ponds were far from optimal, which was not surprising as they were all quite small and situated between arable fields.

Maybe I was trying to make a silk purse out of a sow's ear, but on a brighter note, the farm manager's house was roughly midway between these three fenland ponds and the fish ponds – and the garden contained an old field pond (Figure 9.7). When I counted this pond by torchlight in 1990 it was little more than a marshy depression and I saw a single crested newt. However, it was renovated in 1997, and when I visited again in 2001 the count was 25 crested newts.

There was more good news in 2021. When I revisited the receptor and central ponds after a gap of 11 years, I found that, although the latter pond was dry, crested newts had bred in the receptor pond, indicating that the population still survived 30 years after I began moving their ancestors from the donor site in the town. Renovating old farm ponds has been shown to work elsewhere: in Suffolk, for instance, the Wildlife Trust found that the incidence of breeding by crested newts more than doubled in the space of one year in 50 restored ponds (Sayer *et al.* 2022).

9.5 The early years of the Peterborough brick pits

9.5.1 Background

And now for something completely different: a landscape where ponds number in the hundreds and the crested newt population has probably run into several tens of thousands.

I well remember the request. Sarah Lambert of NCC's East Midlands Region wrote to me on 24 April 1990 asking if I would undertake a survey of crested newts in Orton brick pits to the south of Peterborough (Figure 1.5). The Secretary of State for the Environment had granted permission for Hanson Trust Ltd to develop the area as the city's southern township, but crested newts had been reported. Please could I check it out? The next evening, I followed directions to the works' buildings and my jaw dropped: the site covered an area of several hundred hectares crammed with waterbodies (Figure 9.10). I knew immediately that this was not a task that I could accomplish in a couple of evenings, but I had gathered from Sarah's note that there was some urgency about obtaining information on the status of the newts. So what could be done?

In order to provide background to this request, some historical detail needs sketching out. An in-depth account of the early years of the township has been provided by Tom Denniford (2019), who was project manager during 1989–1998.

Bricks were made for many years in the Peterborough area by utilising the Lower Oxford Clay. This made good cheap bricks – and the adjacent railway line to London and elsewhere simplified and reduced the costs of transport. During the first few decades after the Second World War, London Brick Company dominated brick production in this country and was able to take advantage of the boom in house building. By the 1970s, however, local bricks were starting to be out-competed by better or cheaper products. Another issue was the large holes in the ground that were beginning to accumulate. The latter problem was solved by a huge deal with the Central Electricity Generating Board in which pulverised fly ash from coal-fired power stations in the east Midlands was sent by train to be disposed of in pits in the eastern part of the Orton site. I recall being taken on a trip around this area in 1977 to marvel at the scale of the operation, which for a while solved a significant headache for both of these industrial giants.

Much of the rest of the site remained unaffected, being left as it was after the brick clay had been extracted. The process involved removing the top unwanted

Figure 9.10 Orton brick pits: general view of area surveyed, May 1990 – the pools extended as far as the trees in the background.

layer of the soil and dumping it in ridges that might approach a kilometre in length. The landscape resembled a gigantic version of an old ridge-and-furrow field. Water collected in the furrows, and over the decades the ridges became increasingly vegetated. By 1990, newly worked areas had much open water but were terrestrially barren, whereas the oldest areas had well-balanced waterbodies and thorn scrub was established along the weathered ridges.

Hanson took over London Brick in 1984 and in the later part of the decade explored the possibility of developing parts of the landscape for housing, employing the engineering company Ove Arup to manage technical issues. This had eventually led to Sarah Lambert being contacted by Ove Arup in April 1990 to ask for specific information on crested newts on land at Orton earmarked for Peterborough's southern township.

9.5.2 *The preliminary survey*

I contacted Ove Arup on 26 April 1990 and explained that while I was happy to organise a survey of the area, there was no way that I could do it alone in my spare time: I could find someone who would undertake a brief survey if Ove Arup was willing to support them financially. This could not be a definitive survey, but it should be sufficient to allow basic conservation and planning decisions to be made. They agreed, and Roy Bradley (Figure 12.4b) was duly signed up to do the bulk of the work. At the time, Roy was semi-retired and frequently accompanied me on evening surveys.

Fieldwork at the Orton brick pits was carried out between 2 and 24 May 1990. The standard technique in use at the time was counting newts in the ponds at night, but torch counting hundreds of waterbodies in such a short time was out of the question – Roy and I needed to be selective. We mainly concentrated on the north-eastern part of the site, where the best habitat seemed to occur; and I wanted to choose the better ponds in that area for night counting. I knew from surveying at other local ponds that daytime observations made during warm, calm May weather had a better than evens chance of detecting crested newts (e.g. see Section 10.2). So Roy's first task was to walk a defined area of roughly 100 ha during daytime and note ponds where newts could be seen. We were fortunate in this respect that the water in the ponds tended to be clear and most of the period was blessed with extreme anticyclonic conditions. Sites where newts had been seen were then revisited by Roy during daytime: newts were counted and observations were made on dimensions, water clarity and vegetation cover of each pond. Roy and I then counted newts in all ponds at night where he had recorded them during the day.

Crested newts were recorded at night and/or during the day in 38 waterbodies, with combined totals of 109 during daytime and 608 counted by torchlight. In one pond, 112 crested newts were counted at night. We noted that ponds with fish had few or no newts. Crested newts were especially well established in the older workings where trees and scrub grew on the ridges: counts tended to be higher in the ponds in those areas. This was a staggeringly large number of newts to count. By that time, Rob Oldham and his colleagues at Leicester Polytechnic had collected and collated data on newt populations for seven years, and I was informed that the only night count greater than that at Orton was 705 counted in the old moat at Kirk Deighton in Yorkshire.

9.5.3 *Repercussions*

The developers had some difficulty in understanding that such a count was an index of abundance and not like a human census, because not all newts were in the water at that time and only a small proportion of those in the ponds were likely to have been counted. It was clear that further survey was needed and that it required doing both intensively and extensively. This led to Herpetofauna Consultants International (HCI), which was run by Tom Langton, being contracted to undertake surveys (Figure 12.4a). Within a couple of years, HCI had recorded a total night count of 3,316 crested newts for waterbodies across the entire township site. As a result, an 'estimate' of the actual size of the population was suggested of about 30,000. Tom Langton has explained to me that this estimate was based both on my observations at Shillow Hill in 1984 (Section 10.2) and on knowledge of newt densities in high-quality habitat. This figure was useful in discussions by providing a total that allowed participants to visualise the magnitude of the issue with which we were dealing; and to judge from the number of newts later caught across the site (HCI 2000; see below), it was not wholly inaccurate.

During the early 1990s, NCC was split into the constituent country agencies and the JNCC, and Tony Gent became English Nature's herpetological adviser. My informal involvement came to a halt in the spring of 1993 when I was seconded away from the Peterborough office for a year to study deer damage in woodland. Discussions over the newts remained intense for most of the 1990s for a number of reasons:

- During 1990–1992, crested newts were being discovered in increasingly unexpected numbers in unexpected places, leading to an escalating number of concessions being sought.
- In addition to the newts, the site was notable for holding 10 species of stoneworts, including the fully protected bearded stonewort (*Chara canescens*). It soon became clear that at least part of the site should be designated as an SSSI and as an SAC, but how much should be designated and when?
- Some conservation organisations considered that English Nature should oppose the development scheme, which might then have resulted in massive claims for compensation.
- Without large-scale monitoring and management, successional changes at the site could lead to a reduction in wildlife interest if the site remained undeveloped.

The final outcome was that a reserve of 145 ha was created for the newts in which they would be confined by perimeter barriers. Over a period of years, newts and other amphibians from elsewhere on the township site were caught and transferred to the reserve by HCI: 54,000 adult amphibians and 66,000 juveniles were transferred during the 1990s, including 24,000 adult crested newts and 9,000 adult smooth newts (HCI 2000; Figure 12.4). Within the reserve, management for newt conservation was undertaken, including fish eradication from some ponds. Habitat succession within the newer areas was speeded up by transferring topsoil from the older ridges so as to render the terrestrial habitat more suitable for newts. The township came to be known as Hampton and the reserve as Hampton Nature Reserve. Since 2003, the reserve has been managed by Froglife with funding from O. & H. Hampton Ltd; and from 2006, Froglife organised an annual survey of the newts to ensure the population was thriving.

Figure 9.11 A corner of King's Dyke Nature Reserve in 2022, with chimneys of the still-functioning brick pits behind. This is the oldest part of the reserve, with new areas being added. The site is managed to have a range of waterbodies at different successional stages.

According to Tom Denniford (2019), more than £3.2 million was spent on the newt project up to 1998. More recently, different types of tunnel have been constructed in and under roads in order to prevent newts being killed and to enable genetic mixing to occur, as the reserve is in two parts (Figure 9.6b). The movement of newts through three tunnels under one of the main roads serving the township was studied by Matos *et al.* (2017).

The geology, fauna and flora of the Peterborough clay pits were described in a review by Mark Crick, Pete Kirby and Tom Langton in 2005. Crested newts have been an important conservation feature elsewhere within this pit system. Thus King's Dyke Nature Reserve has been established to the east of Peterborough less than 10 km from Hampton Nature Reserve and dozens of ponds have been created for a range of species (Parker *et al.* 2011; Figure 9.11). The reserve newsletter reported in 2018 that night counts of crested newts had exceeded 600; a single count for the brick-pit complex was 1,207 in March that year.

9.6 Reminiscences of the dangers of survey

Writing about the brick pits has reminded me of some of the dangers associated with amphibian survey, especially counting newts at night. Once, when walking back up one of the higher and steeper ridges at Orton, I toppled backwards and was only prevented from ending up in the water by landing in a thorn bush. Luckily, I was wearing a backpack and was unscathed. Roy Bradley was not so fortunate when we were returning in darkness along a track at the brick pits: one moment he was beside me and the next he was gone. I shone my torch back in his direction and he was literally in a tangled heap on the ground. The loop of one bootlace had hooked itself over a metal lace hook on the other boot – and he fell heavily and hurt his arm. As luck would have it, he was booked to visit the United States the following day. He went, but the arm troubled him for a long time afterwards.

On another night, Roy and I had met at a small village called Water Newton, which, with such a name, was surely destined to have a good newt population. The ancient breeding pond lies immediately to the east of the A1. We completed the count, got in our cars and I drove away first. I was unaware that Roy had parked his Volvo on the verge in such a way that none of its wheels was in firm contact with the ground. This was in the days before anyone carried a mobile phone. Realising his predicament, he went to a nearby cottage to ask if he could use their phone. The gentleman kindly offered to take him home and Roy gratefully accepted. He returned the following day with his son: they jacked the car up and, using bricks and planks, managed to reverse it onto the road. Some years later, I did something similar when carrying out deer surveys for the Woburn Estate; fortunately a passing motorist stopped and was able to tow my car off the verge.

My wife Rosemarie has been unlucky on a couple of occasions when helping me at local newt ponds. Once, she slid down the bank of a dry pond and badly damaged her thumb. On another night, she was carrying a bucket of newts which I had netted in order to measure their lengths when she tripped over wire on a broken fence and landed on the bucket. What really annoyed her was that I appeared to be more concerned about the newts than about her. Sometime later at the same site, we were interrupted by a Land Rover full of locals with guns out after rabbits – they were not too pleased to find that pensioners with torches had already scared their quarry away.

I remember various encounters with the police, such as returning to my car to find them checking the number plate with their records and/or wanting to know what I was doing prowling round at night. Over the years, I found that I was meeting the police less frequently, perhaps because they were becoming more accustomed to people out at night, or maybe because there were simply fewer of them.

The scariest incident I had with the police was sometime in the early 1980s. For a change, it was daytime, and I was looking for toads. While surveying for smooth snakes on the Dorset heathlands, I had been told of a toad pond at the base of the cliffs at Portland – and thought I would drive over and try to find it. As I drove onto Portland, I saw the police stopping and checking cars driving back towards Weymouth, but thought little of it at the time. I parked and walked down the long zigzag path to the bottom of the cliff, but failed to find the pool, so began trudging slowly back. About halfway up, I heard shouting from below, and looking down saw a group of men staring up and pointing. When I looked up the cliff to see what they were pointing at, I saw a similar group of men looking down. It was then that I realised that I was the object of their interest, and as the group from below came closer, I recognised them as police with dogs. I stood very still and tried to look as unthreatening and as unconcerned as possible. The dogs seemed disappointed to find they were not allowed to play with me. A similarly grumpy policeman explained they were looking for a youth who had escaped from the local borstal, and I had been spotted behaving suspiciously by the harbour coastguard. What on earth was an adult doing with a pond net on the cliffs? Fortunately, I carried NCC identification with me at that time.

9.7 Conclusions

What then have I learnt from my field observations of crested newts over the decades? Finding the remains of ponds beside arable fields on Stocking Fen outside

Ramsey in the 1980s was a poignant reminder of the amount of freshwater and terrestrial newt habitat that must have been lost or modified earlier in the twentieth century. The ponds probably did not all have breeding crested newts, but the potential was there for the animals to use them. And the same was true of ponds in the town. I know of six ponds in the town near the fenland edge that had crested newts during the second half of the century: four ponds no longer exist and the other two no longer have breeding newts. I have tried to take an approach on the fenland that was as proactive and strategic as circumstances would allow, but my modest successes do not compensate for those losses.

On the other hand, the size of the population discovered in the Peterborough brick pits is a reminder of the contribution of waterbodies that have been inadvertently created by the extraction industries. Roy Bradley has told me about a number of amphibian ponds in Peterborough which have been destroyed since the 1950s. But the numbers of newts lost were presumably dwarfed by the population of tens of thousands that we now know developed over the same period because of the brick-making process. The success of newts there demonstrates how well the species responds to someone digging a large number of fairly shallow holes and waiting for them to fill up with water and for vegetation and newts to colonise. It can sometimes be that simple.

When I became involved with the newt population at Stanground, the process of safeguarding newts within a housing development was something of a novelty. But gradually such schemes have become much more commonplace. Unfortunately, monitoring has often been deficient, especially during the years leading up to development occurring. In such cases, there has been lack of information to judge the success of the operation objectively and also to aid management of the site and the population. Adequate monitoring before development was often impossible if newts were discovered too late – if past data do not exist, there is nothing anyone can do about it. And I am well aware that professional consultants have been unable to guarantee or justify the duration of post-development monitoring that I have managed to undertake. I have often found it difficult to stop monitoring at a site because local surveys are usually relatively easy and interesting to do and continuing for one more year might provide that final crucial piece of information.

The more monitoring I did, the more I realised that 'apparent' outcome was dependent on the duration of the observation period. This was very evident at the Stanground newt ponds, the Worlick Farm fish ponds and the ponds on Stocking Fen. At Stanground, the first five years of monitoring, straddling the period from two seasons before building started to one year after completion of the estate, indicated impending failure with night counts down and total loss of tadpoles in two of the three seasons that reproduction was studied. After monitoring for 13 years, however, it was clear that the population had been maintained; as it was again after 25 years when Tony Gent and Jim Foster continued the monitoring.

At Worlick Farm, where newts were introduced in 1985, I was reasonably certain that the translocation had failed after seven years. I could so easily have given up monitoring then, but I continued and was pleased to conclude it had succeeded after 14 years. But unless you or someone else keeps monitoring, how do you know the newts are still there? I have not returned to the fish ponds in recent years because I no longer feel fit enough to monitor such a difficult site. And it is not possible to monitor forever, even with other recorders taking over. There must be some reliance on indicators of the likelihood of long-term success, such as the newts having a mixture of breeding sites within easy reach and the guarantee of positive and sympathetic habitat management. Egg searches and/or eDNA analysis of water may be the answer.

Comparing the state of the population in the donor pond in Ramsey in 1989 with the population in the receptor pond on Stocking Fen in 1998, the rescue seemed to be heading for success after 10 years of monitoring. This looked even more certain in 2002, when newts also began to colonise the new central pond on Stocking Fen. By 2007, however, the populations in the two ponds on the fenland were no better than the remnant population had been in the donor pond 18 years before. However, had nothing been done, the population in the donor pond would almost certainly have died out during the early 1990s. By 2010, with fish in the receptor pond, I was pessimistic about the eventual outcome, but I told myself that at worst the translocation had extended the life of the population by a decade or two. Gratifyingly, observations in 2021 revealed that crested newts still bred in the receptor pond.

If a breeding site is monitored, the best that can happen is that numbers increase, but the worst is that the population dies out. And as long as monitoring continues, so the possibility of extinction remains. Thus, a monitoring programme that is focused solely on presence at a breeding site or sites is biased towards failure and loss. Jim Foster reviewed what was known up to 2009 on the success of mitigation operations in Britain and reported that persistence of populations was 'substantially more common' than local extinction (Jehle *et al.* 2011). On the other hand, many populations may persist but with lower numbers of newts (Lewis *et al.* 2017).

Ten years ago, many newt conservationists may have felt that it was desperately difficult to make progress, but as someone who was involved from the 1970s, I believed positive steps had already been taken towards improving the situation. The species had been protected by law since the early 1980s. We knew that not only did we have to be aware of the potential destruction of breeding sites, but we also needed to consider the threats of fish, successional changes and loss of connectivity and metapopulation opportunities, as well as reductions in the quantity and quality of terrestrial habitat. Thanks to the activity of an army of professional ecologists and volunteers, we had an improving knowledge base of where crested newts occurred and what threats existed or might arise. And we had ever-increasing experience of what remedial and creative management might be required.

While this progress was hopeful and reassuring, some problems persisted a decade ago, or at best were only being slowly rectified (Beebee 2015). In particular, as intimated in the introduction to this chapter, the planning process cost developers and the conservation agencies much time, money and angst, while the media largely focused on the perceived folly and cost of mitigation attempts. Even the SAP had not been effectively implemented and relied on 'uncoordinated efforts by miscellaneous organisations and landowners' (Beebee 2015). And we also still lacked a monitoring scheme that was sufficiently robust to detect changes in the national population (Section 2.7).

However, it seems that a corner has been turned in the last few years with broader-scale, often multidisciplinary, approaches. An example of this type of modern conservation is what has just been achieved in the Scottish Highlands

(O'Brien *et al.* 2021). Genetically distinct populations at the extreme north-western edge of the species' range were declining primarily because of habitat succession, desiccation or the introduction of fish. This situation was turned around by identifying where to create or restore breeding ponds in order both to aid natural colonisation and to provide better connections between areas where the newts were genetically similar. Engagement with local stakeholders was an important part of the process, as it is with all initiatives of this type.

It has become possible to command resources to undertake the practical conservation of newts and their habitat on such a scale that makes my own local attempts seem insignificant. At least I feel I managed to demonstrate what might or might not be achievable at a very local level. A District Level Licensing Scheme announced by Natural England in 2016 was a significant step towards correcting some of the former problems with developments. The scheme is based on defining zones of risk for an area and using compensation to fund a net gain in status. Impacts of developments are more than compensated for by a district plan of action for conservation of the species funded via charges paid by the developers. For every pond lost to development four ponds and terrestrial habitat are created or restored. Most of the expense for developers now goes on habitat creation, management and monitoring rather than on mitigation; and newts are colonising the new habitat (Almond 2020).

There are, however, conservation concerns, particularly over the emphasis on compensating for impact, rather than also focusing on avoidance or mitigation at good newt sites. A modified scheme in the South Midlands includes a key element of identifying necessary mitigation and planning conditions at sites where impacts are likely, and seems to have been better accepted. This venture, a three-way partnership between the private sector, conservation organisations and local planning authorities (LPAs), has been described by Tom Tew of NatureSpace Partnership, Jeremy Biggs of the Freshwater Habitats Trust (FHT) and Tony Gent of ARC (2018) as 'a landscape-scale conservation project delivered through groups of adjacent LPAs working together'. Tom Tew referred to their approach as a 'paradigm shift' on the NatureSpace website. A significant additional innovation is that the three bodies formed the Newt Conservation Partnership (NCP) in 2018 to create and restore high-quality freshwater and terrestrial habitat to support this scheme. Monitoring includes determining regional trends in newt status and the part played by both newly created habitat and the development sites. Benefits have already resulted, and these are not restricted to crested newts but help conserve other amphibian species and pond life in general (FHT 2021a, NCP 2021). Monitoring and management are intended to continue in perpetuity. Importantly, developers benefit too because of a faster, more certain process.

The map of risk zones for the South Midlands district in Tew *et al.* (2018) reminded me of the ancient maps referred to at the beginning of the chapter, except that the old maps mentioned dragons incidentally whereas the modern map focused on them specifically. Hopefully, we will continue to build on the progress being made by also taking advantage of new agri-environment schemes, rewilding and other initiatives (Section 12.7), and the zones will expand where 'dragons be likely'.

CHAPTER 10

The newts of Shillow Hill

10.1 Introduction

Shillow Hill is the name given to the area where the B1040 between the villages of Bury and Warboys ascends about 15 m up an incline from the Cambridgeshire fenland to the higher clay-lands (Figure 1.5). This may not sound much of a hill in most of the country, but in this flat region it is a feature worthy of a name. The newt site is on the hillside immediately to the west of the road (Figure 10.1). It extends to about 2 ha and contains two properties: The Spinney on the top of the hill and Windyridge lower down. Apart from the road, the site is completely surrounded by arable fields. In the 1980s, the site was a mixture of woodland, scrub, hedgerow, grassland, marsh, ponds, orchard, gardens and buildings. After that, the woodland matured, and much of the habitat could best be described as wilderness. The site has dried out over the decades, and several small, wet depressions no longer hold water for long enough to attract newts. The main pond is believed to have been dug for clay to make bricks many years ago. In addition to the vast brickworks around Peterborough (Section 9.5), there were other local brickmaking enterprises of various sizes.

In 1983, I was attempting to discover and visit as many local breeding sites of the great crested newt as possible as part of the national effort to gather more information on the species following its inclusion on Schedule 5 of the Wildlife and Countryside Act 1981 (Section 9.2). Bob Stebbings, a national authority on bats, lived at The Spinney with his wife Angela and their children. I knew Bob well from our time together at Monks Wood, and he told me that crested newts bred at the site, particularly in a pond on the other property, which was unoccupied and for sale at the time. As a result of this information, I visited the site and counted 25 crested newts at night in that pond and another 11 in a smaller pond on Bob's property that he had recently created. So this was clearly a reasonable place for the species.

Henry Arnold was another friend working at Monks Wood. Part of his role in the national BRC was to collate and edit records and distribution maps for the amphibians and reptiles; and I had collaborated with Henry on several projects over the years (e.g. see Section 2.5.1). His main interest, however, was mammals, and he worked with Bob on several projects on bats. A short while after I had counted the

Figure 10.1 A view of Shillow Hill from the north in early summer, 1984. The newt site is on the hillside immediately to the right of the road, and the roof of Windyridge can be seen further to the right.

newts, Henry bought Windyridge and moved in with his mother. This presented me with a situation where a population of crested newts occurred on a site owned by conservationist friends, who were pleased that I was keen to find out more about their newts – and I literally waded into action in 1984 (Figure 1.4a). As with many of my studies, it was open-ended since funding was not required, and it rambled on for decades. Throughout this period, I lived no more than 3 km away, so it was easy to visit and I was made to feel very welcome. After a few years Bob moved away, but most of the newt interest was on Henry's property and he continued to help throughout, as did my son, Steven, for the first 10 years or so.

The study was intended to provide information about crested newts. However, as with the vast majority of crested newt sites, smooth newts also occurred and were recorded. The bulk of this chapter is about crested newts, but the last section (10.8) reports observations on the smaller species.

10.2 The early years, 1984–1986

Starting with a clean slate in 1984, I wanted to make observations on the ecology of crested newts at the site that would also be of relevance to conserving their populations in general. This included basic work on survey techniques, asking questions such as:

- What proportion of the newt population was I seeing by torchlight at night?
- Was torch counting over a period of years a reliable guide to how numbers of newts changed in a population?

At that stage, I knew that I would not be able, or indeed would want, to commit as much time to the project in the longer term as I hoped to commit in the early years. I could see, however, that, if the situation permitted, this could turn into long-term surveillance. This was what had already happened with my waterfowl studies begun in 1970 at Grafham Water and deer observations started in 1976 at the local NNRs.

Henry and I named the two semi-natural ponds at Windyridge and The Spinney as Top Pond (Figure 10.2a) and Wood Pond (Figure 10.2b) respectively. Crested newts were much more numerous in the former, and the difference in usage increased as Wood Pond desiccated progressively over the years (Table 9.1 and Appendix 5). Counts of crested newts in Wood Pond had decreased to zero by 1990; a small sump was dug in the pond that autumn, but crested newts were only subsequently recorded there in the springs of 2002–2004, when up to seven were seen at night. (Smooth newts were last recorded in Wood Pond in 2001.) All other counts of newts in the water given in this chapter relate solely to Top Pond, unless specified otherwise.

When full, Top Pond had an edge of about 48 m, an area of 130–140 m² and was flat-bottomed with a maximum depth of about 0.7 m. In 1984, I had considerable practical help and advice about ecological techniques from Rob Oldham and Mary Swan at Leicester Polytechnic (Figure 2.7). That year, standardised bottle trapping (64 for one night or 32 for two nights) was used in both ponds during five alternate weekends to catch newts (Figure 10.2b). Bottle trapping served two purposes. First, it meant that large numbers of newts could be caught and marked, so that population size might be estimated after they were toe-clipped to identify both the weekend trapping session and the pond (Figure 1.4a). Second, bottle-trapping totals could be compared with torch counts to see if they indicated similar trends in numbers (further bottle trapping was done in 1985 and 1986). As with netting in the pond and catching on land, all animals handled throughout the study were released on site as soon as possible.

Torch counting during the early years usually involved nine midweek visits from the second week of April until the first week of June, but in 1986 it extended from March until July. Any problems with water turbidity due to heavy rain were temporary. Counts of crested newts in Top Pond were highest in May each year, with peaks of 50 in 1984, 223 in 1985 and 215 in 1986; average counts in the usual spring to early-summer period were 33, 150 and 145 respectively (Appendix 2). So these counts demonstrated that peak activity occurred consistently in May and that counts could vary significantly between years. The average number of crested newts estimated to be in Top Pond during the trapping sessions in 1984 was 141, indicating that 23% of those in the water were counted at night on average (Cooke 1985c). The number of crested newts on the site that year was estimated to be about 500, and the average night count at both ponds detected 7% of the population. The highest night count of the nine undertaken at Top Pond in 1984 was 50, which was 10% of the estimated population. Although, it may be tempting to add a zero to a night count to derive a figure for number of newts in a population, the percentage will vary considerably from situation to situation – and anyway few people are keen enough to undertake nine counts in a season! As the combined average counts for both ponds was equivalent to 7% of the total population, for most field populations where a single night count is made, the number of newts in the population is likely to be higher than 10 times the count.

The two trapping methods of 64 traps for one night or 32 for two gave comparable results, and there was no significant difference between numbers caught on the two nights when 32 traps were deployed. In all, there were nine trapping sessions during the period 1984–1986; and there was a significant positive relationship between numbers trapped and the average number counted at night during the previous and

Figure 10.2 Newt ponds at Shillow Hill: (a) Top Pond viewed from its north edge, spring 1984 (the pole and old door provided access to deeper water at the southern end but were removed soon after this photograph was taken); (b) Wood Pond viewed from the south, spring 1984, with bottle traps in place.

following week (Cooke 1995a). In other words, trends observed during night counting were reflected in the trap totals. However, there were indications that, when night counts were low, trapping could be relatively more effective, perhaps because individuals might be more active at lower density.

Netting and counting in daylight were compared with torch counting throughout the breeding season of 1986. The netting method involved making a 2 m sweep with a pond net every 1 m of edge. Netting demonstrated a similar seasonal pattern to night counting, but numbers were generally reduced by an order of magnitude or greater (Cooke 1995a). Day counting was less sensitive and reliable, but was rather better at detecting crested newts later in the season. Overall, crested newts were seen on 10 (59%) of the 17 counts.

Top Pond had a mixture of vegetation around, in and over the water. A study was made of where the adults and tadpoles occurred by dividing the pond into 16 sections, each with 3 m of shoreline (Cooke 1986a), using poles cut from the wood to demarcate boundaries between sections. Sections were of three broad habitat types:

- Some had tree canopy cover and consequently had little aquatic vegetation.
- Others were unshaded and had much aquatic vegetation.
- A block in the north-east corner consisted of sections dominated by common reed.

Numbers of crested newts counted at night were higher in the shaded sections than in the unshaded ones (e.g. see Figure 10.3a). Numbers counted in the reedy area were high up until May but were lower when the reed became tall. These observations should not be interpreted as newts always liking shade, because completely overgrown ponds are very poor newt sites – but they did demonstrate that in a pond with extensive areas of aquatic vegetation, courting newts are likely to be attracted to less weedy areas at night (see also Section 8.4). Another feature that appeared to govern where crested newts were seen at night was the profile of the edge. Half of the sections were graded as shallow (water depth less than 20 cm at 50 cm from the edge) and half as deeper (water depth at least 20 cm at 50 cm from the edge). Figure 10.3a shows (1) the locations of the deeper sections, which were mainly along the western side where the original clay pit had been cut into the hillside, and (2) how these sections were more attractive to newts in 1985. In 1986, the deeper sections held 60% of the crested newts counted at night, 85% of those counted during daytime, and 71% of those netted during daytime.

More tadpoles tended to be netted during daytime in late summer 1986 in the unshaded, well-vegetated sections (Figure 10.3b). When tadpoles were counted at night during July and August 1987, they were found to favour both those sections with a tree canopy and those without, but the reedy area was largely avoided.

Pitfall trapping and also examination of refuges in 1984 and 1985 (Cooke 1985c, 1986b) indicated that crested newts were in the following size ranges: young of the year, up to 69 mm total length; immatures, 70–89 mm; and adults, at least 90 mm. However, these groupings were later refined (Section 10.5). There seemed to be no marked preference for newts to be found in any particular terrestrial habitat type within the site. In 1985, a line of traps was installed 25 m out into a wheat field next to the site, roughly 50–75 m from Top Pond. Prior to harvest, more adult newts were caught in the wheat field than in a trap line within the site equidistant from Top

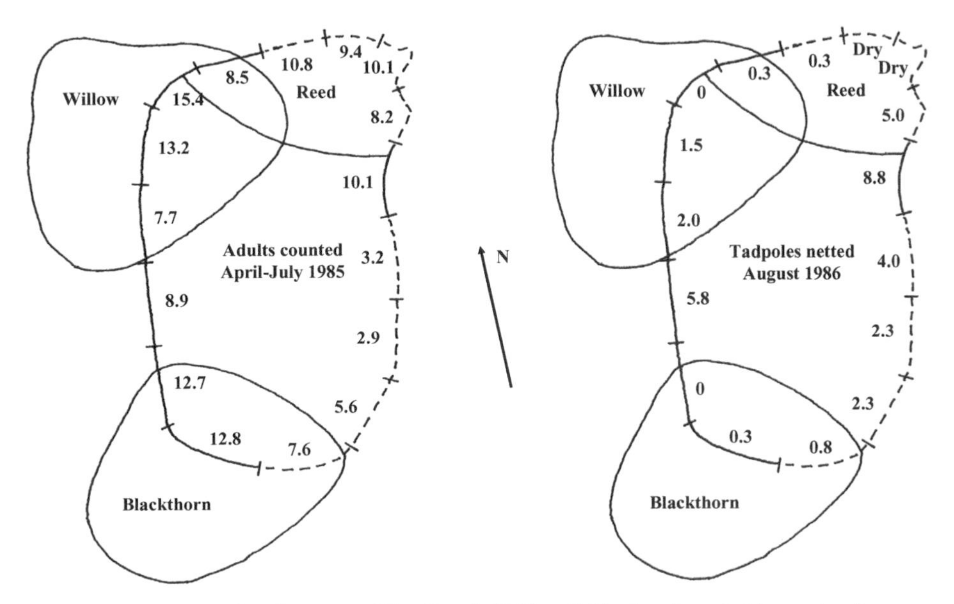

Figure 10.3 Sketch maps of Top Pond, Shillow Hill, showing night counts of adults and catches of tadpoles in the mid-1980s: (a) shows the average number of adults counted in each of the 16 sections from early April to early July 1985 (based on 12 counts); (b) shows the average number of tadpoles netted in those sections during August 1986 (four visits). The solid outline indicates the deeper sections and the broken outline the shallower ones. Also drawn are the reedy area and the extent of tree canopies: willow (*Salix fragilis*) and blackthorn (*Prunus spinosa*).

Pond, and on the opposite side of it. After harvest, however, few adults were trapped on the arable land. No newly emerged young newts were trapped on the arable field. Regular checking of the main road, less than 100 m from Top Pond, failed to detect any road casualties. Thus, although studies were limited, there was no evidence of newts attempting to cross the road or of young newts dispersing.

Newt activity on land was related to rainfall. Traps were checked after intervals of 4–9 days: if rainfall during the intervening period was at least 5 mm then more than twice as many newts were caught. During prolonged drought, numbers caught were reduced to less than a quarter of the level recorded during wet weather.

Today, these observations seem extremely basic and simple, but they were made more than 35 years ago when I – and just about everyone else – had very little knowledge of the species.

10.3 Surveillance using night counting

10.3.1 The method

By 1986, I knew that torch counting showed the same trends in numbers as bottle trapping and had some idea of its effectiveness at detecting both newts in the water and newts in the population. And I had learnt that perhaps less than half of the newts in the population might be in the water at any one time, even at the peak of the breeding season. Nevertheless, night counting seemed to be the best option for providing a low-input index of population size over a potentially long period of time. I knew of no crested newt breeding ponds within 500 m of Shillow Hill, so significant immigration seemed unlikely; but, despite lack of evidence, I have never ruled out emigration occurring.

Detecting newts at night in Top Pond will depend on many factors. The proportion of newts in the pond will be influenced by time of year, seasonal weather conditions, ease of reaching the water and pond conditions. Size of the population may also affect the proportion of newts in the water. Whether a newt is visible in the pond will depend in part on time of year and day; and its behaviour may also be influenced by weather conditions at the time and by density of adults in the pond. Detectability of a newt in open water will depend on pond size, extent of access to the shore, water clarity, amounts of aquatic vegetation, the power of torches used and the skill and care of observers. As I was involved with surveying this pond from early middle age to pensioner, I was also aware of my faculties, particularly eyesight and agility, declining with age!

This may seem a daunting list of issues, but most of them were controlled for:

- Counts were undertaken between 50 and 120 minutes after sunset from the second week of April until the first week of June throughout the period 1983–2009.
- On each count, a single circuit of the pond was made, and the entire edge of the pond could always be scanned.
- The averages of nine weekly counts during 1984–1987 and the averages of five fortnightly counts in 1988–2009 were used to try to counter night-to-night variability in numbers seen.
- Whenever possible, counting visits avoided rainfall or air temperatures of less than 5 °C. Counting was also avoided for several nights after heavy rain, to allow temporary turbidity to clear.
- Filamentous algae and debris from surrounding trees were cleared from the surface of the pond in early April each year, and a stick was taken on each night visit to move aside surface or submerged cover than might hide newts. The stick was also used to prod objects that might have been newts or willow leaves – and to ensure I did not fall in or become stuck in the mud (Figure 9.9b).
- All night counting was done by me with help from Henry and Steven, using torches of up to 70,000 candlepower.

There was, however, not much that we could do about pond conditions – water level, turbidity and development of vegetation through the summer. These variables were recorded on each visit: turbidity was judged on a scale of 0–3; vegetation was estimated as percentage cover at the surface; water level was categorised as 'full' (if water covered the bed of the pond), 'sump' (after the sump was dug in 1991, see below) or 'dry' in each spring, summer and autumn (Appendix 1). The idea was that recording the variables would allow later judgement about whether changes in them had affected numbers of newts counted.

10.3.2 *Trends in counts over the years*

Average annual night counts for 1983–2009 are shown in Figure 10.4. Crested newts occurred in Top Pond each year, but there were considerable fluctuations between

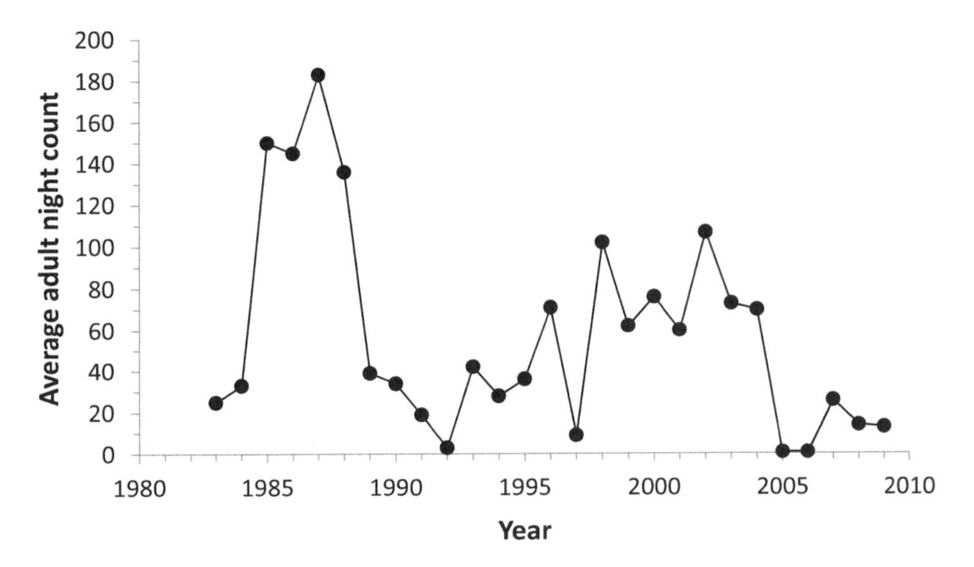

Figure 10.4 Average (mean) adult night counts of crested newts in Top Pond, 1983–2009.

years. Some of the fluctuation was due to pond conditions rather than to changes in population level. The most obvious effect was due to the amount of water in the pond. The sump had been created in January 1991 (Figure 10.5a) because it was clear that the whole site was becoming generally drier and Top Pond had dried out in the summer of 1990 with total loss of tadpoles. When the sump dried out in the summer of 1992, it was deepened further (Cooke and Arnold 2001). But that was not the end of desiccation problems. In September 2004, Henry and I removed 'new' sediment from the centre of the sump and then deepened our original excavation by about 25 cm; accumulation of sediment amounted to about 1 cm per annum over the 13 years between our excavations. The deepest part of the pond was then 1.5 m below top water level, about 80 cm deeper than it had been originally. This was an illustration of the extent to which the site had dried out over the previous 20 years.

The aim of digging a sump was to provide water to allow metamorphosis to occur in moderately dry summers, rather than to provide water for the adults in spring. Because of progressive desiccation, water in Top Pond was confined to the sump in the springs of 1991 (Figure 10.5b), 1992, 1997, 2005, 2006 and 2008. Our efforts at excavation in 2004 did not help much. Average night counts of crested newts during those six springs with water confined to the sump ranged between 0.8 and 19, whereas average counts in the remaining 21 springs ranged between 13 and 183. When only the sump was available, numbers of crested newts were considerably reduced. The extent to which this was due to few newts 'bothering' to return to the water at all or to individuals spending much less time in the water was not investigated.

It helps, therefore, to understand fluctuations of numbers in the population (rather than in the pond) if data for those 'sump years' are omitted (Figure 10.6). In addition, data for 1999 and 2001 can be omitted as average turbidity was unusually high (1.6 and 1.8 respectively), which may have reduced detectability. The value for 1983 has been removed in Figure 10.6 because it was based on a single count.

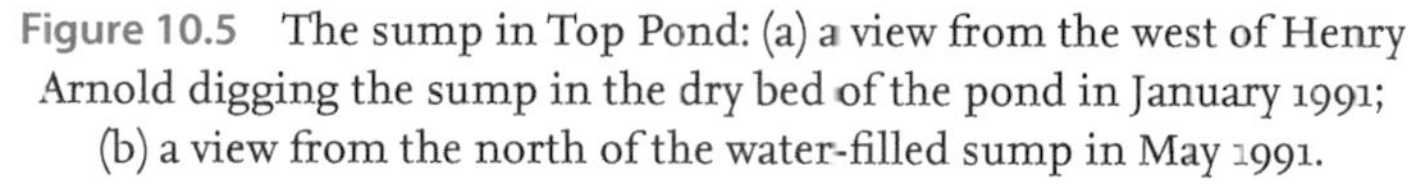

Figure 10.5 The sump in Top Pond: (a) a view from the west of Henry Arnold digging the sump in the dry bed of the pond in January 1991; (b) a view from the north of the water-filled sump in May 1991.

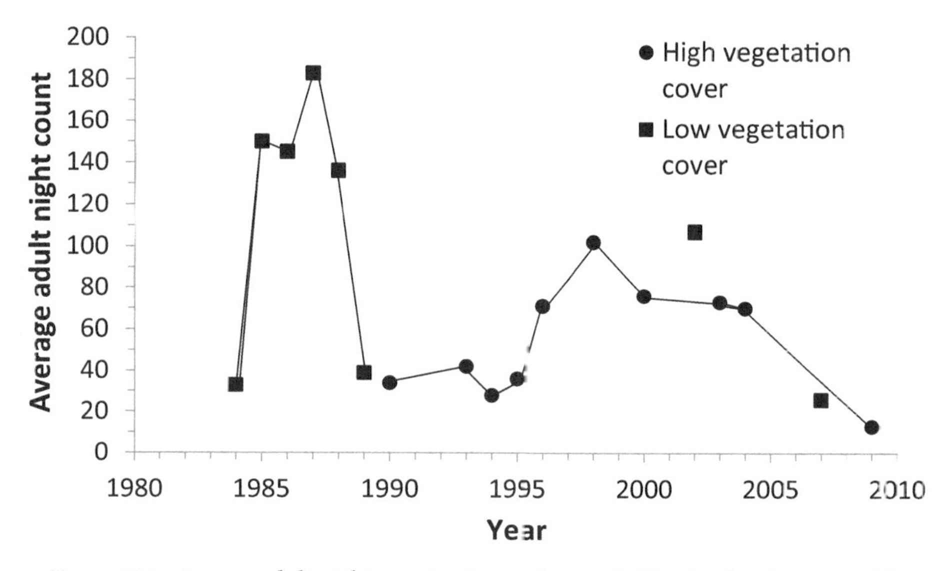

Figure 10.6 Average adult night counts of crested newts in Top Pond, 1983–2009: with data omitted for years when only the sump was available in spring or when the water was unusually turbid or when the average was based on less than five counts. Data are divided into years with high vegetation cover (at least 60%) and low cover (30–55%).

In Figure 10.6, average counts have been divided into those summers when maximum vegetation cover reached at least 60% (high cover) and those when it failed to do so (low cover). Surface cover tended to be greater in the later years. Although there may be relatively little influence of vegetation cover on the counts, they are separated in the graph into low-vegetation counts during 1984–1989 and high-vegetation counts in 1990–2009. This reduces the data from a single group of 27 annual values to two groups of six and ten, but gives a more credible account of changes over the period. There was a major increase in numbers in 1985 – indeed the *average count* of newts in the water in 1985–1987 exceeded the *estimated number* in the water in 1984. This was later followed by a dramatic reduction between 1988 and 1989, and the population then remained low until 1996. Numbers recovered during 1996–2004, but had declined to a low level again by 2009. Factors that might have caused those changes are explored in the rest of the chapter, but first I will explain how and why the study stopped.

10.3.3 Winding up surveillance

By autumn 2009, it was clear that the condition of the pond was not enabling the population to thrive. In three of the previous five springs, only the sump had been available and night counts were low; and in three of the previous five summers, the pond had dried before any emergence of efts (Figure 10.10b). In only one year were water levels high through to summer. Henry and I realised that more serious excavation was needed, but considered that we were physically incapable of manually undertaking what was necessary. Out of the blue, Francesca Barker of Froglife offered mechanical assistance, which we gratefully accepted. What I had in mind was a general deepening of the pond while retaining its existing edges. When the small digger arrived in November, however, I was unable to be present; and the pond received what would be regarded as re-profiling, designed to make it suitable for breeding for a number of years. Total depth was increased to 2.8 m, including an enlarged sump of 1.4 m depth. A ledge along the west side of the pond, which had previously provided safe access for observation, had been removed, so it was now necessary to walk along the sloping side at the angle of the original clay pit in order to observe the western edge of the pond.

Nevertheless, I was optimistic about the future when spring arrived in 2010. By the second week of April, the pond had filled nicely to 2.4 m (the original pond had been dry during the previous summer), but the water was the maximum of three on the turbidity scale (Figure 10.7). The pond remained turbid for the first four counts, when the average number of crested newts seen was nine. Then the water cleared for the last count and 23 newts were recorded, thereby demonstrating both the effects of turbidity and the fact that reasonable numbers of newts remained in the pond in early June. During the next four years, observations were scaled down: single night counts were attempted each year, but turbidity remained a problem and counts ranged from zero to 16. The only plant life noticed in the pond through this period was filamentous algae and duckweed (*Lemna* spp.). I tried daytime netting in 2014 and 2015 instead of torch counting, but, although newts were caught, numbers were not sufficiently high to be able to use them with confidence for monitoring. I rejected bottle trapping as an option because of safety issues for the newts and myself. The 10 newts caught in the pond in 2014 and 2015 ranged in size from 99 to 131 mm, which separated into the following sex/size classes (see Section 10.6.2): two immature females, one small adult female, three large females and four large males, indicating the presence of some younger newts. By 2015, however, it was evident that surveillance using the former methods was no longer feasible and attempts were finally abandoned – 32 years after the newts were first counted. Data from beyond 2009 are not considered in the remainder of this chapter.

It was another six years before I returned to look in the pond again. On 8 July 2021, as Henry was selling the property and moving out, I found Top Pond to be 'full', and the water was sufficiently clear to confirm that crested newts were still breeding. Although there were no large mats of water plants and the area of reed

Figure 10.7 A view from the south of Top Pond in April 2010 after re-excavation the previous autumn. High turbidity of the water was a problem for torch counting for the next five years.

no longer occurred in the north-east corner, aquatic invertebrates were also clearly doing well. Surrounding terrestrial vegetation was denser and lusher than I had previously known. The deepening of the pond in 2009 made further monitoring difficult, but ensured that the population had been able to survive.

10.4 Tadpoles and metamorphosis

Early on in the study, I realised that in order to understand factors underlying population changes, it was essential to have a relative measure of annual output from Top Pond. This was achieved from 1986 onwards by netting the pond for tadpoles on four occasions during July–August each summer when the pond held water. This covered the period from before any efts emerged until after emergence began (e.g. Table 10.3). Successful breeding probably never occurred in Wood Pond because of desiccation. For each 1 m of Top Pond's edge, a 2 m sweep was made towards the edge through the water and any aquatic vegetation (Figure 10.8). When the water level was down to the sump, a sweep was made through each 2 m² of surface area. Tadpoles were identified, counted (Appendices 2 and 5) and released. The assumption was made that the number of tadpoles netted in July and August reflected the extent of emergence of efts.

In six of the 24 summers from 1986 to 2009, Top Pond dried out and no tadpoles survived. An analysis of the period 1993–2002 revealed that number of tadpoles netted was positively related to the number of adults counted in spring, but negatively

Figure 10.8 A view from the reedy corner of Top Pond with the author sweeping for tadpoles in July 2007, when the shoreline had contracted from 48 m to 25 m.

affected in summers when high numbers of adults were incidentally caught along with the tadpoles (Cooke and Arnold 2003). More adults returning to the pond in spring is likely to increase reproduction, but if adults tend to remain in the water for longer this could mean higher levels of tadpole predation. The relationship between average number of tadpoles netted in a year and average adult night count is shown in Figure 10.9 for the 18 years, 1986–2009, when water persisted through to the summer. The years when adult numbers remained high in summer (1987, 1988, 2001 and 2002) are indicated by open circles; in 1988 tadpole catch was good despite adults staying on late in the pond. There was a positive relationship between tadpole catch and count of adults for the 14 years when adult numbers were low in summer.

Night counts of adults and tadpoles were made until late August in 1987, so it is possible to look back at that year in more detail with regard to adult predation on tadpoles. Tadpoles become more conspicuous during the later stages of development. The first night count of pelagic tadpoles was on 7 July, when 55 were seen. Numbers peaked at 109 on 24 July, when 59 adults and three immatures were also counted despite it being high summer. The volume of water in Top Pond at that time will have been in the region of 50 cubic metres, meaning the adult *count* was roughly one per cubic metre, while the actual *density* was probably several times higher. Night counts of both adults and tadpoles had decreased by August when four daytime visits were made primarily to assess the relative abundance of tadpoles by netting. Taking into account the volume of water swept with the net, the average number of crested newts caught equated to densities of about one adult and five

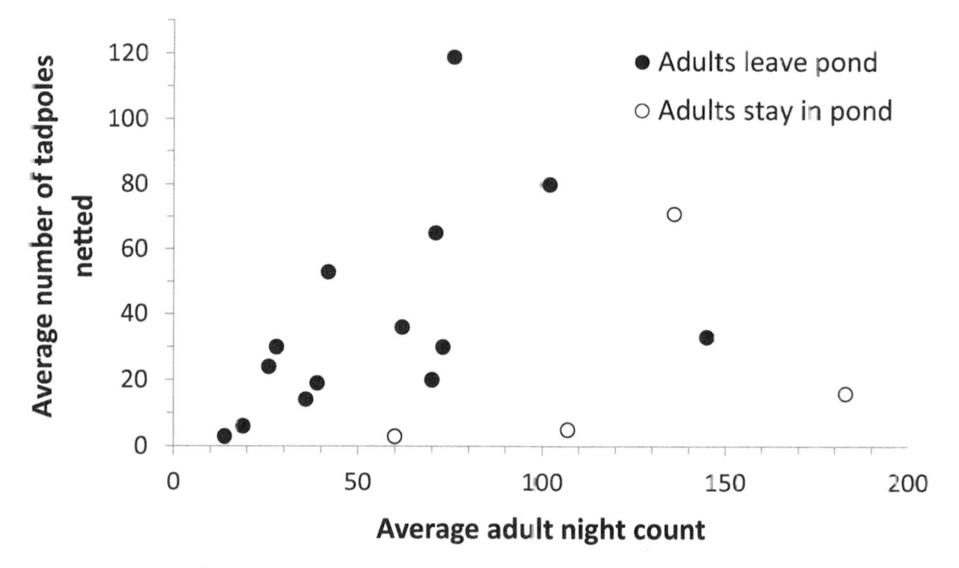

Figure 10.9 The relationship between average number of tadpoles netted in summer and average night count of adult crested newts in spring for the 18 years when water persisted in Top Pond during spring and summer. The four years when adults stayed on in greater numbers later into the summer are shown as open circles.

tadpoles per cubic metre. Depth was about 50 cm that summer so there will have been close and continuous contact between tadpoles and adults despite the former being pelagic and the latter usually showing benthic behaviour. And presumably a certain amount of predation will have occurred throughout the summer.

An analysis of tadpole numbers netted in relation to conditions is summarised for all 24 summers in Table 10.1. Providing some water persisted through until summer, some tadpoles always survived. Average tadpole catch was 23 in the 10 summers when only the sump remained and/or high numbers of adults occurred in the pond, about half the average number when there were favourable pond conditions for the tadpoles. Nonetheless, many tadpoles were netted in 1996 when only the sump held water, so other factors must also be important in some years.

In order to determine whether there might be a link between low numbers of tadpoles and high numbers of predatory invertebrates, the following were counted while netting for newt tadpoles from 1996: larval and adult greater water-boatmen, dragonfly larvae, beetle larvae and adults at least 5 mm in length (Table 10.2). As can be seen, there was no obvious association between high numbers of invertebrates and low numbers of newt tadpoles. And no evidence of invertebrates attacking newt tadpoles was observed during the study. Indeed, the highest counts for crested newt tadpoles and for each type of invertebrate occurred in the year 2000. The trend between number of crested newt tadpoles and number of water-boatmen, for example, was positive, although it was not statistically significant. It should, however, be pointed out that the pond dried out during most years in this period of observation (Appendix 1) so that populations of predatory invertebrates had to keep recolonising. They had little opportunity to realise the density they might have reached had the pond been (more) permanent. However, comparing the four years when the pond

Table 10.1 Numbers of crested newt tadpoles netted in Top Pond, 1986–2009, in relation to different conditions.

Breeding conditions	Number of summers	Average tadpole catch (range)
Pond dried out	6	0
Water in sump only in summer	6	23 (3–65)
High number of adults in water in summer	3	30 (3–71)
Water in sump + high numbers of adults in water in summer	1	5
Favourable conditions	8	49 (19–119)

Table 10.2 Average numbers of crested newt tadpoles and predatory invertebrates netted in Top Pond in years when the pond held water through to the summer, 1996–2008.

Year	Crested newt tadpoles	Greater water-boatmen	Dragonflies	Beetles
1996	65	13	0.3	4
1998	80	11	0	10
1999	36	9	1	2
2000	119	30	8	18
2001	3	19	1	0
2002	5	0.8	2	0.8
2003	30	27	2	5
2004	20	2	0.3	0.5
2007	24	54	0	6
2008	3	5	0	1

had not dried out the previous year with the six when it did dry out, the average numbers of water-boatmen were 16 and 18 respectively, and numbers of crested newt tadpoles were 48 and 32. The differences between the two sets of conditions were not significant for either species. In 1996, 2000 and 2004, some tadpoles had bite marks on their tails, believed to have been caused by them attacking one another; in each of these years, there were unusually high densities of tadpoles in the pond (Appendix 2). Such injuries were also observed in cage trials undertaken by Griffiths *et al.* (1994).

'Walkers' are newt tadpoles that are starting to use their legs, although they are still in the water (Cooke and Cooke 1993; Figure 10.10a). The earliest date on which walkers were recorded in the pond was 10 July, and how development and emergence unfolded in late summer can be illustrated by results from the year 2000, during which the greatest number of tadpoles was recorded (Table 10.3). The edge of the pond decreased from 45 m on the first netting visit to 25 m on the fourth visit, with the number of sweeps being adjusted accordingly. Total number of tadpoles caught decreased progressively, while the number and percentage of walkers increased. No

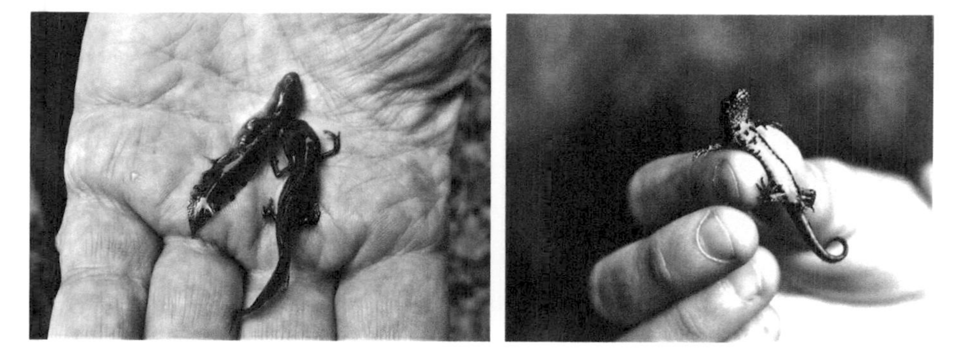

Figure 10.10 Some early life stages of the crested newt. (a) Two tadpoles netted in Top Pond, July 2001. The animal on the right is more developed and has reached the walking stage in which it can right itself onto its belly and walk. The less developed tadpole on the left cannot yet do this and lies on its side. (b) An eft, caught on land in August 1984, with the spot pattern on its underside just beginning.

Table 10.3 Catches of crested newts in Top Pond in summer 2000, indicating numbers of tadpoles changing into the walking stage (and later leaving the pond as efts). Number of sweeps decreased as the pond contracted in size.

Date	Number of sweeps	Number of crested newt tadpoles netted			Number of adults netted
		Walkers	Non-walkers	Total	
08.07.2000	45	0	159	159	0
15.07.2000	42	3	120	123	0
22.07.2000	37	11	99	110	0
06.08.2000	25	33	49	82	0

potentially predatory adults were netted in 149 sweeps of the pond, so it is likely that the reduction in overall numbers of tadpoles was largely due to efts leaving the pond (Figure 10.10b).

After Steven and I had identified the existence of a walking stage (Cooke and Cooke 1993), I measured the length of walkers that were netted, beginning in the summer of 1993. It was noticeable that in years when higher numbers of tadpoles were netted per sweep, walkers were small, and vice versa (Figure 10.11). In other words, there was a trade-off between density and size. The smallest walkers occurred in the sump in 1996, when more than seven were netted per sweep on average, while the largest were in 2001, when cannibalism was believed to have reduced density to one per 10 sweeps. Before 1993, the greatest density of tadpoles was in the desiccating sump in 1991; lengths were not measured then, but tadpoles were recorded as being unusually small.

The connection between good and bad years for tadpoles and population change will be examined in more detail later, but at this stage I will point out that the poor periods for tadpole numbers, 1990–1992 and 2001–2009 (Appendix 2), were associated with low adult counts in the early 1990s and a decline after 2004 respectively (Figure 10.6).

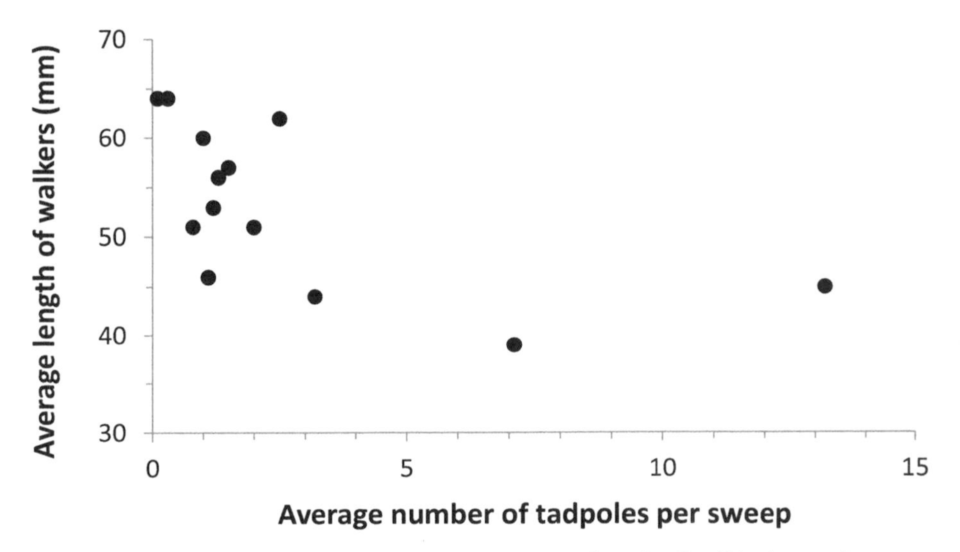

Figure 10.11 The relationship between average length of 'walking' crested newt tadpoles and average number of tadpoles netted per sweep for all years when water persisted in Top Pond until summer, 1993–2008.

10.5 Immature newts

An immature crested newt is an animal that has come through at least one winter, but is not yet ready for breeding. It was, however, difficult to determine whether or not some newts were immatures, and to define the upper limit to their size range. Towards the end of Section 10.2, I described nominal length ranges for efts (up to 69 mm in their first autumn) and adults (at least 90 mm); and I rather lazily termed everything between these lengths as an immature. This assumption is partially correct in that crested newts through their first winter, but less than 90 mm, will almost certainly be immatures. However, there will have been immature females in particular that were longer than 90 mm.

During spring, 1993–2009, I netted and measured 115 breeding females and the shortest was 107 mm; the shortest of 150 males in breeding condition was 93 mm (Section 10.6 and Appendix 3). The size distributions of these newts are shown in Table 10.4 and are compared with the size distributions of newts of length at least 90 mm caught and measured on land during the summer and autumn of 1984 and 1985. The newts caught in pitfall traps or refuges on land showed similar size distribution for the two sexes, with animals being measured in all of the size classes. They were taken at a time when the population was increasing dramatically; many newts were also examined which were less than 90 mm. Males netted in Top Pond had a similar size distribution to the males taken on land, although there were relatively fewer small males, in line with there being generally less recruitment during that period. On the other hand, size distribution of females in the pond was skewed towards larger size. There were none in the range 90–99 mm and fewer measuring 100–109 mm. The total lack of females measuring 90–106 mm suggested that the vast majority in this size range remained on land in spring. In this country, males breed for the first time at two or three years of age and females one year later (Beebee and

Table 10.4 Size distributions of crested newts at least 90 mm in length caught on land in 1984–1985 and in Top Pond in 1993–2009.

	Percentage in each length range (mm)					Total number
	90–99	100–109	110–119	120–129	130–139	
Caught on land, summer and autumn 1984 and 1985						
Males	23	29	35	10	3	31
Females	25	21	33	15	6	52
Netted in Top Pond, spring 1993–2009						
Males	9	25	49	17	0	150
Females	0	5	25	50	19	115

Griffiths 2000). Thus, at Shillow Hill, newly mature males occurred in Top Pond in spring, whereas females of comparable size (and age) were still immature and tended to stay on land. But I had been catching crested newts at Shillow Hill for many years before I had evidence that males started to breed at two years of age and females at three (Section 10.6).

A few newts in the size range 70–89 mm were netted at night with the adults and were recorded during torch counts: these were categorised as immatures. However, no newts of this size were seen during the five springs when only the sump was available or in 2007 when the pond was full. Years with average night counts of immatures of greater than two were: 1985 (7.3), 1986 (4.7), 1987 (3.9), 1988 (3.6), 1994 (5.2), 1996 (2.4), 1999 (3.4), 2000 (2.2) and 2002 (2.2). These were occasions when the population was buoyant and/or recovering, yet numbers of immatures counted in the breeding pond at night were very low when compared with numbers of adults. The ratios of immatures to adults in torch counts were also very different to the catch totals on land in the early years of the study: the ratios of immatures (70–89 mm) to adults on land (excluding females less than 107 mm) were 1:1.2 in 1984 and 1:2.8 in 1985, compared with 1:21 for immatures to adults in the pond in 1985. In 1986, continuing night counting until late July revealed that numbers of immature newts were highest from late May until mid-July (average 14 for seven weekly counts), so my standard night counting 'season' of mid-April to early June will have missed much of this period. In the 20 ponds at Woodwalton Fen, there was an exceptional torch count of 113 adults and 66 immatures in May 1987 (Table 8.3), the most immatures I have recorded during a single mid-season count by some distance.

10.6 The 1990s and beyond

10.6.1 *Introduction*

By the start of 1993, I had been working at the site in my spare time for 10 seasons. Much had been discovered that I hoped was relevant to newt conservation and survey. Torch counting had been shown, within limits, to be of value in monitoring, but I had little or no hard data on why counts had increased so spectacularly between 1984 and 1985 and then decreased equally dramatically between 1988 and 1989. By then, Henry and I knew that the site was progressively drying out, which was further exacerbated by a lengthy regional drought at the beginning of the 1990s. Top Pond had dried out in summer for the first time in 1990 with the loss of all tadpoles. One long-term aim had been to determine to extent to which the population could flourish unaided. But conservation intervention was not ruled out, and we responded to the situation by digging the small sump in January 1991, and that was all that was available for breeding in 1991 and 1992 (Figure 10.5) – success in those two years was minimal.

It seemed as if I was standing at a fork in the road. NCC was in the throes of turning itself in English Nature and the other agencies, and I had passed on responsibility for amphibians to Tony Gent. In May 1993, I was starting a year's secondment with ITE to undertake research on deer: for the first time in my formal career, I was looking forward to a lengthy spell of virtually continuous fieldwork. I remember writing on my office door on the last day of April 1993, 'A. Cooke is out. Please call back in May 1994'. That might have been a time to stop visiting Shillow Hill, and some of my other amphibian monitoring sites too, but I was still enjoying gathering information on newts – and I saw a new opportunity at Shillow Hill.

The failure of the crested newt population to have more than minimal recruitment during the three breeding seasons 1990–1992 provided a marker to see how size frequency of individuals changed over the years as, hopefully, the population recovered. I had already started to deal with a particularly senescent population of large individuals in Ramsey (Section 9.4.2), so perhaps studying what happened at Shillow Hill could provide information to aid in recognising senescent populations and helping them recover.

10.6.2 Separating adults into small and large sizes

In late April 1993 I visited Top Pond one night and did a couple of circuits, netting by torchlight. I managed to catch 15 newts: 10 adult males and 5 females. These were taken into Henry's house, where we weighed them and measured their total lengths. Samples were then netted at night in late April to early May once each year until 2009. I first began measuring the length of newts at Shillow Hill when we trapped them on land in autumn 1984, but some years later I realised that other researchers were measuring snout–cloaca length. I have to admit that this is the better measure, in part because tail tips can get damaged (such newts were not measured). By then, however, I had several years of data on total length so thought it was best to continue with that method. In 2001 I also measured a small number of newts from the tip of the snout to the rear edge of the cloaca in case measurements were needed for comparison with other researchers' data: average length was 70 mm for large males, 72 mm for large females and 69 mm for small females (there were no small males in the sample; see below for definitions of 'large' and 'small').

Jehle *et al.* (2011) stressed the conservation importance of assessing the age structure of crested newt populations and discussed the merits and difficulties of counting growth rings in the bones of digits or limbs for determining the age of newts. Crested newts may continue to grow throughout their lives, but growth rate slows as they mature. These authors pointed out that crested newts were known to live for as long as 17 years, but cautioned against using size alone to estimate age. More recently, there has been a detailed study of newts in 13 populations in Lancashire (Orchard *et al.* 2019). This work showed that growth might continue for 10 years or more, but there was considerable overlap in size ranges at each age – so, for example, a female of snout–cloaca length 70 mm might be aged anywhere between three and six years and exceptionally might be even older. By 1993, I was already wary of assuming age could be estimated from size in toads (Section 4.4.3), but what I set out to achieve at Shillow Hill was to record changes in the size structure of the population over a period of years and relate those changes to breeding success in previous years (Cooke 2002).

I knew from catching and measuring crested newts on land that some males in the length range 90–99 mm had secondary sexual characteristics, and that a few large females occurred in the size range 130–139 mm. Therefore, adults spanned the range 90–139 mm and possibly beyond (Table 10.4). As reproductive success during 1990–1992 had been minimal, it was reasonable to assume that any adult newt netted in 1993 was likely to be at least four years old. The 10 males netted in April 1993 varied between 107 and 117 mm and the five females were between 121 and 135 mm (Appendix 3). Although these were small samples, males less than 107 mm and females less than 121 mm were missing. That year, water persisted in Top Pond through to summer and there were reasonable numbers of tadpoles. The following year, small adults were again missing from the netted sample: males showed signs of growth since the previous year, but size of females was virtually identical. Tadpole production was again reasonable in 1994. Size structure of the population then changed markedly in 1995 and 1996, with small adults becoming increasingly evident. 'Small' adults were defined as males of length less than 107 mm and females less than 121 mm, based on observations on the sample netted in 1993. In the sample netted in 1995, 46% of the males were small but only 13% of the females, a fact that was consistent with females being slower to mature. This told me that the tadpoles from 1993 had started to mature after two years, but I strongly suspected from the increase in small newts in 1996, as well as from the early work of Torkel Hagström (1979) in Sweden on age at first breeding, that some newts were maturing at three years of age – and perhaps even later. In 1996, 75% of the males were small, as were 50% of the females. The small males were three-year-olds from 1993 and/or two-year-olds from 1994, while the small females were likely to be mainly three years old.

The proportion of these small newts each year is shown in Appendix 3. In 2003 the number of small newts counted at night was estimated for each year, 1993–2002, apart from 1997 when only the sump was available, by multiplying the average night count by the proportion of small newts in the netted sample (Appendix 4). Then these figures were compared with the number of tadpoles netted in previous seasons (Cooke and Arnold 2003). The most significant relationship was with the average number of tadpoles caught two and three years earlier. I updated this analysis to 2009, again excluding any years when the newts had only the sump in which to breed, and still found that if high numbers of tadpoles were netted in one year and/or the next one, then the estimated number of small adults counted at night tended to be high two years later (Figure 10.12). Similarly, estimated numbers of large adults (males of at least 107 mm and females of at least 121 mm) counted in non-sump years were related to the average number of tadpoles netted four and five years before (Appendix 4). The relationships between numbers of adults estimated to be in different size classes and tadpole numbers in previous years were perhaps the most surprising results to come from this study. They demonstrated that the annual level of reproduction was reflected in the future size structure of the

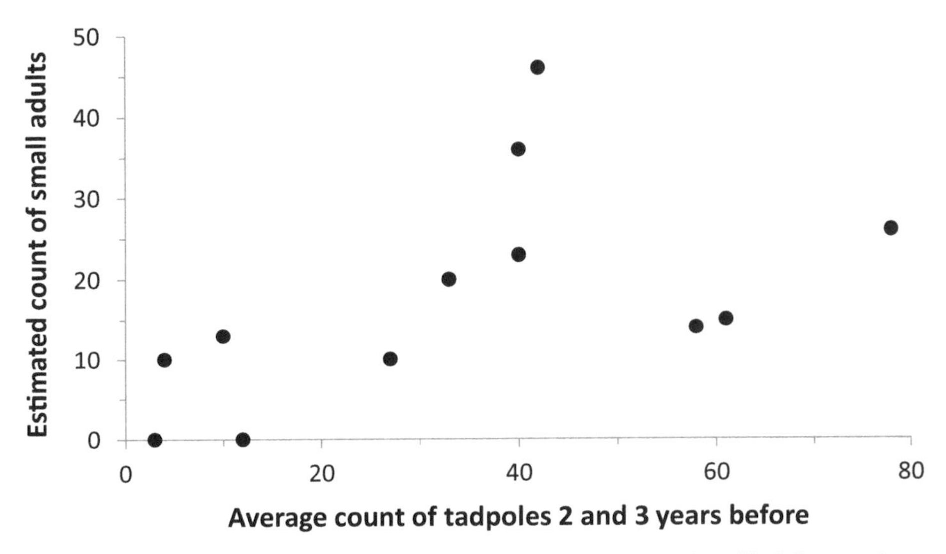

Figure 10.12 The relationship between the estimated count of 'small' adult crested newts in a year, 1993–2009, and the average catch of tadpoles two and three years before in Top Pond. Years when only the sump was available in spring are excluded. There were two years with the values tadpole count three, estimate of small adults zero.

population. I thought there might be hints of such relationships, but did not expect them to be statistically significant, bearing in mind the various factors that could reduce the size of each cohort, the relatively poor relationship between age and size of adults, and the simple techniques used the gather the data.

Although there was a general relationship over a 17-year period (Figure 10.12), a tadpole was more likely to return as a small adult during the first part of the period than later. Thus, as the population recovered in the mid-1990s, estimated night counts of small newts were higher in 1996 (46) and 1998 (36) than when the population had stabilised in 2001 (14), 2002 (26) and 2003 (15). This was despite tadpole counts two and three years before being lower during 1993–1996 (40–42) than during 1998–2001 (58–78). After 1996, the population became progressively more inclined towards larger newts, as reflected by the change in the proportion of small newts in the netted samples (Appendix 3). What factors might have been involved with this change? Tadpole numbers were zero or low in six of the years from 2001 to 2009, with site desiccation being important in four years. However, density-dependent factors were also suspected to be operating, such as cannibalism of tadpoles by adults in 2001 and 2002, and perhaps increased pressure on efts and immatures to leave the site when adult numbers became high. By 2009, the population was dominated by large newts, so it had become senescent again – and low in numbers, just as it was in 1993.

10.6.3 Adult survival

For some pairs of years, it was possible to derive tentative estimates of adult survival using average night counts and the proportion of small newts in the netted samples. This technique focused on those four pairs of years when amount of water, turbidity and plant cover were within the same limits (Figure 10.6):

- There were no small newts in 1993 or 1994 and the drop in average count was taken to indicate loss of newts, survival between 1993 and 1994 being 28/42 = 67%.
- The estimation of survival between 1994 and 1995 was more complicated because the average night count in 1995 was calculated to be 10 small and 26 large newts. If all newts breeding for the first time were small, then annual survival was 26/28 = 93%. However, if some bred for the first time as large adults, then this will be an overestimate. From the size distributions of both sexes, this was considered unlikely as there were noticeable gaps in length between small and large newts: for example, 95–106 mm for small males and 118–126 mm for large males.
- Later estimates of survival were affected by another issue in that it was not known whether newts might remain 'small' for more than one year. If this happened then the derived figure would underestimate real survival. Two more estimates were possible with this caveat: 69% for 1995–1996 and 82% for 2003–2004. The four estimates were in the range 67–93% with an average of 78%.

Most studies reporting annual survival data for adult newts are based on capture–mark–recapture studies (von Bülow and Kupfer 2019). There are, however, difficulties with deriving information on survival from such studies, including that some animals in a population will probably not be recaptured (Jehle *et al.* 2011). Probability of (re)capture is usually unknown, and survival estimates are typically viewed as minimum values. Reported rates of annual survival have been highly variable but are generally in the region of 50–90%. I am aware that I have used unconventional techniques, in part because I have rarely had the time to undertake detailed trapping and marking. I have, however, had the commitment and good fortune to be able to undertake straightforward observational studies regularly over a long period of time. This has enabled me to compare observations made in different years but under virtually identical sets of conditions, which should mean similar detectability. I would argue that a technique such as counting newts at night is simply sampling without catching them, but I have often felt that many scientists do not accept such techniques as being valid. Similarly, the long-term counting of newts in the two ponds at Stanground (Jehle *et al.* 2011; Section 9.3.2) seems to have been largely ignored by other authors.

To return to the point, my estimates of annual survival are consistent with an animal that is capable of surviving and breeding for a number of years. Survival of adults should have been good if they stayed within the boundaries of the Shillow Hill site. During the period reported here, food and natural refuges will have been plentiful and harmful disturbance was minimal. However, venturing outside the site will have left them exposed to a variety of hazards on the road and the arable land. There was no evidence of them foraging across the road but checks were limited for reasons of personal safety. Adults foraged on the arable land prior to harvest and may have been exposed to agricultural vehicles spraying and combining. Although several pitfall traps were destroyed during ploughing in 1985, no dead newts were found.

10.6.4 *Body condition*

Higher numbers of newts were caught on land in 1984 when there had been rain. This implied that they foraged more when the weather tended to be wet, and I wondered if there might be a relationship between rainfall during their terrestrial phase in one year and body condition the next spring. I defined a Body Condition Index (BCI) as the weight in milligrams divided by the cube of the length in centimetres. A significant relationship existed between average BCI of newts netted each spring and rainfall during June to October of the previous year (Figure 10.13). The years that followed wet summers and autumns were generally productive from the point of view of breeding. Of the years covered in Figure 10.13, Top Pond remained full throughout spring and summer in only four years – and these were among the five that followed the wettest June–October periods. These four years were among the best seven in terms of numbers of tadpoles in summer, and three of them were the most productive in terms of tadpoles surviving per breeding adult. The summer that failed following especially wet weather during June to October was 2002: exceptionally high numbers returned to breed but water retreated to the sump and adults remained in the pond until summer, resulting in low numbers of tadpoles surviving to metamorphosis. Laurence Jarvis (2015) caught newts from four ponds in Epping Forest during 2007–2011 and determined their BCI using their snout–cloaca length. There were changes in BCI from year to year, but it was not found to be affected by winter temperature.

Figure 10.13 is an example of a number of significant relationships apparent in this study. So many years of data are available that it is possible to be rigorously selective over which to use and still have enough to test in a statistical sense. This type of result does not always imply a causal relationship. It does not mean, for instance, that rainfall in one year has necessarily caused the newts to be in better condition during the following spring – although in this case there appears to be a logical connection between rainfall and body condition. Proof of the connection would need further testing, perhaps under controlled conditions using captive newts that were

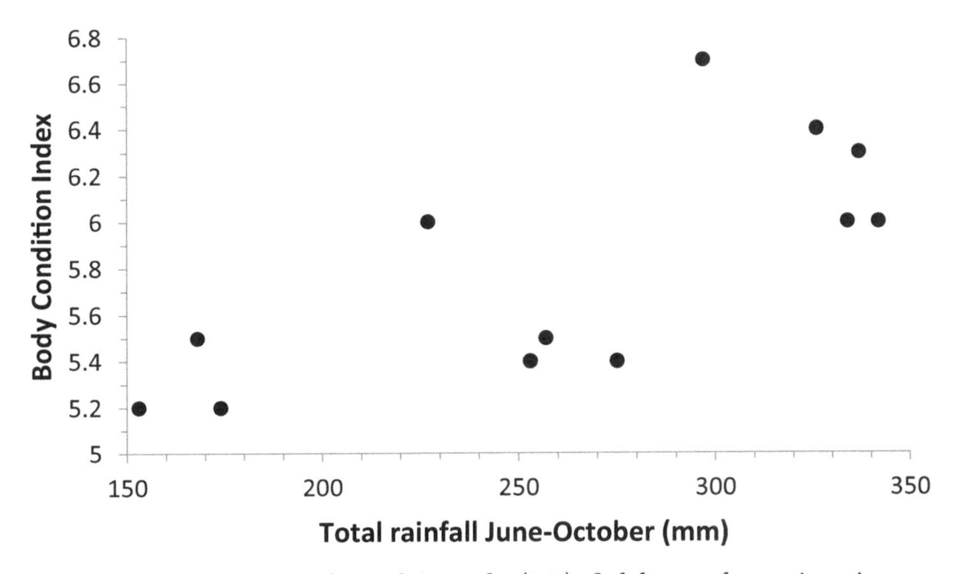

Figure 10.13 Average Body Condition Index (BCI) of adult crested newts in spring plotted against rainfall during June to October in the previous year, 1993–2008. Years when fewer than 10 adults were weighed and measured are excluded.

regularly monitored. As a further caution, the more tests that are done, the more likely it is that one or two will show statistical significance by chance. Nevertheless, it is hoped that the results of the study will give pointers to more detailed and controlled work that can be undertaken in the future.

10.7 Conclusions from the study of crested newts

10.7.1 Fluctuations in numbers counted

From 1984, night counts of crested newts in Wood Pond were no greater than eight and most were zero, so information on overall change came from night counts in Top Pond. The 27 years from 1983 to 2009 meant this was by some distance the longest continuous study of a crested newt population. I am not aware of a longer study to this day. The next longest seems to be an ongoing investigation on a population in experimental ponds at the University of Kent: Zakaria (2017) provided data for 2001–2015 and so far there are another six years of data (Richard Griffiths personal communication). There was also a 19-year investigation at a breeding site in Germany based on recognising individuals by their belly patterns (von Bülow and Kupfer 2019). The data from Shillow Hill have sometimes been used to illustrate how much a population can vary over time (e.g. Baker 1999), but using the counts unselectively can be misleading because exceptionally low counts were occasionally caused by pond conditions rather than by the population being very low. Low counts were associated with only the sump being available (and, after 2009, with higher than normal turbidity). Using Figure 10.6, numbers counted at night fluctuated between 33 and 183 for counts with 'low' vegetation cover during 1984–1989 and between 13 and 102 for 'high' vegetation counts during 1990–2009. Fluctuations were roughly six-fold and eight-fold respectively for the two time periods.

What then has been learnt about why the population fluctuated? A count of 25 crested newts in 1983 and an average count of 33 in 1984 suggested it was a colony of fairly typical size. English Nature (2001) classed a population with a maximum count between 11 and 100 as being 'medium-sized'. In 1985, however, my view on the relative population size changed as average count escalated to 150 adults, and it remained greater than 100 until 1988. Night counts suggested a four- to five-fold rise in number of adults, whereas bottle trapping indicated about a three-fold increase. Significant recruitment was suspected of being at least part of the cause of the increase in breeding numbers. Adult lengths were not studied in the breeding site in spring 1985, but newts had been collected on land and measured in the summer and autumn of 1984. If this sample is analysed and interpreted as was done for adults netted in spring 1993–2009 (Section 10.6), 55% were 'small' (i.e. females measuring 107–120 mm or males of 93–106 mm). Moreover, there were large numbers of immatures: females of 70–106 mm and males of 70–92 mm. Results were very similar for newts caught on land in summer and autumn 1985. So the population was buoyant at that time and its composition resembled that in the mid-1990s when night counts increased roughly three-fold between 1995 and 1998, despite dry conditions in 1997.

However, the fact that the count increased as much as it did in one year from 1984 to 1985 suggests that other factors were also involved. Any movement from Wood Pond and smaller, even more transient, on-site waterbodies to Top Pond did not appear to be important between those two years (Cooke 1995a). It is though possible that there was a behavioural change in the mid-1980s, with newts remaining for longer in the pond so that higher proportions of the population were counted. Post-breeding, more adults than usual were netted during the tadpole assessments in July and August 1987 and 1988. Although no netting was undertaken for tadpoles in 1985, torch counts continued until late July that year, with more than 100 adults being counted each week from the second week of April until the first week of July: 1985 had the longest span of counts of more than 100.

No other crested newt ponds are known within 500 m of the Shillow Hill site. There may be movement further than this between ponds serving a metapopulation, but it seems that this involves the movement of first year newts rather than adults. Thus, in an agricultural landscape in Germany, some young newts were found to walk further than 800 m to neighbouring ponds, but the vast majority of adults remained faithful to their breeding site (Kupfer and Kneitz 2000). Similarly, after a 12-year study of a metapopulation in Kent, Richard Griffiths, David Sewell and Rachel McCrea (2010) concluded that increases in adult population sizes were driven by increases in recruitment rather than migration of adults between subpopulations. Nevertheless, adult movements of greater than 500 m have been documented (e.g. Haubrock and Altrichter 2016), so some immigration may have occurred at Shillow Hill. At the site, 1985 was the only year in which the magnitude of a population increase was difficult to explain.

Despite high numbers of adults being counted during 1986–1988, catches of tadpoles were not unusually high: predation by adults and presumably to a lesser extent by immatures was suspected of lowering tadpole production. The failure of recruitment to sustain such a high population in the longer term might have contributed to the sudden and dramatic decrease in night counts between 1988 and 1989. However, the size of the reduction (71% in terms of reduction in average counts of adults) far exceeded annual losses of adults estimated in later years (7–33% for four years, Section 10.6.3). This fact suggested that another important factor was unusual mortality between those two breeding seasons, especially if significant emigration of adults or a shortening of time in the pond did not occur. John Baker (1999) studied a newt population in a pond in Milton Keynes from 1988 until 1995: average annual survival of adults was 69%, but rates were on two occasions surprisingly low at 31% and 37%, suggesting that dramatic, unexplained losses can occasionally occur. Richard Griffiths and his colleagues (2010) found that survival was low in their metapopulation in Kent when winters were wet and mild. The winter of 1988/89 was fairly mild at Shillow Hill, but both the previous winter and the next one had fewer days with air frost; and rainfall was lower than average that winter. There was no obvious physical change at the site which might have caused high mortality. Possibly, catastrophic losses occurred on the farmland or the road, or maybe there was an outbreak of disease while the population was living at high density. At this stage, more than 30 years later, one can only speculate about the cause.

The population remained low for several more years. The pond dried out in summer 1990 and a small sump was created during the following winter to try to safeguard the tadpoles in future dry summers. In the 19 subsequent summers up until 2009, the pond was 'full' in seven, but dried out completely in five. In the other seven years, the sump provided tadpoles with a chance of survival; average number of tadpoles in 'sump summers' amounted to 41% of the average in summers when the pond remained full. Digging the sump was therefore judged to be fairly successful. Its early years, however, were not auspicious as an average of only six tadpoles was recorded in the desiccating sump in 1991 and it was totally dry in the summer of 1992. The story from 1993 onwards has already been described in some detail (Section 10.6). Apart from the dry year of 1997, tadpole numbers were good more or less throughout 1993 until 2000. This resulted in significant recruitment to the adult pool from 1995 until 1997, followed by a progressive decrease in the proportion of small adults. In the Lancashire study of Orchard *et al.* (2019), although some males first became sexually active at two years and females at three years, few individuals of this age were captured at the breeding site, leading the authors to suggest that the young adults did not 'partake in reproduction' until they were older.

In Top Pond, there was also a fairly steady reduction in the proportion of immature newts counted at night over the period 1994–2004. Similarly, few immatures were encountered in a French population when the adult numbers became high (Arntzen and Teunis 1993). Initial declines in immatures and small adults at Shillow Hill were considered likely to be due to density-dependent effects such as competition between tadpoles at high densities leading to smaller and perhaps less robust individuals (in 1996, 2000 and 2004) and higher than usual predation by adults on tadpoles (in the summers of 2001 and 2002). In addition, it is possible that there was increased pressure on young newts to disperse away from the site when adult night count was relatively high, from 1996 to 2004. Although, there was no firm evidence of immigration at Shillow Hill, emigration of young newts cannot be ruled out and it is conceivable some found suitable breeding sites beyond 500 m. After 2004, the population at Shillow Hill appeared to decline because there were desiccation issues in four of the five years 2005–2009.

10.7.2 *General considerations*

NCC's guidelines of 1989 recommended that crested newt breeding sites eligible for recommendation as SSSIs should have night counts of at least 100 over a three-year period, in part to confirm population stability (Section 9.2). By 1989, I regarded the Shillow Hill site as a candidate for notification because average counts exceeded 100 during 1985–1988, but I was concerned that counts in 1983, 1984 and 1989 were much lower, suggesting high counts might not be sustainable in the longer term. Later counts only very rarely attained the threshold and the site was never notified. In the light of current knowledge, if I was formulating SSSI guidelines now I would propose something that also embraced the characteristics and attributes of the whole area and the population.

In 2004, I calculated the HSI for the site as 0.71 using the method of Oldham *et al.* (2000). This value was pulled down by the fact that Top Pond was relatively small and there were no other known breeding ponds in the vicinity. The survey undertaken by Rob Oldham and his co-workers found that a site with an index of about 0.7 typically had a night count of approximately 10, whereas the average night count for Top Pond in 2004 was 70. This suggested that while the population appeared strong at the time, it was vulnerable because it was dependent on a single, fairly small pond.

In 2000, Richard Griffiths and Clair Williams constructed population models for the crested newt using published data for different aspects of life history. The aim of modelling was to test the effects of isolation, drought, habitat fragmentation and dispersal on the risk of extinction. Isolation and drought were found to increase the risk of extinction considerably. The study at Shillow Hill was not a 'hands-off' exercise: I wanted to know what management was required for the population to survive and, if possible, thrive. Top Pond dried out in summer six times in 26 years (roughly one year in four), but without the sump the pond would have dried out in 13 years (one year in two). Had the sump not been dug, the population would still have survived the difficult dry period in the early 1990s, but recovery would probably not have been so swift. As density-dependent factors appeared to operate, the breeding population might well have reached the same level by around 2001, but it would then have faced a severe struggle to remain extant, as Top Pond would have been dry without the sump in seven of the eight years up to 2009 (Appendix 1). The 12-year study on the Kent metapopulation (Griffiths *et al.* 2010) demonstrated that three of the four sub-populations suffered reproductive failure in most years, and recruitment within the metapopulation was reliant on one sub-population. By the end of the 12 years, the estimate for number of newts in the metapopulation was roughly 100, down from more than 400, and at least one of the sub-populations was extinct. The study site of Bernd von Bülow and Alexander Kupfer (2019) in Germany dried out during nine of 19 summers, including one period of four summers in succession; the population survived but declined more or less throughout the whole period.

This discussion may make it seem as if pond desiccation is bad news for newt populations, but this is not necessarily so. In Chapters 8 and 9, there are several references to predatory fish detrimentally affecting local newt populations, and many such cases have been reported from elsewhere (e.g. Beebee 1997, Baker 1999). If a pond dries out in autumn or winter, then that is unlikely to be a problem for the newts but it will eliminate fish. Similarly, in a pond that rarely dries out, invertebrate predators such as diving beetle larvae and dragonfly nymphs may have an impact on survival of newt tadpoles, which more frequent desiccation could remedy. In Top Pond, predatory invertebrates did not seem to be a problem for newt tadpoles between 1996 and 2008, but the pond dried out in most years. Fish were never recorded in Top Pond, so were not a consideration, but the newt population would probably have done better had the pond been 'full' each spring and dried out less frequently in summer; this is especially true for the period after 2004. A complicating factor, however, was that an elevated incidence of predation by adult newts on their own offspring was suspected in some years when the pond was full and spring counts of adults were high, such as in 1987, 2001 and 2002.

With the benefit of hindsight, it would have been better not to have depended totally on Top Pond for recruitment, but to have had one or more other ponds that were at least occasionally attractive and productive. Wood Pond was never more than 50 cm deep and suffered from siltation issues and from the general desiccation of the site. It was deepened in 1990 and again in 1992, but did not hold any water until heavy rains occurred in late 1992. After that, crested newts were only recorded in three springs and the pond remained stubbornly dry from 2004 onwards. Access by machine to that part of the site would have been difficult. When Top Pond was re-excavated in 2009, a small pool was dug by machine in an area that had been marshy 20 years before. This held water in the spring but dried out in summer – and no newts were ever recorded in it.

In conclusion, this was a straightforward, relatively low-input study on crested newts undertaken in my spare time. It was very much a product of the 1980s and contrasts with a number of recent studies on the species regarding complexity and input. Some of its strengths were that it permitted timely dissemination of information on certain aspects during the 1980s and early 1990s, including at meetings of practitioners; and annually repeated observations yielded useful datasets, including on numbers of tadpoles, which have been largely ignored by other workers.

10.7.3 A summary for the population at Shillow Hill

To summarise conclusions for adult crested newts:

- They foraged on land throughout the site.
- Most may have been on land at any point of the breeding season.
- They occurred on the adjacent arable land prior to harvest.
- Adults were most active on land during and after rain, and body condition in spring was related to rainfall during the previous summer and autumn.
- Annual survival was estimated during four years to be 78%, but much lower survival was suggested for one other year.
- Variation in numbers counted at night was affected by pond condition as well as by population size. Behaviour was also a factor, with high numbers sometimes remaining in the pond well into the summer.
- Average night count in 1984 was 23% of the total estimated to be in the pond; and average counts in Top Pond and Wood Pond that year represented 6% and 1% respectively of the estimated adult population.

Conclusions for tadpoles:

- Number of tadpoles surviving was affected by a range of factors including pond condition, number of breeding adults and the density of predatory adults remaining in the water in the summer.
- Size of tadpoles just prior to emergence was inversely related to their density in the water.

Conclusions for immatures and recruitment:

- Immatures occurred in the breeding pond in small numbers at 1–2 years of age.
- Many newts took 2–3 years to become sexually mature, with males typically maturing one year earlier than females.
- An individual was less likely to remain or survive to breed if adult numbers were high.

10.8 Smooth newts at Shillow Hill

The legal protection afforded to crested newts in 1981 not only helped the species directly but indirectly aided the conservation and knowledge-base of ponds and pond life. Had I not been interested in learning about crested newts, I would most certainly not have recorded smooth newts for 27 years at Shillow Hill.

In this section, information is given on night counts and on tadpoles caught in summer which was obtained at the same time as the information on crested newts. A positive relationship was found between numbers of smooth newt adults counted at night in Top Pond and numbers caught in bottle traps (Cooke

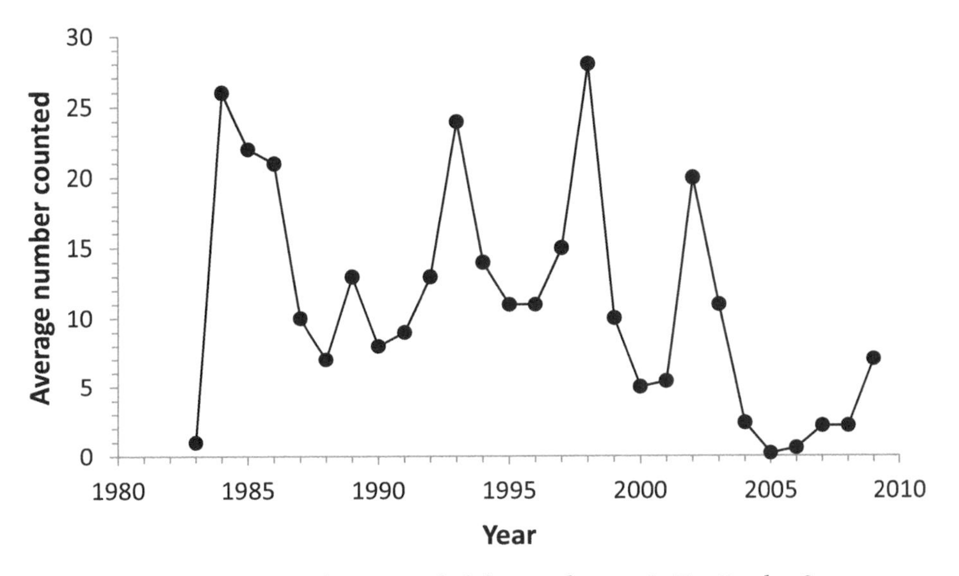

Figure 10.14 Average night counts of adult smooth newts in Top Pond, 1983–2009.

1995a). Both methods indicated that fewer smooth newts than crested newts bred in the pond. Average number counted at night for 1983–2009 (Figure 10.14 and Appendix 5) demonstrated fluctuations between 28 in 1998 and 0.2 in 2005. There were broad similarities between night counts of the two species: high numbers in some years in the 1980s, low numbers during the drought of 1990–1992, recovery during the mid- and late 1990s, and generally poor numbers from the mid-2000s. On the other hand, there was a decline in smooth newts from 1984 to 1988 that might have been in response to a boom in numbers of adult crested newts from 1985 to 1988. Crested newts were not averse to swallowing adult smooth newts when confined together in bottle traps, and predation may have occurred in Top Pond particularly when the crested newt population was very high. In the six springs when only the sump held water, average counts of adult smooth newts ranged from 0.2 in 2005 to 15 in 1997; the sump years of 1992 and 1997 were the only occasions when numbers of smooth newts were higher than those of crested newts. The former species appeared rather less inconvenienced by the reduced amount of water.

Catches of smooth newt tadpoles are shown in Figure 10.15. There was complete loss of tadpoles when Top Pond was dry in summer in 1990, 1997, 2005, 2006 and 2009. It also dried out in summer 1992, but a small plastic sheet was placed in the puddle before total desiccation and this saved a few smooth newt tadpoles; no crested newt tadpoles were recorded that year. The general failure of tadpoles during 2001–2006 is an obvious feature of the graph, and the pond only dried out in two of those years, 2005 and 2006. In 2001 and 2002, adult crested newts stayed on in the pond until well into the summer and were suspected of predating their own tadpoles; it seems that, not surprisingly, they preyed on tadpoles of the smaller species too. In both 2003 and 2004, the pond contracted to the sump, but crested newt tadpoles were relatively numerous and in 2004 were present at exceptionally high density (Appendix 2). Bitten tails were very noticeable on crested newt tadpoles that

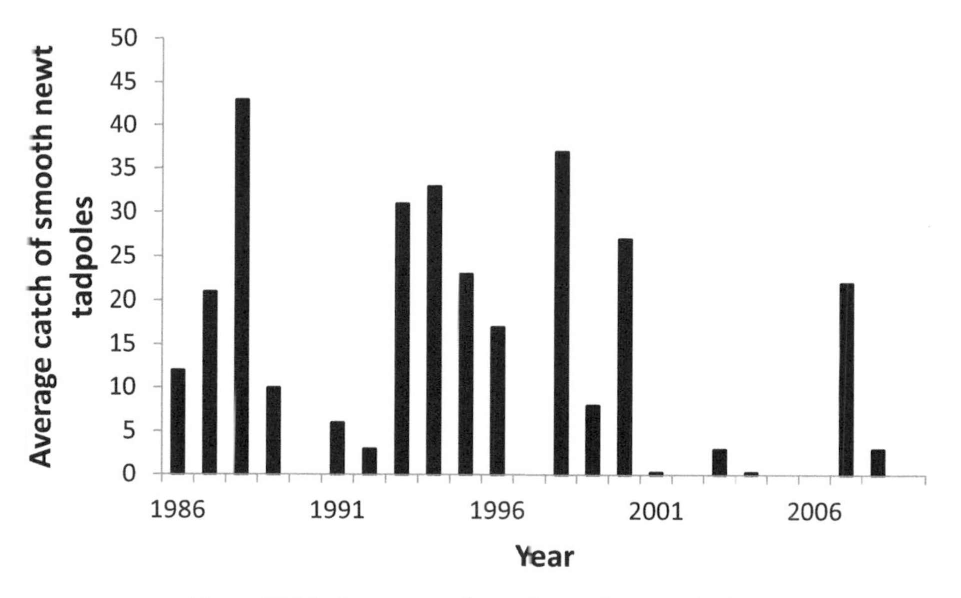

Figure 10.15 Average numbers of smooth newt tadpoles netted in summer in Top Pond, 1986–2009.

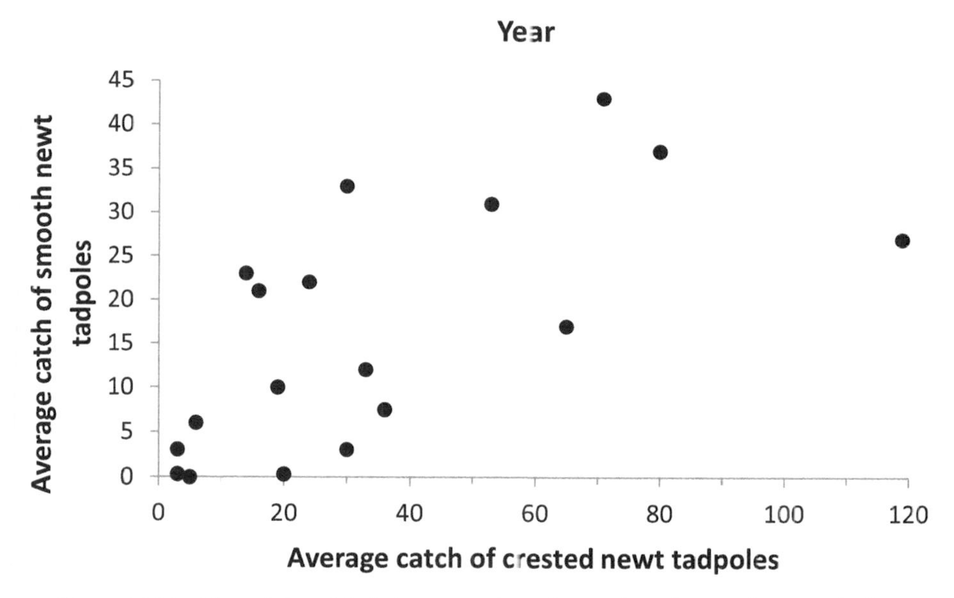

Figure 10.16 The relationship between catches of smooth and crested newt tadpoles in Top Pond, 1986–2009, omitting years in which the pond was dry by summer.

year, and it appears likely that smooth newt tadpoles were predated by tadpoles of the larger species. Such predation has been seen under experimental conditions (Griffiths *et al.* 1994).

Despite the suspected interspecific predation, there was a general relationship between numbers of tadpoles of the two species (Figure 10.16) showing that what was good for one species was generally good for the other too. In the five best years for smooth newt tadpoles, when the catch exceeded 25, the pond remained full into the summer. Even in the year 2000, when the catch of crested newt tadpoles was greater than 100 and many had bitten tails, there was sufficient dilution of their density for a reasonable number of smooth newt tadpoles to survive (Figure 10.15). The comparatively small size of crested newt tadpoles that year may also have helped to restrict interspecific predation.

The overall conclusion regarding smooth newts was that the population functioned 'smoothly' until the early 2000s when a combination of pond desiccation and crested newt predation combined to reduce tadpole survival. In the years following the re-profiling of the pond in November 2009, smooth newt tadpoles were netted in each summer 2010–2014 except 2011; the highest numbers were 13 in 2012 and 17 in 2013. Thus, breeding success was rather better than during 2001–2009.

CHAPTER 11

Long-term studies and climate change

11.1 Introduction

Simple long-term field studies have been an important part of my work on amphibians from the early 1970s to 2010. They have produced results throughout this period, but now they have been completed it is useful to look back at them holistically to reflect on how they fitted in to the ever-expanding knowledge of amphibians in Britain.

In the 1970s, I realised that repeating straightforward surveillance on an annual basis on frogs and toads at St Neots and Ramsey need not require a huge amount of time, but few herpetologists seemed to be doing it, particularly on the widespread species. On the other hand, some of our natterjack sites have now been monitored annually for conservation reasons for many decades. For instance, spawn strings at Woolmer in Hampshire have been counted each year since 1972 (e.g. Beebee and Buckley 2014a).

A particular focus of research and survey by British herpetologists has been on whether amphibian populations have declined at a regional or national scale, resulting in projects that originated in the simple postal enquiries of the 1970s. And from the 1980s, people began to appreciate that declines were happening in many other parts of the world. Long-term population studies have been undertaken in an attempt to understand better why declines occur and how to remedy them. Other studies delving into ecological processes have also progressed for many years. Sometimes these various long-term datasets have turned out to have relevance to issues that were not envisaged at the outset, such as the effects of climate change. Indeed, our current knowledge of the impacts of climate change leans heavily on long-term data compiled for other reasons.

In Switzerland, Meyer *et al.* (1998) considered that collecting long-term information was necessary to determine whether observed changes were simply natural fluctuations that occur in amphibian populations or could indicate a definite change in population size over a defined period (see also Atkins 1998). In three Swiss populations of common frogs recorded for between 23 and 28 years, one demonstrated a significant long-term decrease, thought to be the result of introduction of goldfish, whereas the other two did not. The populations did, however, provide empirical evidence of short-term boom and longer-term decline. If this is often the norm, shorter-term studies of such populations might conclude that a decline had occurred when it was simply natural fluctuation (Pechmann and Wilbur 1994). In two of the Swiss cases there was evidence of density-dependent regulation of the population. Where density-dependent factors operate, population growth decreases if density increases, and vice versa.

Another issue as regards determining population trends concerns quality of data. In 2002, at a meeting of the Society of Conservation Biology in Canterbury on global amphibian declines (Griffiths and Halliday 2004; Figure 11.1), Benedikt Schmidt argued that most of the data used to determine trends were unreliable counts unadjusted for detection probabilities (Schmidt 2004). While this was certainly true, I tried to control counting conditions for newts at night or toads dead on roads, and for frog spawn I attempted to count every clump (i.e. perform a census). Hopefully, anyone who has been counting to monitor change has at least considered and tried to control for changes in detection probabilities. I acknowledged the concern especially if datasets were being used for a purpose for which they were not intended. By 2004, however, I had been counting amphibians for 30 years and it was too late to make changes, even if I had been able to do so.

Much of my early long-term fieldwork was open-ended surveillance, because I often started recording without any defined end point. My primary intention was

Figure 11.1 Contributors to the Society of Conservation Biology meeting in 2002: from the left, Richard Griffiths, Tim Halliday, Benedikt Schmidt, Peter Daszak, Jim Collins, Jean-Marc Hero, Jim Foster, Per Sjögren-Gulve. Photo courtesy of Richard Griffiths.

sometimes simply to see what happened rather than conserve that particular population. Few people probably set out to record the same thing for many years – life is too uncertain for that. Repeating observations annually does though have a habit of becoming increasingly interesting, even addictive, and helps to shape and focus a project in later years. Because much of it was in my own time and did not need extra funding, I was able to stop or change direction without any problem. As I have lived in the same area since 1968 and my sites have tended to be in nature reserves or on private land with sympathetic owners, I have often been able to continue with my simple recording – and the years have ticked by.

Since the 1980s, situations have arisen which required this type of input to answer a specific question – usually, did management or intervention work? At sites that have involved monitoring translocations or the impacts of house building, I attempted to record for as long as was needed to determine success or level of impact. Thus, at Stanground on the edge of Peterborough, where housing was to be constructed beside a great crested newt breeding site, I considered that monitoring should continue for long enough to be able to determine whether the population had at least been maintained and whether the newts and their progeny had bred (Section 9.3). I monitored for 10 years, including two years pre-development, before I felt sufficiently confident to report in print that the venture had succeeded (Cooke 1997b). With some projects, I found it difficult to draw a line under the recording unless forced to do so by circumstances. At Stanground, I monitored for another couple of years before stopping when I no longer worked in the city.

The next section concentrates mainly on my datasets for the widespread species in sites in Cambridgeshire. Here I am more concerned with patterns of change and whether populations survived rather than reasons for change, which were discussed in previous chapters. That is then followed by a section on the natterjack, and the chapter ends with sections on climate change and phenology.

11.2 Long-term studies on populations of widespread species

11.2.1 Frogs and toads

Information from frog and toad sites with more than 10 years of observation is presented in Table 11.1.

The aims at St Neots Common were to understand more about frogs during the breeding season and to monitor how the population changed over whatever time period I was able to record. I began annual monitoring in 1971 and reluctantly finished in 1984, by which time the frog population was buoyant (Figure 5.4). However, numbers of spawn clumps were much reduced in 2005 and then completely absent in 2021, 50 years after I had first started. As the common remained more or less intact throughout that time and as frogs had bred in many different waterbodies, the demise of the population was both a disappointment and a surprise. It did though finally make me realise that nothing can be taken for granted as far as the permanence of frog populations is concerned.

As frogs became more widespread and abundant in Ramsey and Bury, so they colonised at least two of the local toad ponds, but with rather different long-term results. They were recorded as a by-product of my interest in the local toads. At Bury Pond, they began breeding in 1990. Spawn was recorded every year up to 2011

Table 11.1 Summary of information on frog and toad breeding sites for which observations spanned more than 10 years. Counts were of clumps of frog spawn, live toads in the breeding ponds or dead toads on roads. At the start, populations were established, colonising naturally, or had just been introduced.

Location	Period (span of years)	Recording interval	Population at start	Recorded at end?	Section in book
Common frog (spawn clumps)					
St Neots Common	1971–1984 (13)	Annual	Established	Yes	5.2.2
	1971–2005 (34)	Irregular	Established	Yes, but fewer	5.2.4
	1971–2021 (50)	Irregular	Established	No	5.2.4
Bury Pond	1990–2011 (21)	Annual	Colonising	Yes, established	5.4.1
	1990–2021 (31)	Irregular	Colonising	Yes, established	5.4.1
Horse Pond, Ramsey	1992–2010 (18)	Annual	Colonising	Yes, but intermittent	5.4.2
	1992–2021 (29)	Irregular	Colonising	No	5.4.2
Garden Pond, Ramsey	1985–2007 (22)	Annual	Introduced	Yes, established	5.3.3
Common toad (live adults)					
St Neots Common	1971–1984 (13)	Annual	Established	Yes	6.2.1
	1971–2005 (34)	Irregular	Established	No	6.2.4
Bury Pond	1989–2010 (21)	Annual	Established	No, last recorded 2006	6.3.3
Common toad (dead adults on roads)					
St Neots Common	1971–1984 (13)	Annual	Established	Yes	6.2.1
	1971–2005 (34)	Irregular	Established	No	6.2.4
Bury Pond	1989–2010 (21)	Annual	Established	Yes, but rare	6.3.3
Horse Pond, Ramsey	1984–2010 (18)	Annual	Established	No, last recorded 2008	6.3.2
Field Road Pond	1974–2010 (36)	Annual	Established	No, last recorded 2009	6.3.1

(Figure 5.12), so they bred in the same pond for at least 22 consecutive years. Maxwell Savage (1961) apparently never recorded any of his ponds for longer than nine consecutive years, but he did state that he knew of no pond that held spawn every year. Possibly in his day (the late 1920s to the 1950s), suitable ponds were much more common and frogs had many more breeding site options open to them.

At Horse Pond, Ramsey, frog spawn was first recorded in 1992 and was seen in 15 of the 19 years up until 2010 (Figure 5.13); no spawn was found during a search in 2021 and none was noticed during casual visits in the intervening years. The population in Horse Pond was smaller and less successful than that in Bury Pond, perhaps because of having to contend with tame ducks, goldfish and a relatively busy road next to it. Any discussion about whether declines are part of natural fluctuations raises the question of what is 'natural' about these human-created environments.

None of these three sites had any management specifically to encourage or conserve frogs, but the plastic pond in our third garden was designed and maintained with frogs in mind. Spawn was laid in it each year from 1985 up to 2007 (Figure 5.9), with production going through two distinct cycles of increase because of successful breeding and decrease because overcrowding resulted in recruitment failure (Figure 11.2): density-dependent factors were operating. We moved house in 2007, and the pond was destroyed by the new owner in 2008. This was not a total disaster, as other (less attractive) ponds survived in that garden and two adjacent gardens.

In Bury Pond, Horse Pond and the garden pond, I recorded how frog populations attempted to become established: spawn clumps increased to an initial peak after 12, 2 and 9 years respectively. In Bury Pond, a statistically significant decline occurred over the 10-year period from 2002 until 2011; this would have been regarded as of concern if no monitoring had been done before and later, but frogs were still breeding in 2021. At Horse Pond, amount of spawn also declined significantly over the last 10 years of continuous recording, but this time it was a prelude to long-term failure. Nevertheless, monitoring over a couple of years would have been as likely to conclude that the population was increasing as decreasing at all three sites. In other words, there was no clear evidence of short-term boom followed by long-term decline (Meyer *et al.* 1998).

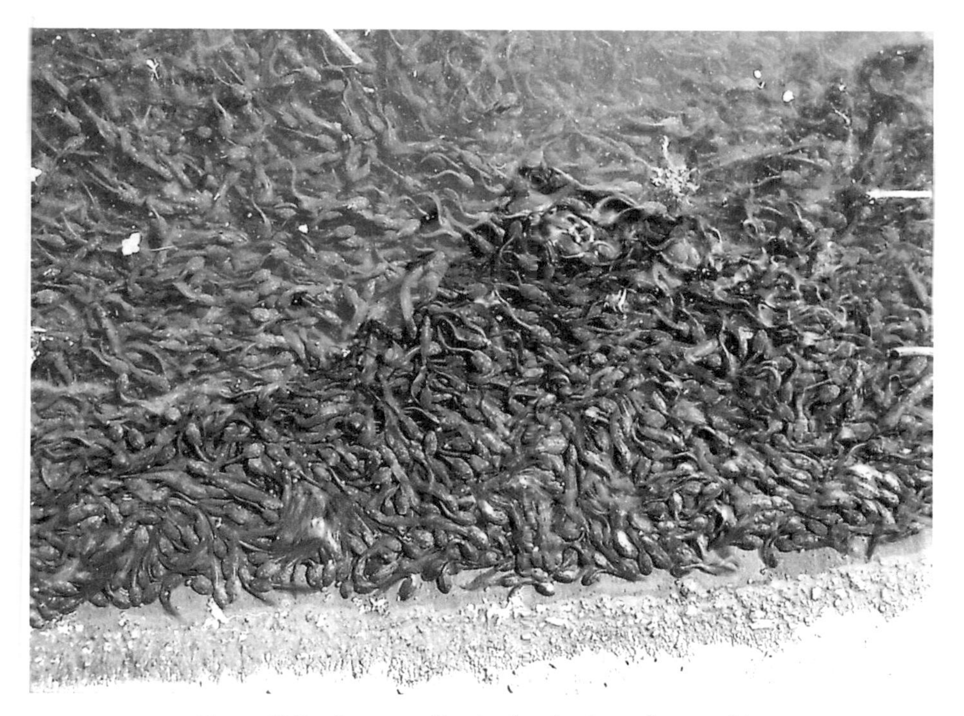

Figure 11.2 Overcrowding in the plastic garden pond in Ramsey, with frog tadpoles massing at the edge.

The frog populations studied at St Neots, Ramsey and Bury seemed rather more robust than their toad counterparts (Table 11.1). On Neots Common, toads were abundant in 1984 when surveillance stopped, but the number killed on the roads was alarming (Figure 6.2, Table 6.1). When the common was visited again in 2005 and 2021, there was no sign of toads alive in the breeding sites or dead on the roads. Declines of toads in Ramsey and Bury started in the early 1990s, and the species was rare at each location by 2007. If the decline at St Neots happened at roughly the same rate, it probably started soon after annual surveillance stopped in 1984.

Although frogs also died out at St Neots Common, they evidently held on for longer. And the frog's long-term success at Bury Pond indicated there was nothing intrinsically wrong with the pond – and pointed again to terrestrial problems causing the toad declines. Observations suggested that frogs were not just more adept at utilising garden ponds, they were better able to survive in larger sites in towns and villages, perhaps mainly because of their more modest migratory behaviour. It is not surprising that the toad continues to be one of the species giving most concern in a regional and national context. Populations tend to suffer when exposed to the twin threats of development and road traffic. The toad's ability to travel long distances to and from its breeding site offers both advantages and threats. In our busy world, the threats outweigh the advantages in ever more places.

Numbers of toads counted in Bury Pond increased in 1991, but then went into terminal decline, and numbers of road casualties showed a similar trend (Figure 6.10). The Field Road dataset of toad casualties on roads is my longest for any site (Figure 6.7), extending from 1974 until 2010. Overall, it indicated an increase to the early 1980s, followed by a long slow decline (Cooke and Sparks 2004) over about 25 years. For runs of 10 years of data, only the first period, 1974–1983, showed a significant increase, and some of the later runs, such as 1995–2004, showed a significant decrease. Inter-year fluctuation in numbers was one reason for the relative paucity of significant trends. Also, rather strangely, the four highest counts occurred at six-year intervals. This was presumably coincidence, with the gaps between minimum counts varying between four and seven years. Casualty counts at Field Road were an indirect measure of population status, and random population fluctuations can occasionally produce the appearance of cycles (Pechmann and Wilbur 1994). Each of these four peaks was the result of a sudden, dramatic increase and a similar decrease, which may indicate poor survival of adults, at least in those years.

Elsewhere, Roger Meek (2022) counted road casualties of the closely related spiny toad (*Bufo spinosus*) in France to determine long-term trends: data for four populations indicated general stability during 2005–2019. Reading and Jofré (2021) found reduced numbers of common toads returning to breed at a site in Dorset over the period 1983–2020. Adult numbers in a population in the Swiss Alps did not change significantly during 1982–2001 despite low reproductive success in the 1990s (Grossenbacher 2002). Some toads of both sexes in the Alps lived longer than 15 years, and it was suggested that extreme longevity masked reproductive failure during the study period.

11.2.2 *Newts*

Information from long-term studies on the three species of newts is summarised in Table 11.2. Long-term observations on the grid of 20 ponds at Woodwalton Fen and 37 ponds dispersed in the New Forest were derived from irregular, not annual, visits. The ponds at Woodwalton Fen were dug in 1961. They were managed for dragonflies, with scrub being occasionally removed from around their edges, but, nevertheless, scrub invasion appeared to be the principal reason for the decline in occupancy for both species of newts (Table 8.2). Some ponds were much less prone to scrub invasion than others, and even if there was no management, it is likely that small numbers of newts would still have been breeding in 2005, at least 35 years after the ponds were first colonised.

Table 11.2 Summary of information on newt breeding sites for which observations spanned more than 10 years. Live adults were counted at night or netted during daytime. At the start, populations were established, colonising naturally, or had just been introduced.

Location	Period (span of years)	Recording interval	Population at start	Recorded at end?	Section
Crested newt					
Woodwalton Fen, 20 ponds	1974–2005 (31)	Irregular	Colonising	Yes, established	8.4.2
Field Road Pond	1986–1997 (11)	Annual	Established	No, last recorded 1993	9.2
Monks Wood, 2 ponds	1986–1997 (11)	Annual	Established	Yes	9.2
	1986–2005 (19)	Irregular	Established	Yes	9.2
Stanground Ponds	1986–1998 (12)	Annual	Established	Yes	9.3.2
Shillow Hill Top Pond	1983–2009 (26)	Annual	Established	Yes	10.3.2
	1983–2021 (38)	Irregular	Established	Yes	10.3.3
Shillow Hill Wood Pond	1983–2009 (26)	Annual	Established	No, last recorded 2004	10.2
Worlick Farm fish ponds	1985–1998 (13)	Annual	Introduced	Yes, established	9.4.1
Stocking Fen receptor pond	1991–2021 (30)	Irregular	Introduced	Yes, but rare	9.4.2
Smooth newt					
Woodwalton Fen, 20 ponds	1974–2005 (31)	Irregular	Established	Yes, but fewer	8.4.2
New Forest ponds	1974–1989 (15)	Irregular	Established	Yes	8.5.3
Shillow Hill Top Pond	1983–2009 (26)	Annual	Established	Yes	10.8
Shillow Hill Wood Pond	1983–2009 (26)	Annual	Established	No, last recorded 2001	10.2
Palmate newt					
New Forest ponds	1974–1989 (15)	Irregular	Established	Yes	8.5.3

In the New Forest, there were few changes between 1974 and 1989 in the number of ponds with the three newt species or in numbers of newts caught, with the palmate newt being numerically dominant. This was my only field study that included palmate newts, and it indicated that this species has probably been better protected against modern human-made changes in freshwater and terrestrial habitat than other widespread amphibian species. Not only can it flourish on acid moorland, but it has a more rural distribution than the smooth newt and is the most common newt in Wales, Scotland and the English uplands (Beebee and Griffiths 2000).

Surveillance of established populations of crested newts was undertaken annually for more than 10 years at Field Road in Ramsey, at Monks Wood and at Shillow Hill (Table 9.1). These studies were all started in the 1980s in response to the increased interest in the species created by its protection under the 1981 Act. When I began at each site there was no reason to believe these populations were especially threatened in any way. I was, however, aware of crested newts breeding in Field Road Pond, mainly because I had seen occasional road casualties since 1974. Night counts during 1986–1997 indicated that the population died out during the 1990s. Stickleback predation of tadpoles took most of the blame, but road deaths cannot have helped. In contrast, night counts at the two small ponds on the edge of Monks Wood suggested small, fairly stable populations during 1986–1997. The ponds are about 400 m apart and one is a similar distance from a third breeding pond in the wood, so there is likely to be some exchange. Newts were recorded in both ponds on the edge of the wood when they were next checked in 2005.

Night counting was used to record fluctuations in the crested and smooth newt populations at Shillow Hill from 1983 up to 2009 (Figures 10.4 and 10.14). Various methods confirmed that crested newts continued to use the site during 2010–2014 and again in 2021, 38 years after the site was first visited. The two breeding ponds suffered desiccation problems, both from the lowering of the water table and from build-up of sediment. Attempts were made to deepen the ponds. The smaller and less important breeding site, Wood Pond, was eventually allowed to dry out, and crested newts were last recorded in it in 2004. The two ponds were less than 100 m apart and exchange of newts between them had been demonstrated (Cooke 1985c), so loss of Wood Pond was not considered to be critical. Monitoring provided a long-term view of how a crested newt population changed and responded to some intervention. Numbers of crested newts in the main breeding site, Top Pond, were exceptionally high during 1985–1988, but crashed in 1989, apparently primarily because of high mortality. Excavating a sump in 1991 led to a population recovery, but desiccation recurred, leading to a major re-profiling of the pond in 2009. Unfortunately this made the pond too turbid for torch counting, so the study was eventually stopped.

In Section 10.7.2, I surmised that crested newts at Shillow Hill would have struggled to survive until 2009 had the sump not been created. Without any intervention, including the re-profiling in 2009, the species would have been even less likely to survive until the present time as conditions at the site generally became drier still (Figure 11.3). The population of crested newts studied by von Bülow and Kupfer (2019) survived reproductive failure in nine years out of 19, including four consecutive years, but declined dramatically. These authors listed 14 sites where long-term studies on the population dynamics of crested newts had been undertaken across its European range: these varied from 2 to 19 years in duration but only two spanned more than 10 years. The study at Shillow Hill was not mentioned, presumably because methodology was unconventional.

Adult smooth newts at Shillow Hill did not appear as affected as crested newts by reduced amounts of water in Top Pond in spring. However, despite the low count in the single visit in 1983, the data indicated a significant overall decrease in number counted between 1983 and 2009 (Figure 10.14). There was considerable inter-year fluctuation in average count, and the highest in 1998 was more than 100 times the lowest in 2005. There were increases during 12 years, decreases in 12 and two when there was no change, so decreases did not outnumber increases. Nevertheless, it is hard to see how the smooth newt population would have survived till now had there been no intervention.

Two translocations of crested newts were carried out at ponds on fenland outside Ramsey. At Worlick Farm fish ponds, crested newts were introduced to the newly renovated ponds in 1985. Torch counting revealed small numbers of what were considered to be translocated individuals from 1986 to 1988, but then no newts were seen during 1989–1991. This worrying period was followed by a steady increase in numbers from 1992 up until 1998, when monitoring stopped. In the second translocation, adult crested newts and eggs were moved during 1991–1993 to another newly renovated pond nearby on Stocking Fen, and surveillance was undertaken up to

Figure 11.3 Crested newts were first seen in Top Pond, Shillow Hill, in 1983 and were still breeding 38 years later in 2021. However, without intervention they would have been unlikely to survive that long. Reduction in water availability was caused by sedimentation and lowering of the water table, probably due to maturation of the trees and climate change. Photo by Mihai Leu.

2010, apart from during 1999–2000. Numbers counted at night increased to 2001 and then decreased. Torch counting tailed off from 2007 because of several years of high turbidity; by that time colonising sticklebacks were threatening to eliminate the newt population. When the site was visited again in 2021, however, no sticklebacks were seen and breeding by crested newts was confirmed. This was 30 years after the initial translocation and despite the site receiving no conservation management during that time.

At Stanground, I first confirmed the presence of crested newts in 1982, and later monitored each year from 1986 to 1998 to determine the impact of houses being built adjacent to the ponds in 1987–1989. Tony Gent and Jim Foster continued the recording up until 2010, although there was a gap in 2002. Wildlife Trust staff, who are responsible for management, last surveyed in 2019 and confirmed breeding. Although this project has succeeded in maintaining a breeding population of newts next to the new development for more than 30 years, there have been issues of lowering of the water table because of a changed drainage system, growth of vegetation in and beside the ponds, and also colonisation by sticklebacks. Had there been no detailed monitoring combined with an ability to undertake essential management over many years, this venture would almost certainly have failed. It is also worth remembering that the site could so easily have been drained prior to house building in the late 1980s before anyone realised that newts were there. Another breeding site, roughly 100 m to the north, is in different ownership and numbers of newts counted at night were very low or zero during 2001–2010 (Jehle *et al.* 2011) because of lack of management. As with Wood Pond at Shillow Hill, unavailability of this breeding pond was not catastrophic: adult newts were likely to be gradually attracted to the two managed ponds. It has though decreased the number of sites in the metapopulation from three to two, which could be a disadvantage, and, unless the northern pond can be renovated, renders the monitoring and management of the two remaining ponds even more critical.

Crested newts have had a tough time in this area over the last 50 years or so, and these losses have been discussed in previous chapters. Of ponds that escaped development, some were filled in because they were no longer needed, some were overgrown by scrub and trees, and others were colonised by fish. Out of the populations monitored in the long term, crested newts disappeared from Field Road Pond 20 years after I first became aware of them. Conversely, they remained in the Monks Wood ponds 20 years after I started monitoring, and a population was still increasing in the Worlick Farm fish ponds 13 years after newts were introduced. Crested newts continued to breed in the receptor pond on Stocking Fen 30 years after the initial translocation. They remained in the Stanground ponds 37 years after I had confirmed their presence despite the adjacent housing development and neglect of another pond nearby. And crested newts were still breeding at Shillow Hill 38 years after I first saw them there. But it has taken considerable effort to achieve even this modest success. It is another illustration of the need for proactive and creative conservation. We cannot just rely on trying to save what we have. In a recent study of scientific and practical importance, Cayuela *et al.* (2020) analysed data from five long-term studies on crested newts in western Europe and concluded that population dynamics depended more closely on local factors, such as habitat characteristics, than on wider environmental fluctuation, such as changes in regional climate.

11.2.3 *Connections*

The essence of most of my long-term studies was that measures or indices of population size, demography or reproductive output were recorded over more than a decade. In addition to monitoring population success or failure, this approach sometimes generated data for a species at a site that gave pointers to population processes via correlations. Furthermore, relationships were occasionally noted between data for the same species at different sites and for different species at the same site. Examples are given below.

Total counts of clumps of frog spawn and peak counts of adult toads were positively related at St Neots Common (Section 6.2.3). It was not possible to investigate in detail whether this had occurred by chance or whether there were ecological or environmental factors behind the observation, because relatively little was discovered about why either population fluctuated. It was difficult, though, to find common ground between potential reasons for change for the two species.

Looking at situations where frogs or newts colonised without help or were introduced, some similar trends were noted. For example, there was a cycle of increase or spread over a period of years followed by a decline for frogs at Bury Pond and for crested newts at Woodwalton Fen and Stocking Fen. There were signs of two such cycles for smooth newts at Woodwalton Fen and for frogs at Horse Pond and in our third garden. Declines appeared to be caused by a variety of factors such as scrub encroachment, fish predation, road traffic mortality and high population density. At Worlick Farm fish ponds, counts of crested newts were still increasing when monitoring stopped.

The three main breeding sites of common toads in Ramsey and Bury were roughly 1 km apart, with roads and housing between them (Figure 5.11). Trends in casualty counts were related with numbers decreasing from the early 1990s in each site, which suggested there were common factors causing the decreases (Section 6.3). Effects of weather conditions on toad migration and on carcass persistence may have made some contribution to the fine detail of the overall declines. The issue is explored further in Figure 11.4, where casualty counts are presented for roads around Field Road Pond and Horse Pond back to 1984, when annual counting began at the latter site. It is noticeable that the two datasets have a range of features in common: decreases from 1985 to 1988 and from 1991 to 1994, low counts in 2002, and identical counts from 2006 to 2010. Baker *et al.* (2011) stated that toads are able to colonise a new pond 1 km distant, and recommended a distance between ponds of not more than this. It is difficult, however, to believe that there was much or any exchange between the two sites. The straight-line distance between the ponds is 900 m, but they are on opposite sides of the centre of Ramsey, so any toad walking between them would encounter hazards in the form of roads, kerbs, buildings, fences and walls and, if it survived, would need to walk further than 1 km. Few road casualties were found further than 300 m from any of the local ponds. This supports a suggestion that the trends in numbers occurred independently, responding in an almost identical fashion to very similar sets of conditions.

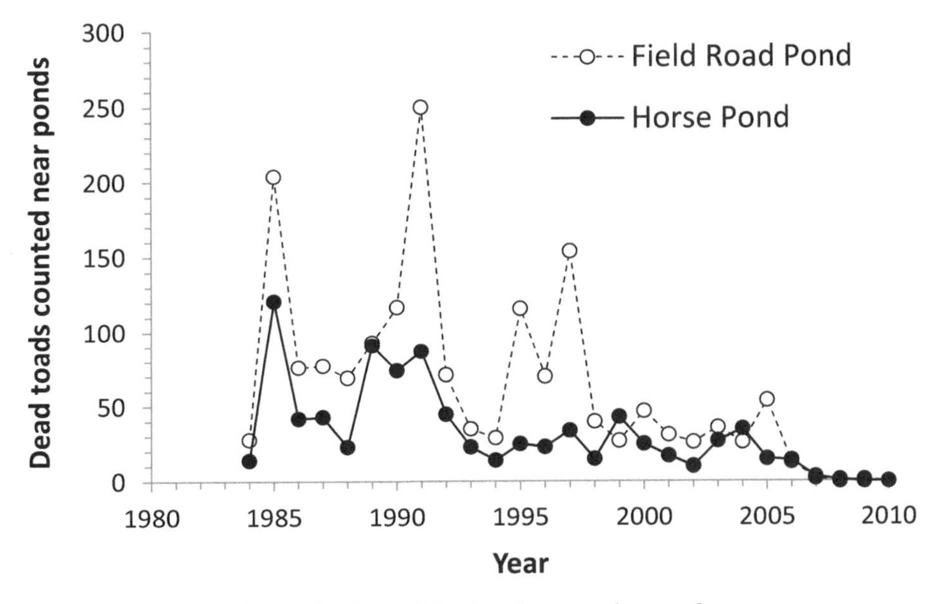

Figure 11.4 Numbers of dead toads counted on roads near Field Road Pond and Horse Pond, Ramsey, 1984–2010.

At Shillow Hill, the fact that annual catches of smooth and crested newt tadpoles were positively related (Figure 10.16) was something of a surprise, as it was considered that high densities of crested newt tadpoles would, given the opportunity, prey on their smaller relatives and depress their numbers (see Griffiths *et al.* 1994). Similarly, there was no evidence of significant predation by invertebrates on crested newt tadpoles (Table 10.2). Indeed, the year 2000 had the best summer for both invertebrate predators and newt tadpoles. In addition, there was no sign of an impact of invertebrates on crested newt tadpoles at the Stanground site (Cooke 2001b). On the other hand, there was evidence of predation on tadpoles at Shillow Hill when greater numbers of adult crested newts stayed on in Top Pond later into the summer (Section 10.4). In years when this did not happen, a positive relationship existed between number of adults in the water in spring and early summer and number of tadpoles caught in July and August (Figure 10.9). There were also indications of a positive relationship between number of tadpoles caught and number of adults counted in future years (e.g. Figure 10.12).

11.3 Natterjack toad

With the wealth of information available on natterjacks it is not surprising that similar relationships to those described for newts immediately above have been investigated for that species. Thus, relationships have been discovered between spawn production and toadlet numbers, both for an area over a long period of years (Smith and Skelcher 2019) and between sites (Beebee and Buckley 2014b). However, long-term datasets did not convincingly reveal a link between toadlet production and numbers of spawn strings 2–5 years later (Beebee and Buckley 2001, 2014b).

The amount of monitoring done on the natterjack toad since 1970 is truly remarkable. In the fourth edition of the natterjack Site Register (Beebee and Buckley 2014a), 196 people were thanked for contributing, and I suspect that most of them (like myself) were field recorders. This input enabled a massive dossier to be constructed, not just detailing all records from every site but also including collation and analysis that could only have been dreamed of 50 years ago. The objective of course was not to produce an impressive document but to describe the past and aid future conservation of the species. As is explained in the introduction to the Register, the populations that existed after 1970 probably only represented a small fraction of those that existed in the early 1900s. The nadir is thought to have occurred in the 1960s. Since 1970, monitoring and conservation action halted this decline until recently (see also Buckley *et al.* 2014, and Section 7.3). This is one of the major achievements of British herpetology during this period.

The recent review by Smith and Skelcher (2019) again highlighted the dedication involved in such monitoring, reporting on 31 years from 1987 to 2017. Phil Smith was the local expert who, in the late 1970s and beyond, advised NCC on the impact of a proposed golf course on the Sefton Coast (Section 7.1; Figure 7.1a). Since then he has continued to monitor and coordinate others in that area, which still has the largest natterjack population in Britain. Because recording was standardised, it was possible to determine how spawn and toadlet totals were correlated with time and environmental variables. Both measures of success decreased over time, and failure was related to lower rainfall in spring, which translated into a long-term reduction in the height of the water table. In addition, increasing stability of the dunes and vegetation succession led to fewer suitable breeding slacks and a more vegetated environment generally. Natterjacks prefer unshaded, shallow slacks and an open terrestrial environment with short vegetation and sandy patches; stable dunes mean an increasingly well-vegetated habitat which favours competitors such as common toads. Problems were occurring despite huge conservation input over the decades, and Smith and Skelcher concluded that more slacks should be created and older ones restored, livestock grazing should be increased, scrub and trees should be removed and coarse vegetation in pools mown. Achieving and maintaining optimal habitat for natterjacks may require even more effort than monitoring.

Even for a species as thoroughly recorded as the natterjack, population estimates and trends are not necessarily clear-cut. The standard survey method is to count spawn strings, but not all females breed every year, so multiplying number of strings by two (based on a sex ratio of parity) is likely to lead to an underestimate of the total adult population (Beebee and Buckley 2014b, Smith and Skelcher 2019), although spawning twice per annum was reported at Woolmer (Banks *et al.* 1993). Trevor Beebee has told me of an incident at Woolmer when natterjacks needed to be removed from a relatively small area because of operations at the Ministry of Defence site and, based on spawn string counts, an unexpectedly large number were caught. Spawn string counts are though used as an index to monitor trends. In the past, some 'educated guesses' of population sizes may have been optimistic, especially at poorly recorded sites (Trevor Beebee personal communication), thereby suggesting changes where none had been suspected. Such an example arises when comparing the national estimate of 15,000 adults in the 1980s (Banks *et al.* 1994) with the later census estimates for the 1990s of 3,000–4,000 based on spawn string counts (Beebee and Buckley 2014a). Moreover, the 'effective population size' of about 600 was lower still than the census estimate because many adults fail to leave progeny (Beebee 2009, Beebee and Buckley 2014a).

11.4 Effects of climate change

Climate change was blamed by Smith and Skelcher (2019) for the worsening trend in the premature desiccation of dune slacks on the Sefton Coast. A warming climate impacting amphibians was certainly not a scenario that will have occurred to many British herpetologists until towards the end of the twentieth century. Trevor Beebee and I first wrote about earlier migration to amphibian breeding sites in the mid-1990s (Beebee 1995, Cooke 1995b). Long-term datasets have proved invaluable in understanding past impacts and predicting future ones. McGrath and Lorenzen (2010) assessed the performance during 1985–2006 of 20 well-spaced natterjack populations around Britain in relation to management history and climate. They showed that while enough rain was needed to allow spawning and toadlet emergence, too much rain might lead to pools becoming more permanent, so permitting populations of invertebrate predators of tadpoles to become better established (see Section 7.2.2). This was another example of the conditions required by natterjacks being frustratingly precise. Two more issues affecting natterjacks were raised by Trevor Beebee in his review of the effects of climate change on British wildlife (Beebee 2018a). First, mild, wet winter weather may drown toadlets in their burrows; and second, rising sea levels could restrict the zone of suitable habitat available to important colonies on the northern coast of the Solway by hemming them in against unsuitable farm fields just inland.

Observations from long-term studies on other species of amphibians also help to inform the debate about climate change. Thus, Griffiths *et al.* (2010), after their 12-year study of a metapopulation of crested newts, reported that mild, wet winters were associated with reduced survival of adults. Conversely, at Shillow Hill, I found that body condition of adult crested newts in spring was positively correlated with rainfall during the previous summer and early autumn (Figure 10.13). These observations provide examples of the timing of rainfall being important. Chris Reading studied one of his breeding sites of common toads in Dorset for 42 years. He used data for the period 1983–2005 to test if interruption of hibernation because of mild weather affected the toads (Reading 2007). Warm weather resulted in reduced body condition of female toads, with their annual survival and egg production declining. Later, Reading and Jofré (2021) examined data for 1983- 2020 and reported that higher summer temperatures were correlated with decreases in body condition and size of toads returning to breed for the first time. These changes indicated a reduction in the intake of energy. This was not the result of increased intraspecific competition, as an abundance index indicated a reduction over time; a reduction in invertebrate prey caused by increasing temperatures was suggested as the likely cause.

11.5 Phenology

Phenology is the study of the timing of events in the life cycles of animals and plants in relation to external factors. Climate is one of those factors, and phenology is part of the process of understanding the effects of climate change. During my lifetime, I have been aware of changes occurring for dates of frogs breeding, bird migrants arriving and wild flowers blooming – and in some cases I have the records to prove it. The story of learning about the effects of conditions on when frogs spawn is a long one.

Recording phenological data for common frogs had evidently been a pastime among literate country folk well before the Royal Meteorological Society began coordinating other phenological observations throughout Britain during the 1800s (Sparks and Carey 1995). The recording of dates when frog spawn was first seen was added to the list of observations in 1926 and the collection of data continued up until 1947. Maxwell Savage wrote a paper in 1935 on when frogs spawned, based on early results from this scheme – and he then revisited the subject in his monograph on the frog (Savage 1961). He found that spawn date depended on a range of factors: it could be delayed by an increase in temperature in the month of spawning and an increase in sunshine in the previous month; however, the date could be brought forward by raised temperature two months before spawning and an increase in rainfall during the previous month. It was also affected by latitude and longitude, indicating that geography had a part to play. Savage was able to construct a famous map showing variation in spawn date across the British Isles. The earliest dates (in January) tended to be in the south-west and the latest were up mountains in the north. The difference between the earliest and latest spawning was 2–3 months.

Half a century later, Carroll *et al.* (2009) reassessed frog spawn dates using nearly 70,000 records from the UK Phenology Network, which had been initiated in 1998. For the first 10 years of data, they reported a significant negative relationship between the average first spawn date for a year and the average January–March temperature in Central England. An increase of 1 °C brought spawning forward by about five days. These authors constructed a UK map of spawning for comparison with Savage's earlier map for the British Isles: there were general similarities but also subtle changes consistent with spawning becoming earlier. For the whole dataset, 0.2% of records occurred before 1 January, another indication of warming conditions (Figure 11.5).

Carroll *et al.* (2009) also noted that examination of long-term data from individual sites failed to detect any significant change, a possible explanation being the considerable inter-year variation that occurs in such datasets. Certainly, my longest continuous series of spawning, from the plastic pond in my third garden, showed only the slightest hint of earlier spawning over the period 1985–2006. When I published data for my ponds in this garden in 2003, I also reported that spawn was laid in the warmer plastic pond before the concrete one, with an average gap of 18 days.

I collected data on first spawning for frogs from 1971 until 2016 in the former county of Huntingdonshire, and those records did show a significant trend towards earlier spawning (Figure 11.6). No attempt was made to find ponds with early spawn; this was a passive exercise, involving collating information from the few ponds in which I was most interested for other reasons. Up to 1985, spawning almost

Figure 11.5 Common frogs paired up in a pond in southern England, late December 2015. Photo by Marc Baldwin.

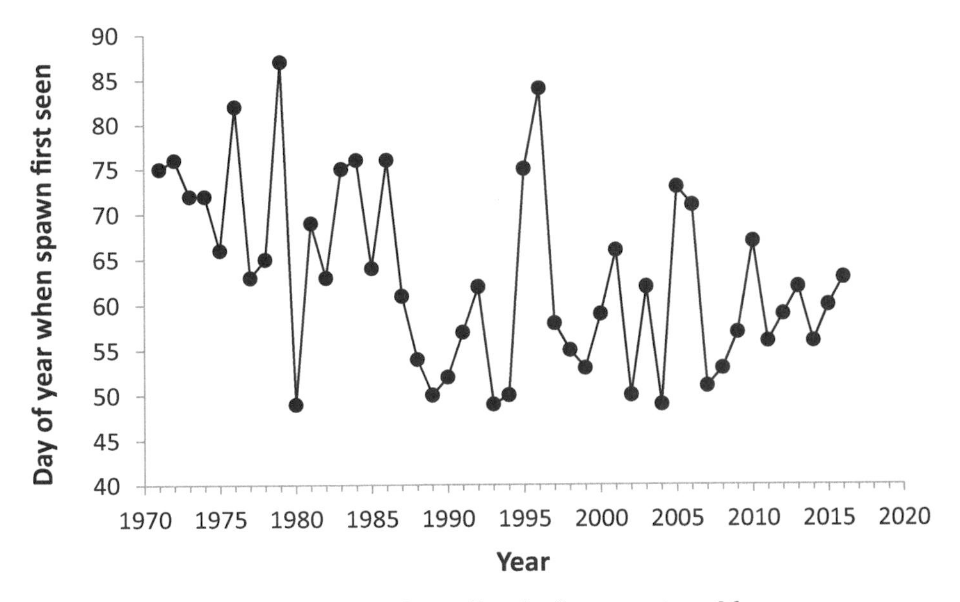

Figure 11.6 Date of recording the first spawning of the common frog in Huntingdonshire, 1971–2016.

invariably began in March, whereas from the year 2000 spawning tended to start in late February or early March. I do not doubt that temperature affects frog spawning date, but I suspect that these data for Huntingdonshire over-emphasised that effect. The early records in the data series were all from St Neots Common, because initially it was the only site I knew and I was keeping a close watch on it; but as frogs spread so there were more sites from which to choose, and some of these were warm sheltered ponds in gardens where frogs spawned relatively early irrespective of the weather that year. The plastic pond in my garden was one such breeding site.

There have been several papers and fragments of information on the phenology of common toads in Britain. Thanks to Maxwell Savage, Malcolm Smith (1969) was able to devote several pages to the phenology of the frog, but his account for the toad was only part of one paragraph and implied that the main spawning season was during the second half of March and the month of April. I referred in Section 4.2.3 to data on the breeding of frogs and toads that was collected during 1952–1959 by Deryk Frazer, who kindly donated the material to me in the early 1970s. I realised that there was not enough information on toad sites to be able to construct a map as Savage had done for the frog, but it might be possible to compare dates for the toad with those for the frog in some parts of the country (Cooke 1976). Sample sizes were small, but analysis suggested that in 12 English counties with reasonable datasets, toads usually started to spawn from early March to mid-April. In counties where frogs spawned early, such as Cornwall, then toads were about a month behind; in counties where frogs started in late March, toads varied from three weeks later to spawning at around the same time. Armed with this information and Savage's map, it was possible to decide roughly when toads might begin to spawn in any part of the country. At that time, of course, there was no suggestion that spawn dates might change in the fairly near future.

Moving on by a couple of decades, those of us with appropriate long-term datasets on toads began to examine them for signs of warmer weather in late winter and early spring leading to evidence of earlier breeding activity. In 1995, I published an article about road deaths of toads at Field Road in Ramsey (Cooke 1995b). Casualty counts were made just after the main inward migration to the pond, so the date of this stage of the breeding cycle was recorded each year. By then I had 21 years of observation (1974–1994), which revealed that peak migration occurred significantly earlier in the year during the 1990s. Soon after, Chris Reading (1998) reported that records for first toad spawn did not become earlier during 1980–1998 for his study colony in a Dorset clay pit. He did, though, find that the time of peak arrival at the breeding site was highly correlated with the average temperature over the previous 40 days. Field Road data were updated by Sparks *et al.* (2007) in a comparison of phenological data in the UK and Poland. The same time period of 1978–2005 was studied for both countries: the trend for earlier migration at Field Road was highly significant, and the date was negatively related to average January–March temperature.

Phenological changes for other species were first reported by Trevor Beebee (1995): natterjack toads spawned earlier at Woolmer in Hampshire over the period 1979–1994, and all three species of newts began to appear earlier in his garden ponds near Brighton in Sussex between 1978 and 1994. Populations of smooth and palmate newts were trapped at Cardiff University's study pond at Llysdinam in mid-Wales during inward and outward migration in 1981–1987 and 1997–2005 (Chadwick *et al.* 2006) in order to test whether increases in spring temperatures had changed the phenology of the two species. Differences in average median arrival dates between the two periods ranged from four days earlier for female smooth newts to 18 days earlier for male palmates. Average median departure dates differed by 11–14 days, with the aquatic phase being 3–4 weeks longer during 1997–2005. For every 1 °C rise in minimum temperatures in February, the inward migration was advanced by 2–4 days.

As regards crested newts, I have checked to see whether the date of the peak count in the breeding season at Shillow Hill changed during 1984–2009. The trend was for the date to become earlier, but the relationship was not statistically significant. The same applied to data for smooth newts at the site. Dates were also checked for when the percentage of 'walkers' in the sample of netted crested newt tadpoles first exceeded 20% (Figure 10.10a); again, the trend was for the date to become earlier, but it was not statistically significant.

In conclusion, there was evidence that aspects of breeding behaviour in all six of these species responded to changes in temperature, but sometimes variability between years was too great for even long-term studies to demonstrate a change over time at a single site.

CHAPTER 12

A recent history of amphibian conservation

12.1 Introduction

The indefinite article has been deliberately included in the chapter heading because this is my version of recent history, with inevitable gaps, plus my slant on what happened. Other reviews have been published that cover events in earlier decades (Beebee and Griffiths 2000; Beebee 2014). This final chapter has been included because it brings together the various historical threads in the book. It focuses on people, organisations, research, conservation events and publications in approximately chronological order. Rather than repeating statements and accounts from earlier in the book, the section or chapter is indicated in which further information can be found. At the end is an attempt to suggest the direction in which amphibian conservation and research might proceed in the few years following late summer 2022 when I stopped writing. This was the most difficult part of the book to write because of the level of uncertainty in UK conservation and in the world in general – and because I am now viewing the herpetological scene from the outside.

The chapter begins with a section on journals and books. These are the bedrock of herpetological information but during recent decades online versions have totally or partially replaced paper copies, and social media seems to have taken over the lives of many people. In addition, newsletters and handbooks have appeared (Section 12.5). Analysing the contents of such material can reveal its quantity and quality, as well as monitoring progress and concerns.

12.2 Journals and books

The BHS was formed in 1947 and began to publish the *British Journal of Herpetology* in 1948. The publication became larger and shinier over the years. In December 1985 it changed its name to *The Herpetological Journal* and became even thicker, going primarily online in 2006. The issue in December 1985 had 40 pages, while my last paper copy in July 2006 had 104. With the loss of 'British' from the title, there seemed to be progressively fewer papers on British species in those two issues, for example, there were eight British papers in 1985 and just one in 2006. When an interesting paper does turn up nowadays, however, it is often particularly pertinent. As a pensioner working from home, a big advantage for me of online journals is that I have far better access to literature generally than I would have had in the pre-digital era. Issues of the *Journal* are freely available online to BHS members back to December 1985, with those older than two years open-access to non-members, but I am not aware of any issues of the *British Journal of Herpetology*, 1948–1985, being online.

The BHS Bulletin was originally called the *Newsletter*, and first appeared with its new name in 1980, being edited by John Pickett and Simon Townson. Like the *Journal*, it became larger and glossier over time, and changed its name to *The Herpetological Bulletin* in the year 2000, before going online in 2007. Now all issues are available online, and those older than one year are open-access. The first issue in June 1980 contained, in addition to a number of articles, a broad collection of information including members' advertisements for captive animals. By 2007, contents were restricted to an editorial, articles and a book review. The role of newsletter was taken over by *The Natterjack*.

According to my records, I had seven articles published in the *Journal* in the 1970s, five in the 1980s, three in 1990s, but none during this century. In the *Bulletin*, I had four in the 1980s, three in the 1990s and eight in the early 2000s. The trend in the *Journal* reflected changes in the nature and quality of the articles it accepted rather than changes in what I was writing.

In order to see how the flavour of the subject matter of papers published in the *Journal* might have changed during my time as a herpetologist, I compared the first 10 British papers on the six British amphibians (i.e. those I have covered in previous chapters) beginning in 1969 with the last 10 published up to summer 2021. Nine of those in the older group were on the natural history or behaviour of the five widespread species, while the remaining one was a policy statement by BHS's newly formed Conservation Committee. The last paper was the only one to refer to natterjacks. In addition, there was a paper on British species from outside this country: a description from Sweden of a photographic method of recording belly patterns in newts. Turning to the recent group of 10 papers, eight were on natterjacks, great crested newts, genetics or combinations of these subjects, with all being driven by conservation. The other two were on diseases and frogs in pre-industrial Britain. There were also four articles that were of some interest but did not meet my criteria: two on the pool frog, one on genetic variation of smooth newts in Hungary and one on the effects of fishing on common frogs in the Pyrenees. Not surprisingly, the older group were mainly straightforward descriptive accounts of the most common species, whereas the newer group tended to be geared towards conservation of the rarer species and were more complex pieces of work.

My comments on publications of the BHS illustrate changes that have occurred over the years, but in the earlier decades of my career in particular I followed a range of journals. I have deliberately focused on the British scene in this book, but the other herpetological journals consulted regularly were based beyond these shores. *Amphibia–Reptilia* was the journal of the Societas Europaea Herpetologica (SEH), while three came from North America: the *Journal of Herpetology*, *Herpetologica* and *Copeia*. The last one was named after Edward Drinker Cope, the famous nineteenth-century palaeontologist and naturalist; however, because of his views on race and women, this journal changed its name in 2021 to *Ichthyology and Herpetology*. Herpetological articles can of course be found in a range of journals devoted to different specialisms such as ecology, zoology, conservation, wildlife management and phenology. Because of my earlier formal interest in pesticide impacts, I needed to be aware of what was being published in journals specialising in pollution and toxicological issues. The number of different journals appears to have increased steadily throughout the last 50 years, particularly since the advent of the Internet.

Turning briefly to books, when Malcolm Smith published the first edition of his New Naturalist in 1951, he was able to include some of the early papers from the *British Journal of Herpetology*. Many of his references did, though, date back to the 1800s. One well-used source was *The Zoologist*, a monthly periodical about natural history which was published between 1843 and 1916. Moving into the early twentieth century, his list of sources became more recognisable, ranging from *The Field* to the *Proceedings of the Zoological Society of London*. Because of his herpetological experience and his skill at writing and presentation, Smith managed to distil a century and a half of diverse literature into a coherent and readable account that is still of interest today. I believe it is fair to say that no one produced a broad-spectrum text on British amphibians and reptiles that took us significantly further forward until the third New Naturalist on the subject by Trevor Beebee and Richard Griffiths (2000), almost half a century after Smith's book appeared.

The problem with all books of this type is that they are out of date by the time they are published. Nowadays, websites and social media fill an information vacuum for many people: they should be up to date, but usually they are not very detailed. Trevor Beebee (2013b) contributed another book on *Amphibians and Reptiles*. It was produced by Pelagic Publishing, a small publisher that has been particularly supportive of books on amphibians, including the book you are currently reading. Trevor's book and that by Howard Inns (2009) should provide ample information to identify animals that are seen or caught. Ease of identification has improved hugely since I had to use cigarette cards to try to determine the species of newts I saw in the 1950s (Section 1.2.1). In addition, to these broad-based books, there are monographs: *The Natterjack Toad* (Beebee 1983) and *The Crested Newt* (Jehle *et al.* 2011). I am surprised that no one has yet published a monograph on the common toad, as there is now much information on the species and it is of conservation concern.

12.3 The Monks Wood years, 1968–1978

There is a limited amount that can be learnt about wildlife by reading books. It is essential to be able to observe animals closely in the field, to think about them, to talk about them and to ask questions. Fifty years ago, I was fortunate to have older colleagues in the Nature Conservancy, such as Frank Perring, John Heath and Ian Prestt, who were willing to accommodate my queries and requests. They were specialists in other areas, but they were a generation older and had grown up when frogs and toads were more common. John came with me on one of my very first field trips to see what had been described as a colony of frogs. They turned out to be toads. One of the first things I did was to be sure I could tell the two species apart!

Despite encouragement from colleagues, I felt initially that I was working on my own – not that I minded. At that time I did not personally know any active herpetologists. My entry into the field of herpetology was as someone who was starting out as a researcher on conservation issues and had stumbled onto amphibians. I knew little about them, but soon realised that frogs were in a bad place – or, as regards Huntingdonshire, not in many places at all.

Although I spent much of my time during the 1970s working on the possible effects of pesticides on amphibians, most of the work being done more generally by the few British herpetological researchers was basic ecological observation or an enquiry into (changes in) status. In 1970, I carried out my first national enquiries on whether and why frogs and toads had changed in status in the British Isles (Section 2.2). A bonus of this type of work was that it put me in personal contact with a large number of people who had an interest in these species. It was probably a minor interest for most correspondents, but a few were passionate herpetologists who wanted to know what I was discovering – and I wanted to know what else they could tell me. I corresponded with some for years and still do with several, such as Michael Clark in Hertfordshire. Suddenly I felt less isolated, even if these were people I rarely, if ever, met.

Of those herpetologists that I managed to meet, some were stalwarts of the BHS, such as Deryk Frazer, who was 58 years old when we first surveyed newts in the New Forest in 1974 (Section 8.3), and Maxwell Savage, who was in his mid-seventies when he helped with my survey of potentially polluted frog sites in 1975 (Section 3.4.2; Figure 1.1). Both men continued to be dedicated to learning about amphibians and enjoyed being in the field. I also met people of roughly my own age who already had a much broader and deeper knowledge of the species and their acute conservation needs. Even at that time, Trevor Beebee (Figure 12.1) and Keith Corbett (Figure 12.3)

Figure 12.1 Trevor Beebee at a frog pond in the grounds of Aldershot Crematorium, Hampshire, in 1975.

were among those who stood out as those most likely to make a difference. The BHS Conservation Committee had been formed in 1969, and Keith and Trevor were members. I started going to meetings and met others of a similar age and determination. One or two, such as John Buckley (Figure 12.5), leant towards amphibians, but others, including Michael Lambert and Jon Webster, focused primarily on reptiles. In this part of the book it becomes difficult to divorce amphibians from reptiles.

During the mid- to late 1970s, most of the conservation concern for amphibians was aimed at the natterjack, but there was a gathering gloom about the decline of widespread amphibians and of the crested newt in particular. Because of the reorganisation of the Nature Conservancy in 1973, I was then with ITE and undertaking work of relevance to conservation of the widespread amphibians – and I should have been viewed as a natural ally by other people on the Conservation Committee.

12.4 The years with NCC and English Nature, 1978–1998

This section takes my review up until 1998. Although Tony Gent took over amphibian duties for NCC in 1990/91 (Figure 12.4b), I stayed with NCC and English Nature until 1998 and remained active on the herpetological scene, especially locally.

In summary, this period saw a surge in ecological and conservation-based research, plus data gathering on sites following the introduction of the Wildlife and Countryside Act 1981. The still relatively common, but legally protected, crested newt had started to cause issues for developers as well as for conservationists (Chapter 9). Conservation action stabilised the national natterjack population despite initial problems over not fully understanding the precise conditions required for freshwater and terrestrial habitat (Sections 7.3, 11.3 and 11.4). Events at the end of the 1980s included the First World Congress of Herpetology at Canterbury (Figure 3.8), realisation that amphibians were declining drastically in many parts of the world (Sections 3.5 and 3.7.2) and the inception of the HCT. Some of the 1990s was taken up with trying to discover whether various environmental hazards that were of concern elsewhere might be causing problems in Britain (Section 3.6). However, this was also a time when volunteers were forming groups and becoming involved in more local survey and conservation. Action plans were put in place for species such as the natterjack and crested newt (Section 2.7), and we discovered that we should be conserving things called metapopulations (e.g. Section 8.6) and should have conserved our pool frogs (Section 12.6.1).

My situation had shifted in 1978 when I joined the Chief Scientist Team in NCC. Team members had different specialisms and were responsible for giving advice within and outside the organisation and for setting up and running relevant conservation research programmes. Prior to me joining, Don Jefferies was responsible for herpetofauna in addition to his weightier responsibilities for mammals and toxic chemicals. The team was expanding: I was to take over toxic chemicals and herpetofauna from Don, leaving him to concentrate solely on mammals. The aim was for amphibians and reptiles to receive more staff time, but in reality my pollution remit widened over time.

By the late 1970s, relations between BHS's Conservation Committee and NCC were not good, as is revealed by the Committee's annual report for 1979 (Beebee 1980b). Members of the Conservation Committee rightly considered that it was one of their duties to lobby for better conservation for amphibians and reptiles. Problems were identified in three English regions where the endangered species of herpetofauna occurred; the issues concerned slow scheduling of SSSIs, inadequate conservation management, and Conservation Committee members not being consulted over development proposals. This was the situation into which I had dropped. Because of my past and current jobs, both sides probably viewed me with a degree of mistrust. I could see why problems arose. Conservation Committee members were generally totally committed to the conservation of one or a small number of species at a site or in a region and resented what they may have seen as a lack of commitment. On the other side, NCC staff were often overstretched and had to prioritise within their area, so resented being told they were not doing their jobs well enough or fast enough. Some rifts never healed, but I continually preached the merits of the other side to whoever would listen. Despite such problems, an agreed national plan for the conservation of the natterjack and the two endangered reptiles was finally published (NCC 1983). From the 1970s onwards, other individuals and academics at universities had emerged with herpetological knowledge and conservation skills. The situation was not entirely frictionless as, occasionally, newcomer rubbed up against old hand, and personality clashes occurred. A certain lack of harmony seems to remain today between some parts of the herpetological community.

Nevertheless, the ever-expanding range of people dedicated to research and/or conservation of the species was beginning to build a hill, if not yet a mountain, of results and information. I was not aware of Tim Halliday (Figure 11.1) until about 1977 when he and Nick Davies became television celebrities because on their observations on mating strategies and success in the common toad (Davies and Halliday 1977, 1978). The fact that the largest male toads with the deepest croaks were most successful appealed to scientists, media and public alike. In Ramsey, the local biology teacher launched a small boat on Horse Pond (Figure 6.6b) to collect breeding toads to try to duplicate the researchers' results. At the time, Tim Halliday was at the University of Oxford and had already described in minute detail the elaborate courtship behaviour of the smooth newt (Halliday 1974 and later articles). He soon moved to the new Open University at Milton Keynes, and during the 1980s he was particularly instrumental in pulling together researchers for regular meetings and exchanges of information under the banner of the name *Triturus* (which began as the Amphibian Ecology Group in 1982). Some of his research students, such as Paul Verrell and John Baker, would undertake further studies on newts.

In 1969, when I first started collecting the small amount of research information on the common frog in addition to Maxwell Savage's contribution, I found two papers from 1963 on breeding and homing behaviour by R. S. Oldham. It was, however, about another 10 years before I finally met Rob Oldham (Figure 2.7). By that time, he was a lecturer at Leicester Polytechnic, so, like Tim Halliday, he was reasonably close to my base in Cambridgeshire – and that was the beginning of a collaboration that would endure for more than a decade (Sections 2.5.3, 3.6.6, 4.4, 9.2 and 10.2). Clive Cummins was even closer, as he had started at Monks Wood in the early 1980s as a replacement for Al Scorgie (Figure 8.3), who had been ITE's freshwater ecologist. Although he was not a replacement for me, Clive decided to focus on amphibian research. I am somewhat embarrassed to compare the statistical rigour of his studies (e.g. Section 3.6.3.) with my own a decade of two earlier. Later, he would form an amphibian habitat advisory service with his wife, Mary Swan (Figure 2.7). Chris Reading was another ITE researcher; he was based at Furzebrook in Dorset and undertook what turned into extremely long-term studies on common toads (e.g. Section 11.4).

The former University of Wales Institute of Science and Technology came into prominence in the late 1970s and early 1980s because of the activities of Paul Gittins, Fred Slater and their colleagues on toads at the lake at Llandrindod Wells and the pond at Llysdinam Field Centre. Llysdinam was where Richard Griffiths undertook postdoctoral studies on newts and toads (e.g. Griffiths 1985, 1986, Griffiths *et al.* 1988; Figure 12.2).

It is a pertinent reminder for me to look back at who served on the BHS Council and Conservation Committee during the 1980s. Michael Lambert was Chairman of Council for most of that time. Geoff Haslewood chaired the Conservation Committee from 1983 up until 1988 and deserves considerable praise for the calm manner in which he dealt with issues of the time. It probably helped that he was at least a generation older than most of the rest of us. At the other end of the age scale, Tom Langton was listed in 1983 as Conservation Officer, but soon joined the Fauna and Flora Preservation Society (FFPS) as its herpetologist.

Keith Corbett (Figure 12.3), who had been so good at acquainting me with the endangered herpetofauna species and sites at the end of the 1970s (Section 1.2.2), was

Figure 12.2 Richard Griffiths at Llysdinam in 1983 with a prototype arrangement of bottle traps for monitoring amphibian activity at different depths in the pond (photo courtesy of Richard Griffiths). His paper in 1985 introduced bottle traps, but news had travelled quickly and, by then, they were being used by a number of UK fieldworkers (e.g. Figure 10.2b). I stopped using them in the early 1990s because of concerns for newt safety and my lower back. Dewsbury traps are also now used by many herpetologists and organisations. The image illustrates the general lack of personal risk assessment in the 1980s.

Figure 12.3 Keith Corbett holding a smooth snake in 2003, during his time with HCT. Photo by Jonathan Webster.

the BHS's 'European Observer'. This was an unusual title, but he certainly earned the right to be described as such. Through the 1980s, I fielded comments from him that NCC was not doing anything or enough in Britain when compared with other European countries. For instance, Swiss amphibian breeding sites had been surveyed on a canton basis with a national total of 8,000 sites covered (Grossenbacher 1988); so why was Britain not doing something similar rather than the more limited approach of the contracts with Leicester Polytechnic? The answer was the constraint of scale and resources, as Keith knew well enough. Similar problems have more recently been faced by NARRS (Section 2.7). At the end of the decade, Keith wrote a most useful book summarising the extent of the conservation of reptiles and amphibians in different European countries (Corbett 1989). This provided me with the opportunity to understand better how well we were conserving herpetofauna compared with other countries. I assessed our performance by scoring each country on a scale of 0–5 according to 'knowledge', 'species protection' and 'habitat protection'. A score of 5 represented the best that might reasonably have been achieved rather than absolute perfection. On my scoring of Keith's information, Britain was eighth equal out of 31 countries. I then asked Keith to comment on my scoring and correct it if he disagreed; his scoring demoted Britain to fourteenth equal. My conclusions were that while we certainly were not among the most backward countries, we could learn from others, particularly East and West Germany, the Netherlands, Sweden and Switzerland. My analysis of Keith's information was never published but was circulated to NCC colleagues as well as to him. During the late 1980s, UK legislation was criticised by non-government organisations as focusing on species protection rather than habitat protection. That, however, was what we had to live and work with at the time.

Contracts with Leicester Polytechnic during the 1980s and 1990s grew out of the need for more basic information on the crested newt and other widespread species because of increased protection given by the Wildlife and Countryside Act 1981. The initial NCC contract began in 1983 and was handled by Mark Nicholson under Rob Oldham's supervision. Its aims were to learn more about the freshwater and terrestrial requirements of the crested newt and compile a site dossier. This was followed in 1986 by the Amphibian Communities Project, which attempted to expand information gathering to any sites with amphibians, especially those sites with several species. Field survey by volunteers was an important part of the work, with an emphasis placed on blanket cover of areas of varying sizes; by 1987, response was encouraging but patchy (Griffiths 1987, Swan 1987).

In 1985, FFPS and the national BRC organised a seminar for herpetofauna recorders and workers separate from the annual meetings for researchers which had taken place for several years (Anon 1985). At the same time, FFPS published a provisional list of county recorders. The importance of holding such meetings for fieldworkers was recognised, and Leicester Polytechnic hosted a series of annual recorders' meetings in January or February, 1987–1992, as components of the contracts on Amphibian Communities and Herptile Sites. In 1990, a number of county group representatives met to establish an umbrella group, first called Herpetofauna Groups of Great Britain (Langton 1990), then known as Herpetofauna Groups of Britain and Ireland (HGBI) later in the 1990s (but see also Section 12.5). Herpetofauna Conservation International Ltd (HCIL) acted as secretariat; this was a not-for-profit company run by Tom and Catherine Langton involved with publishing and raising funds for conservation, and was distinct from the similarly named Herpetofauna Consultants International (HCI), run by Tom Langton. When the final contract between Leicester Polytechnic and NCC finished, the annual meetings were organised by HCIL and continued in other places, for example at Birkbeck College, London, in 1993 and at Manchester Metropolitan University in 1994. Annual herpetofauna workers' meetings still take place, organised by ARC and Amphibian and Reptile Groups of the UK (ARG UK), as do annual science meetings organised by ARC and the BHS.

The late 1980s witnessed several important events relating to the conservation of amphibians and reptiles. To me, it seemed as if Ian Swingland had appeared out of thin air in around 1986 and proceeded to (re)organise our herpetological world. He had taken a post at the University of Kent at Canterbury in 1979 and, in 1989, set up the Durrell Institute of Conservation and Ecology, which was named in honour of the famous naturalist Gerald Durrell. The same year, Ian Swingland was Director of the First World Congress of Herpetology at Canterbury (Figure 3.8) and founded the

HCT with Vincent Weir. By that time, I was still NCC's adviser on amphibian conservation, but Malcolm Vincent had taken over the role of adviser on reptiles. For several years leading up to the First World Congress, I was involved with planning meetings as I was convenor of one of the 27 symposia (Section 3.5). I also contributed at meetings to discuss the role of the new HCT. The Trust's first employee was Keith Corbett (Figure 12.3), and by 2009, when the name was changed to Amphibian and Reptile Conservation (ARC), its staff complement had grown to about 20. This charity has continued to grow and undertakes a range of activities aimed at conserving the species and saving their habitats; currently it manages about 80 reserves with a combined area of more than 2,000 ha. The fact that ARC has permanent staff means it is now far better placed to safeguard the species than the BHS Conservation Committee members, who had more limited time and funds for such activities during the final decades of the twentieth century. The BHS has a conservation officer, who is responsible for liaising with other relevant bodies and generally supporting the conservation of our amphibians and reptiles.

In 1995, HCIL changed its name to Froglife and was soon able to employ several members of staff to aid conservation of widespread species of amphibians and reptiles, and run an enquiry service; the organisation was based at Bramfield in Suffolk (Langton 1995 and personal communication). Froglife expanded into offices in nearby Halesworth in 2000 and then moved to Peterborough in 2003. Tom Langton had long been involved with the area of Orton brick pits that became Hampton Nature Reserve, south of Peterborough (Section 9.5; Figure 12.4a); and from 2003 Froglife took over its management. Because of this commitment, the organisation maintained a base in the city and this became its home. Tom and Catherine Langton handed full control of Froglife to new trustees and a chief executive in 2007. Froglife was then, and still is, an independent charity committed to conserving amphibians and reptiles and their habitats throughout Britain. It has a particular emphasis on educating and enthusing people about our native species.

12.5 Newsletters, handbooks and atlases

As knowledge and activity built up, so tailored advice and information on identifying, surveying and conserving amphibians began to appear. From the 1980s, there was a steady stream of newsletters and other sheets and notes with advisory material, in addition to more formal handbooks. Although publication of such material covers nearly 40 years, this seems an appropriate place to slip in this section. The first newsletter of this type was probably *Herpetofauna News*, edited by Tom Langton. Twelve issues, published by FFPS, appeared between 1985 and 1988; and these were followed by at least eight more published by HCIL from 1989 up to 1995.

Among species-specific material was *Great Crested News*, which ran from 2001 until at least 2007 and was funded by combinations of English Nature, Froglife, HCT and the Countryside Council for Wales. HCIL produced a few slim advice sheets in the 1990s, and these were revisited and enlarged by Froglife in 2002 and 2003. By then it seemed that printed material was starting to run out of steam, and more up-to-date and briefer bites of information online were taking its place.

Many local amphibian and reptile groups (ARGs) were forming in the 1990s, and *Herp-Line* was a newsletter devoted to these groups; it was produced by Froglife

Figure 12.4 Herpetologists at Orton brick pits, Peterborough: (a) Tom Langton during the initial stages of HCI involvement with the crested newt population, 1990; (b) Tony Gent (left) discusses practicalities with Roy Bradley in 1991, soon after the former joined English Nature. Tony Gent was succeeded by Jim Foster in 1999. Roy Bradley was then working for HCI.

until at least 2003. The umbrella organisation HGBI provided a forum and source of advice for the ARGs. In 1998, advice notes were produced on (1) evaluating mitigation/translocation programmes: maintaining best practice and lawful standards and (2) commercial consultancy work: guidelines on ARG involvement. HGBI was rebranded as ARG UK in 2005, and now covers the British Isles (ARG UK 2013). Angie Julian has been employed to run the organisation and represent ARG UK in professional circles. Advice sheets have been produced and adopted as national standards: for example on a modified approach to the calculation and interpretation of crested newt HSIs (ARG UK 2010; see also Buxton *et al.* 2021). Editions 1–12 of the newsletter *ARG Today* (2006–2013) were available on the website in 2022, as were 10 advice notes (2010–2018) and news items dating back to 2012. Froglife's newsletter, *Natterchat*, and ARC's *e-Newsletter* can also be found online.

The first issue of *FrogLog* was in March 1992. This was the newsletter of the Declining Amphibian Populations Task Force (DAPTF), which had been set up by the International Union for the Conservation of Nature and the Species Survival Commission. DAPTF's mission was to organise a monitoring programme to determine the status of amphibians worldwide, assess the implications of declines, study potential factors and make policy recommendations. In 2011, the Amphibian Survival Alliance (ASA) was launched to develop a better world for amphibians through coordinated action. Since then, *FrogLog* has been ASA's main way of communicating with herpetologists. *FrogLog* went digital in 2005 and became, in the words of ASA, 'the world's number one amphibian conservation digital magazine'. It is free to read or download and is certainly a mine of information.

In addition to newsletters and advisory sheets, there were many pamphlets produced to aid conservationists, planners and developers. Eventually, sufficient information had accumulated for a number of important handbooks to be produced, including the *Natterjack Toad Conservation Handbook* (Beebee and Denton 1996), the *Herpetofauna Workers' Manual* (Gent and Gibson 1998), the *Great Crested Newt Conservation Handbook* (Langton *et al.* 2001), *Great Crested Newt Mitigation Guidelines* (English Nature 2001), *Common Toads and Roads: Guidance for Planners and Highways Engineers* (Barker 2009), the *Amphibian Habitat Management Handbook* (Baker *et al.* 2011) and the *Amphibian Survey and Monitoring Handbook* (Wilkinson 2015). There were also relevant handbooks from other conservation bodies such as *The Pond Book: a Guide to the Management and Creation of Ponds* (Williams *et al.* 2010), which was published by the Freshwater Habitats Trust (FHT). These handbooks contained a wealth of information, very little of which was available when I began working on amphibians in 1969. Many of these publications have been updated.

As regards atlases, national distributions progressed from being presented in paper form on a vice-county basis (Taylor 1948), then dot maps of locations (Taylor 1963), before ending up as presence in 10 km squares on the National Grid (e.g. Arnold 1973, 1998); now national distributions are available online via the NBN Gateway. Several county atlases have been produced by ARGs, such as *Amphibians and Reptiles of Surrey* (Wycherley and Anstis 2001). Over the decades, knowledge of distribution at a coarse scale has changed relatively little, unlike knowledge of site-scale distribution.

12.6 Events and developments since 1998

12.6.1 Other amphibian species

One of the most important developments of the late twentieth and early twenty-first centuries was the gathering of evidence that indicated the pool frog was a native species in Britain (Snell 1994, 2016, Kelly 2004, Beebee *et al.* 2005, Raye 2017). Three species of water frogs, the pool frog, marsh frog (*Pelophylax ridibundus*, formerly *Rana ridibunda*) and their hybrid, the edible frog, were known to have been introduced since 1800 to various parts of Britain. Perhaps not unreasonably, it was assumed by most conservationists that any water frogs found here were descended from introduced individuals. But then, in the 1990s, Charles Snell pointed out that such evidence was lacking to explain the existence of some populations of pool frogs. This stimulated a multidisciplinary investigation which found genetic, acoustic, archaeological and archival evidence for the species to be considered as native to this country, being most closely related to Scandinavian populations. Sadly, the wake-up call came too late to prevent the last population at Thompson Common in Norfolk from dying out. Habitat management was undertaken in Norfolk to 'redress the presumed reasons for extinction'. Yearly introductions of different life stages of Swedish pool frogs were made to an unnamed site in Norfolk between 2005 and 2008 (Figure 12.5). Annual peak counts of frogs indicated a relatively stable population during 2010–2014 (Figure 12.6), and another introduction was started in 2015 to Thompson Common. Currently, there are breeding populations at these two sites (John Baker personal communication).

Figure 12.5 John Buckley of HCT collecting pool frog tadpoles in Sweden in 2006 for translocation to Norfolk. Photo courtesy of John Buckley.

Although pool frogs were lost from this country after I had handed over responsibility for amphibian advice in NCC, I have regretted not recognising the species' predicament earlier when I might have been able to do something about it. I was, however, interested in a population of European tree frogs (*Hyla arborea*) breeding in a pond in the New Forest, Hampshire, after I had been introduced to the pond and the frogs by Deryk Frazer in 1974. As the species breeds as close as the Atlantic coast of northern France, I did wonder whether it could conceivably be native to southern England. On my subsequent surveys of the New Forest ponds I made a point of checking whether tree frogs were still present: they were in 1982, but not in 1989. I later found out that this fitted in with Charles Snell's observations and conclusions that extinction occurred between 1984 and 1987, after the species had been present in the pond since at least the early years of the twentieth century (Snell 2006). In 1989, I was told that Forestry Commission staff had recently reported tree frogs in three places in the New Forest, so I was not as disappointed at finding none in the usual pond as I otherwise would have been. However, things literally seem to have gone silent since then. An intensive search of old literature revealed likely references to tree frogs from the 1500s, whereas there were references to water frogs back to Saxon times (Raye 2017). This led the author to suggest that they may have been introduced to England during the sixteenth century, as they were used in traditional medicine. There was, however, evidence of tree frogs in Britain during interglacial periods of the Pleistocene (Holman 1998).

Howard Inns (2009) listed the tree frog among a number of non-native species that have occurred in the British Isles as a result of animals escaping from captivity

Figure 12.6 Pool frogs mating in Norfolk in 2013. Photo by John Baker.

or being deliberately released – and described it as extinct. He mentioned the agile frog (*Rana dalmatina*) because it occurs naturally on Jersey in the Channel Islands. Agile frogs declined to very low numbers on the island, but intensive conservation has improved the situation (Wilkinson and Buckley 2012, Ward *et al.* 2016). Subfossil remains of this species and the moor frog (*Rana arvalis*) were discovered in Saxon deposits in eastern England (Gleed-Owen 2000). Both species breed in northern France. There is no mention of them in historical records in Britain, perhaps because they were thought to be varieties of common frogs owing to their close resemblance to this species (Raye 2017). Agile frogs and moor frogs evidently died out in England during the last 1,000 years or so. In contrast, twenty-first-century science added another species to the list of amphibians occurring in the British Isles, the variously known Jersey, western common or spiny toad (Arntzen *et al.* 2014). This had previously been considered a subspecies of the common toad, but morphological and genetic studies revealed it to be a distinct species, *Bufo spinosus*. It breeds in that part of France closest to Jersey, and the island was the last of the Channel Islands to be connected to mainland Europe; none of the other islands in the group has native populations of toads.

Amphibians and reptiles that have become extinct in Britain are being bred in captivity with a view to releasing them into areas of rewilding, particularly where beavers (*Castor fiber*) have been reintroduced (Barkham 2021). Amphibians mentioned in this interesting enterprise include tree, agile and moor frogs. This is an example of a modern issue that raises modern questions. The appearance of information on the occurrence of these species is relatively recent, as is the concept of rewilding. But what about the disease risk, and will such activity deflect attention and resources away from helping those native species that have had such a hard time since the 1930s? Experience with the reintroduction of the pool frog demonstrates the level of thought, planning and commitment that such a project requires.

12.6.2 *Other events and developments since 1998*

Undoubtedly my most well-thumbed and scribbled-on hard copy of the *Herpetological Journal* is Volume 10, number 4, from October 2000. This contains papers from a one-day symposium on 5 December 1998 that was organised by the BHS Research Committee and was entitled 'Scientific Studies of the Great Crested Newt: its Ecology and Management'. As Clive Cummins and Richard Griffiths explained in the editorial, there was a feeling at the time that the academic community and practitioners who were engaged in mitigation work were not communicating sufficiently well. The meeting was an attempt to improve the exchange of information and views, and was seen as a contribution to the crested newt's SAP. In addition, that issue of the *Journal* is a particularly poignant reminder for me of subsequent events, as a month after attending the meeting I was diagnosed with a serious illness, which severely impacted my herpetological input, in some instances permanently.

Sadly, several major contributors to herpetological conservation died in the early years of the new century. Michael Lambert (1941–2004) worked in pest control and visited many warmer parts of the world, where he was able to spend much time studying and conserving tortoises. He was active in various herpetological societies and groups including the BHS. In 2005, I was able to give a presentation on 'Long-term surveillance' at a meeting entitled 'Herpetofauna Ecology and Conservation' which was organised jointly by the BHS and the HCT in his memory. Deryk Frazer (1916–2008) qualified and worked in medicine but switched to conservation when he joined the Nature Conservancy in 1959. He too was heavily involved with the BHS, being its president from the late 1950s up until 1981, and a life member thereafter.

Tim Halliday (1945–2019) was introduced above in connection with his seminal work on the sex lives of our amphibians and initiating *Triturus* research meetings (Section 12.4). These gatherings later spread internationally, and he became known personally by herpetologists in many countries. Tim helped to organise the First World Congress in 1989 and, because of concern at amphibian declines across the globe, he was involved in setting up the DAPTF (Section 12.5) – and became its International Director from 1994 until 2006. After his death, no less than 16 appreciations appeared in *FrogLog* issue 120.

My illness and its treatment had long-term effects which severely disrupted any plans for continuing with various aspects of my previous formal and informal work. Participation at meetings was badly affected, but illness did focus my attention on completing fieldwork and writing up observations. My amphibian fieldwork after 1998 centred on concluding monitoring already started on crested and smooth newts, common frogs and common toads. Very little fieldwork was done after 2010 for a variety of reasons. For example, numbers of toads breeding in my three local ponds dwindled to very low levels between 1990 and 2010 (Section 6.3), finally rendering searches of road casualties a fruitless exercise. I kept a hopeful eye open after 2010, but rarely saw a single toad. The early years of this century were spent puzzling over why the toad populations had disappeared and, in particular, whether road traffic might be partially responsible. Conveniently, I was able to present my conclusions at an EU workshop on 'Amphibian mortality on roads' held in Peterborough in March 2012 (Cooke 2011; Figure 12.7a–c). In addition I gave a few talks, mainly to local groups about my work and my views on amphibian conservation. And I was pleased to be consulted by individuals and organisations about planned initiatives, such as setting up an ARG for Cambridgeshire and how NARRS might be undertaken (Section 2.7).

Since the late 1960s, there have been huge changes in how many people are involved in amphibian conservation and the degree of contact and cooperation. Then I may have been the only person who was paid to conserve frogs. However, my national enquiries demonstrated that many individuals knew something about their local amphibian populations, but were not being paid to follow their interests. In the Breeding Sites Survey in 1970, replies were received on how status of frogs and/or toads had changed from more than 300 sites (Section 2.2). There was though very limited contact or coordination between the observers. It was likely that many of them did not know another individual who was interested in frogs or toads. Now the situation is very different. Numbers of at least part-time professional herpetologists have been swelled by the ranks of the herpetological and conservation organisations, by university groups, by specialist consultants and so on. I have recently

Figure 12.7 Some steps in the development of toad protection and recovery: (a) an early road sign in 1979 to warn motorists at Llandrindod Lake, Powys; (b) a tunnel made by ACO Polymer Products Ltd at the Thetford by-pass, Norfolk, in 1992; (c) a close-up of the tunnel mouth and the barriers leading toads into it; (d) a view in 2022 of a pond created in 2015 under the Million Ponds Project, which common toads have colonised naturally, Great Fen, Cambridgeshire (photo by Rosemarie Cooke).

searched online for 'environmental consultant great crested newts' to count how many websites came up – I stopped counting at 50. It is the job of some of these people to try to connect with the rest by organising meetings, compiling websites and generally making information available.

Then there are the genuine amateurs who are prepared to give up some time to make a difference by undertaking surveys, working on habitat conservation and doing a multitude of other tasks. These people include ARG members and those who contribute to citizen science projects. Such distinctions are artificial and misleading to some extent because a number of ARG members are professional consultants. Some of those people may have started out as volunteers but have developed the necessary expertise and qualifications to become consultants. The umbrella group, ARG UK, is able to make rapid collective responses when necessary. Thus, during the 7th Quinquennial Review of the Wildlife and Countryside Act in 2021, the Review Group changed eligibility criteria so that species to be listed on Schedule 5 needed to be in 'imminent danger of extinction'. Such a change would have meant there was no control of sale for the four most widespread species of amphibians, while the widespread reptile species would, additionally, not be protected against being killed. This had implications for conservation, for public perception of the needs of our herpetofauna, and for disease control. Following coordinated responses by ARG UK and other bodies, the proposed changes were reconsidered. This was one of a number of recent examples of UK conservation charities coming together to fight for a common cause.

Pond creation and restoration has continued apace during the early years of this century. For instance, in 2008, the FHT launched the Million Ponds Project, described as a 50-year mission to attempt to ensure that eventually we have at least as many ponds as occurred in 1900 (FHT 2013, ARC 2016b). ARC, a lead partner in this project, reported that 467 ponds were created for crested newts in the first part of the project up to 2012. Benefits of this project for other amphibians have also been reported (e.g. NCP 2021; Figure 12.7d).

Before progressing to ask how things might change in the next decade or two, it is worth recalling some significant observations, achievements, fears and events that occurred and/or ended during the first couple of decades of the twenty-first century. The following topics have special relevance for the future and were summarised in the sections indicated.

- Particular concern was voiced for the crested newt and common toad, and there were initiatives to aid conservation and knowledge of herpetofauna more generally: SAPs, the Habitats Directive, NARRS and other citizen science projects (Section 2.7).
- There was continuing concern over pesticide use, disease and other hazards (Sections 3.6 and 3.7.2).
- Problems caused for toads by road traffic and development were highlighted (Section 6.4).
- Intensive and extensive monitoring confirmed that a programme of dogged monitoring, management action and introduction was required just to maintain natterjack numbers (Sections 7.3 and 11.3).
- The ability of established and undisturbed newt populations to persist for decades was demonstrated, provided management was on hand when needed (Sections 8.6 and 10.7).

- Introductions of crested newts or permitting developments close to breeding sites could be successful, providing sufficient attention and commitment was sustained over many years (Section 9.7).
- New strategic approaches to licensing of developments and conservation of crested newts were described (Section 9.7).
- The value of long-term datasets was illustrated for monitoring specific populations and for providing information more broadly, for example on the effects of climate change (Chapter 11).
- Genetic studies were noted as contributing in an increasing variety of ways (e.g. Beebee 2018b), such as by examining genetic diversity and fitness (Sections 5.5, 7.2.2), resolving issues of phylogeography (Section 9.7) and identification of species (Section 12.6.1).

Overall changes in status since the late 1960s, as described in this book, can be summarised as follows: the common frog has shown some recovery; the smaller newts have just about held their own; the natterjack has suffered recent declines after a period of stability; the common toad and crested newt have given rise to more prolonged concern although their rates of pond occupancy seem to have stabilised; and the pool frog has been reinstated after becoming extinct. Without the determined conservation effort, any negative trends would have been very much worse. Essentially, what we have done so far is more or less stop the rot. Now we need to make things better. Despite the disappointments and concerns, knowledge about our amphibian species and how to conserve them has developed immeasurably during this time and is surely now sufficient for us to be reasonably confident that we are equipped in that respect to provide them with a sustainable future? At least partially rectifying declines that occurred in the middle of the last century should not be beyond present and future generations. But to be successful, we need support and structure in the shape of political and public attitudes, legislation and resources.

12.7 The current conservation climate and an attempt to predict the future

When trying to understand the current and future conservation scene as regards official policies and commitments, it becomes essential to look back in some detail at recent history. The following account is sometimes delivered from an English perspective, but the other UK countries have generally followed suit. The year 2010 saw the publication of the seminal report *Making Space for Nature*. This was a report to Defra by a committee chaired by Professor Sir John Lawton. It presented a long-term vision for England based on large-scale habitat restoration with an emphasis on ecological processes and ecosystem services for the benefit of wildlife and people. Soon after, the UK government published a White Paper and also produced a strategy document for England's wildlife and ecosystem services, *Biodiversity 2020*, which set out how international and EU commitments might be implemented during the decade (Defra 2011). One of the aims was to establish 'a coherent and resilient ecological network capable of responding to the challenges of climate change and other pressures', as had been recommended in the Lawton Report.

In 2016, national Action Plans for species were no longer formally recognised by government. In England, ARC then worked with Natural England under a Back from the Brink project to aid natterjack conservation on the Sefton Coast; and the ARC website also announced that the organisation would revise Action Plans to help all involved with assisting relevant amphibians and reptiles (ARC 2016a). By 2022, however, only the rare reptiles' SAP seemed to have been updated. The SAPs for common toad, crested newt, natterjack and pool frog all dated from 2009.

The Brexit vote in 2016 meant the UK would no longer be obliged to conform to EU rules and regulations, and there was a collective holding of breath. Two years later, the government released its 25-year plan to improve the UK's environment: *A Green Future* (Defra 2018). This looked forward to delivering a 'Green Brexit' by reforming management of agriculture and fisheries, restoring nature and caring for our land, rivers and seas. Targets included restoring 75% of protected sites to favourable condition and creating or restoring half a million hectares of 'wildlife-rich habitat' outside protected sites. The Nature Recovery Network was announced by Defra in 2020: this was to be an important part of the 25-year plan in which Natural England and its partners would work together to improve sites and promote recovery in threatened species.

The UK duly left the EU at the end of January 2020. In September 2020, John Lawton and his committee wrote to the Prime Minister acknowledging that their report 10 years earlier had positively influenced government policy. They noted, however, that there had been little progress towards establishing the 'landscape approach' and requested a commitment to increase spending on rebuilding England's natural infrastructure by £1 billion (Brotherton 2020). At the same time, ARC and many other conservation charities in the UK urged government to begin a 'new era for nature' after failing to halt environmental decline during the previous decade (ARC 2020b). In an interview in April 2021, Tony Gent of ARC explained the importance of more robust land-use policies and funding mechanisms to encourage sympathetic land management, such as via agri-environment schemes (NHBS 2021).

The Agriculture Act received royal assent in November 2020, with a basic principle of farmers being paid to steward the land and deliver public goods. Farmers needed further information on the subject of subsidies, and Defra produced guidance in March 2021 on three Environmental Land Management schemes which related to conservation action on farms:

- The Sustainable Farming Initiative scheme would pay farmers to manage their land sustainably.
- The Local Nature Recovery scheme would pay to support local nature recovery and meet local priorities.
- The Landscape Recovery scheme would pay to support landscape and ecosystem recovery.

In November 2021, the Environment Act was given royal assent. It included a commitment to halt species decline by 2030, and stated that new developments, including major infrastructure projects, would be required to deliver a 10% increase in biodiversity. There would be Local Nature Conservation Strategies covering the whole of England that were intended to help developers avoid the best existing habitat and focus compensatory habitat creation where it was most needed. Species Conservation and Protected Site Strategies would feed into these local strategies for certain species and protected sites respectively. A new green watchdog, the Office for Environmental Protection (OEP), was established. There were some concerns about these proposals, such as over the level of independence to be enjoyed by the OEP, but generally they were welcomed by conservationists.

Also in 2021, the national nature conservation bodies of the UK collaborated with John Lawton to produce a document, *Nature Positive 2030* (Natural England *et al.* 2021). This was in response to the Prime Minister, Boris Johnson, signing up to an initiative at the United Nations General Assembly in 2020 to protect 30% of our land and seas by 2030. The report warned that we are running out of time, and action is needed urgently if we are to have any chance of achieving objectives.

Wildlife does not respond to schemes and strategies alone. Enough coordinated and appropriate action is needed to produce significant effects and benefits. In this respect, there does seem to have been some joined-up thinking in the various ways that agricultural schemes and environmental strategies may work together on the ground to improve the situation. Thus, with nearly three-quarters of the UK being farmed, agricultural policy can have profound effects on the conservation of crested newts (Jehle *et al.* 2011), and on amphibian conservation more generally. These authors found that agri-environment schemes were producing good results but that the extent of uptake was unpredictable. Now if Local Nature Recovery schemes plug into local conservation strategies, farmers may be more encouraged to participate by seeing how their contribution fits in. In addition, 'regenerative agriculture' has become more talked about recently; this is a holistic approach intertwining farming and ecological systems, and wider adoption of such techniques should lead to a general improvement in biodiversity. It is important in any broad prediction of the future not to view it simply from the perspective of an environmentalist. How, for example will our farming community adapt and cope with tariff-free food coming in from new markets overseas, and to what extent will we continue to provide food for ourselves? The Covid-19 pandemic, the Russian invasion of Ukraine in 2022 and a huge rise in the cost of living have all added further significant layers of uncertainty about the future. Food and energy security are of increasing concern. Priorities can change quickly, especially when new demands on the nation's purse arise – and environmental schemes may be the ones to be axed, forgotten or delayed.

Worries over having adequate commitment and resources are of course not new. However, a pertinent recent report in June 2021 from the Environmental Audit Committee of the House of Commons (HCEAC) considered Defra's 25-year plan and the other initiatives to be a 'welcome start' but stated that action needed to be 'stepped up in scale, ambition, pace and detail'. Unfortunately, recent governments have a solid track record of over-promising and under-delivering on environmental issues. The HCEAC proposed a package of recommendations spanning monitoring, funding, policy implementation, economics and education.

In March 2022, a Defra Green Paper launched a consultation on a new tiered approach to 'modernise' species and site legislation post-Brexit. There is widespread concern at how EU laws may be revoked or reformed. The EU's Habitats Directive has provided protection for 30 years, but will it survive and function effectively for much longer (Clarke 2022)? Similarly, there are likely to be threats to the Natural Environment and Rural Communities Act of 2006. Amongst other things, this converted English Nature into Natural England, but it also required local authorities and government departments to have due regard to conserving biodiversity, including by focusing on Priority Species such as the natterjack toad, pool frog and crested newt.

The following statements on what could happen are general and brief. The future is probably more difficult to predict now than at any stage in the last 60 years as the UK and much of the rest of the world faces a string of threats, disturbances and upheavals. By the time this book is published, the way ahead may be somewhat clearer. Nevertheless, with the major, largely beneficial, approaches described or taking place already (e.g. district licensing, agri-environment schemes and rewilding) it does seem that amphibian conservation will become an even more collective effort for groups of people at local and national levels. There is likely to be an increase in some activities, such as responding to official proposals, surveying, advising, planning, producing strategies, creating and maintaining habitats and monitoring whether various initiatives succeed. For those in positions of national or local influence, there will undoubtedly be a continuing struggle to ensure adequate protection for habitats, sites and species. Conversely, there may be less work for consultants on crested newts and development proposals because of new licensing arrangements. Whether there are sufficient people with the dedication and expertise required for all the new work will depend to some extent on funding. If the money is there, then chances of success will increase, but the HCEAC referred in 2021 to a 'severe skills shortage in ecologists' as a result of 'cuts to public spending on biodiversity'.

Those working in amphibian conservation must continue to fight and guard against any perception that the species they especially care about lack the charisma and vulnerability of some other fauna. In Britain, amphibian equivalents of people who tick off lists of species of birds or invertebrate groups are essentially absent. There may be few native amphibian species, but conserving them helps other wildlife that shares their habitats such as birds, butterflies and dragonflies. And amphibians represent local wildlife that can be closely and enjoyably observed, including at night. Legal protection and having agreed knowledge, plans and targets become important when 'competing' with other conservationists for attention and funds. Up-to-date and workable plans and targets plus monitoring schemes that are adequate to detect changes in status are all needed now the 'playing field' is changing and the EU will no longer be keeping a critical eye on (lack of) progress. As regards the natterjack, ARC was, in 2021, in the 'final stages' of compiling an Action Plan designed to 'increase the size and range of natterjack populations in the UK' (ARC 2021a); important elements of future conservation action include restorative work on existing sites, reintroduction and monitoring. The new National Amphibian Survey will hopefully provide good monitoring data on status trends for the widespread species, and for the crested newt in particular (ARC 2021b; Section 2.7). In addition, analysis of eDNA in pond water is already being used to detect presence of crested newts over large areas and to monitor changes on a landscape scale (e.g. FHT 2021b, NCP 2021, Buxton *et al.* 2022).

One prediction that seems certain to come about is that new techniques will to some extent replace more traditional methods. Thus, cameras are being adapted to record amphibians on land (e.g. Hill *et al.* 2019) and in water (Xavier Mestdagh's

Newtrap, described by Simpson 2021a); these are analogous to the camera traps that have proved so useful for the study of larger land animals. With the exceptional quality of the photos now taken on mobile phones, there will hopefully be a better archive of photos than in the past, but this will require thoughtful labelling and storage. Research on innovations will continue, and herpetologists of the future are likely to find that increasing amounts of time are spent in the office or laboratory rather than in the field. Publication processes for research results will continue to evolve online, and hopefully information will become easier and cheaper to produce and access. Associated activities, such out-sourcing the promotion of research results, will continue to expand. As many research topics become more complex, so the trend towards multi-authored papers is likely to endure. In the reference list to this book, all 20 references with six or more authors date from 1990 onwards, 16 having been published this century.

The recent move to request information on amphibians in gardens from contributors to schemes run by the BTO and RSPB is most welcome (Section 2.7). Those organisations are able to muster huge numbers of birdwatchers and naturalists for their monitoring programmes, and perhaps this resource can be tapped into more widely than just in gardens? BTO's BirdTrack now allows users to log observations on amphibians and reptiles in order, for example, to transfer distribution data to the NBN. Concern has been expressed about the extent to which the BTO has strayed into encouraging its members to record non-avian fauna and interpreting the resulting data (Wheeler *et al.* 2019). However, providing recorders are sufficiently accomplished with regard to identification, and conclusions about the data remain objective, then making use of large numbers of observers in the field in the right places and at the right times of day and year has to be a good idea. Personally, I have amassed long-term datasets on deer when I've been primarily recording birds and vice versa. And the series of torchlight counts of crested newts in Monks Wood (Table 9.1) was done after I had undertaken dusk surveillance of deer in the wood.

The UN Climate Change Conference in Glasgow in the autumn of 2021 reaffirmed to the world that climate change will worsen before any recovery may begin. The HCEAC argued in 2021 that the biodiversity crisis should be raised up the political agenda to a level similar to that occupied by climate change: each government department should be required to take into account the impact of its actions on biodiversity. Such a change was given some impetus by the publication of *The Economics of Biodiversity: the Dasgupta Review* (HM Treasury 2021). This insisted we should 'change our measures of economic success to guide us on a more sustainable path'. In the gloom of 2022, however, such advances seem even more distant.

Whatever happens, the need will remain for studies into the effects of climate change on amphibians, including monitoring how impacts change over time. As was shown in Chapter 11, local studies by individuals or small groups can contribute significantly in this area, but this work needs further coordination as amphibians as a group are both at threat from climate change and could serve as a natural barometer of its impacts. And concerns over pollution, disease and other hazards will never disappear – new and old issues will constantly require (re)appraisal.

Although there must be overarching strategic approaches to amphibian conservation and research, opportunities for action will not be restricted to areas of farming and development, and will crop up in a multitude of places at a range of spatial scales, from single garden ponds to nature reserves to landscape-wide projects. In Section 6.4, I referred to the importance of large nature reserves in Cambridgeshire for common toads and how the species has been encouraged to colonise new habitat in the landscape-scale Great Fen (Figure 12.7d). Managers of nature reserves and rewilding programmes may need to be reminded about the potential to conserve the widespread herpetofauna as well as more conspicuous species of wildlife. A recent article on rewilding described how Kent ARG was involved with surveying and monitoring herpetofauna in Blean Woods in relation to the introduction of large herbivores (Simpson 2021b). The number of rewilding projects in Britain seems certain to increase, and one topic that requires more attention is how well amphibian populations might fare in landscapes receiving minimal or no intervention (e.g. see Sections 4.4.4, 6.4 and 11.2.2). Overall, many more sites and areas will nominally be set aside for conservation, but their future success is in the balance and by no means assured. Opportunities to make a difference will continue to arise. Be prepared to grasp them.

Appendix 1

Shillow Hill Top Pond: seasonal water conditions, 1984–2009

Year	Seasonal water conditions in the pond		
	Spring	Summer	Autumn
1984	Full*	Full	Dry*
1985	Full	Full	Full
1986	Full	Full	Full
1987	Full	Full	Full
1988	Full	Full	Full
1989	Full	Full	Dry
1990	Full	Dry	Dry
1991	Sump*	Sump	Dry
1992	Sump	Dry	Full
1993	Full	Full	Full
1994	Full	Full	Full
1995	Full	Sump	Sump
1996	Full	Sump	Dry
1997	Sump	Dry	Dry
1998	Full	Full	Dry
1999	Full	Full	Sump
2000	Full	Full	Sump
2001	Full	Full	Full
2002	Full	Sump	Dry
2003	Full	Sump	Dry
2004	Full	Sump	Dry

(*Continued*)

(*Continued*)

Year	Seasonal water conditions in the pond		
	Spring	Summer	Autumn
2005	Sump	Dry	Dry
2006	Sump	Dry	Dry
2007	Full	Full	Dry
2008	Sump	Sump	Dry
2009	Full	Dry	Dry

[a] Full = water at least covered the base of the pond; Sump = water in sump only; Dry = pond dry.

Appendix 2

Shillow Hill Top Pond: average night counts and tadpole netting results for great crested newts, 1983–2009

Year	Night count		Tadpole netting		
	Adults	Immatures	Number	Number per sweep	Walker length (mm)
1983	25*		—	—	—
1984	33*		—	—	—
1985	150	7.3	—	—	—
1986	145	4.7	33	0.8	—
1987	183	3.9	16	0.4	—
1988	136	3.6	71	1.6	—
1989	39	1.0	19	0.5	—
1990	34	1.0	0 (D**)	—	—
1991	19 (S**)	0 (S)	6 (S)	4.7	—
1992	3 (S)	0 (S)	0 (D)	—	—
1993	42	0.2	53	1.1	46
1994	28	5.2	30	0.8	51
1995	36	1.0	14 (S)	1.2	53
1996	71	2.4	65 (S)	7.1	39
1997	9 (S)	0 (S)	0 (D)	—	—
1998	102	1.2	80	2.0	51
1999	62	3.4	36	1.3	56
2000	76	2.2	119	3.2	44
2001	60	1.2	3	0.1	64

(Continued)

(Continued)

Year	Night count		Tadpole netting		
	Adults	Immatures	Number	Number per sweep	Walker length (mm)
2002	107	2.2	5 (S)	0.3	64
2003	73	0.2	30 (S)	1.5	57
2004	70	0.2	20 (S)	13.2	45
2005	0.8 (S)	0 (S)	0 (D)	—	—
2006	0.8 (S)	0 (S)	0 (D)	—	—
2007	26	0	24	1.0	60
2008	14 (S)	0 (S)	3 (S)	2.5	62
2009	13	1.2	0 (D)	—	—

* Count includes immatures.

** S = water in sump only; D = pond dry.

Appendix 3

Shillow Hill Top Pond: results from annual spring netting of crested newts, 1993–2009

Year	Lengths of adult newts (mm)				Proportion of small adults *	Average Body Condition Index **
	Males		Females			
	No.	Average (range)	No.	Average (range)		
1993	10	112 (107–117)	5	126 (121–135)	0.00	6.3
1994	5	117 (113–126)	6	126 (122–130)	0.00	6.0
1995	13	112 (95–126)	8	127 (113–133)	0.29	6.0
1996	12	102 (93–108)	8	118 (107–130)	0.65	5.5
1997	17	108 (96–124)	8	121 (107–133)	0.60	5.2
1998	8	115 (104–129)	9	121 (113–133)	0.35	6.7
1999	3	113 (106–123)	6	125 (108–137)	0.33	5.4
2000	18	114 (103–127)	12	122 (108–135)	0.30	6.4
2001	6	117 (113–123)	7	122 (117–129)	0.23	5.4
2002	16	114 (105–127)	13	124 (110–136	0.24	6.0
2003	15	111 (97–122)	13	124 (108–139)	0.21	5.5
2004	14	113 (104–119)	6	124 (113–134)	0.14	5.2
2005	0	—	0	—	—	—
2006	1	124	0	—	0.00	5.2
2007	2	124 (118–129)	4	119 (108–131)	0.50	5.5
2008	6	118 (113–126)	5	126 (117–128)	0.09	5.4
2009	4	122 (113–129)	5	127 (121–134)	0.00	5.8

* Small males are shorter than 107 mm, small females are shorter than 121 mm.

** Body Condition Index is the weight in milligrams divided by the cube of the length in centimetres.

Appendix 4

Shillow Hill Top Pond: estimated number of small and large crested newt adults counted at night in springs when the pond was full

Year	Average number of adults counted	Estimated number of small* adults counted	Estimated number of large* adults counted
1993	42	0	42
1994	28	0	28
1995	36	10	26
1996	71	46	25
1998	102	36	66
1999	62	20	42
2000	76	23	53
2001	60	14	46
2002	107	26	81
2003	73	15	58
2004	70	10	60
2007	26	13	13
2009	13	0	13

* Small males are shorter than 107 mm, small females are shorter than 121 mm. Large males are at least 107 mm, large females are at least 121 mm.

Appendix 5

Shillow Hill Top Pond and Wood Pond: average night counts and tadpole netting results for smooth newts, 1983–2009

Years	Average night count of adults		Average number of tadpoles netted in Top Pond
	Top Pond	Wood Pond	
1983	1	11	—
1984	26	8	—
1985	22	—	—
1986	21	4	12
1987	10	2	21
1988	7	0	43
1989	13	1	10
1990	8	0	0 (D*)
1991	9 (S*)	0	6 (S)
1992	13 (S)	0 (D)	3 (D)
1993	24	0	31
1994	14	0	33
1995	11	0	23 (S)
1996	11	2	17 (S)
1997	15 (S)	0 (D)	0 (D)
1998	28	2	37
1999	10	—	8
2000	5	—	27
2001	5	2	0.3
2002	20	0	0 (S)

(Continued)

(*Continued*)

Years	Average night count of adults		Average number of tadpoles netted in Top Pond
	Top Pond	Wood Pond	
2003	11	0	3 (S)
2004	2	0 (D)	0.3 (S)
2005	0.2 (S)	0 (D)	0 (D)
2006	0.6 (S)	0 (D)	0 (D)
2007	2	0 (D)	22
2008	2 (S)	0 (D)	3 (S)
2009	7	0 (D)	0 (D)

[a] S = water in sump only; D = pond dry.

References

Advisory Committee on Pesticides and other Toxic Chemicals (1969) *Further Review of Certain Persistent Organochlorine Pesticides Used in Great Britain.* London: HMSO.

Allain, S.J.R. and Duffus, A.L.J. (2019) Emerging infectious disease threats to European herpetofauna. *Herpetological Journal* 29: 189–206. https://doi.org/10.33256/hj29.4.189-206

Allain, S.J.R. and Goodman, M.J. (2019) Cambridge amphibian survey report 2017. *Nature in Cambridgeshire* 61: 70–75.

Almond, J. (2020) District level licensing for crested newts – by numbers! *Natural England Blog.* https://naturalengland.blog.gov.uk/2020/12/11/district-level-licensing-for-great-crested-newts-by-numbers. Accessed 14 June 2022.

Amphibian and Reptile Conservation (2016a) Species Action Plans (SAPs). https://www.arc-trust.org/species-action-plans-saps. Accessed 28 December 2020.

Amphibian and Reptile Conservation (2016b) Million Ponds Project. https://www.arc-trust.org/million-ponds-project. Accessed 22 January 2022.

Amphibian and Reptile Conservation (2020a) Newts and Project Speed. http://www.arc-trust.org/news/newts-and-project-speed. Accessed 15 December 2020.

Amphibian and Reptile Conservation (2020b) New era for nature conservation needed for UK. https://www.arc-trust.org/news/new-era-for-nature-needed-for-uk. Accessed 15 January 2021.

Amphibian and Reptile Conservation (2021a) Saving species: natterjack toads. http://www.arc-trust.org/saving-species-natterjack-toad. Accessed 15 January 2021.

Amphibian and Reptile Conservation (2021b) National Amphibian Survey. https://amphibian-survey.arc-trust.org. Accessed 21 January 2022.

Amphibian and Reptile Groups of the UK (2010) *Great Crested Newt Habitat Suitability Index.* Advice Note 5. Amphibian and Reptile Groups of the United Kingdom.

Amphibian and Reptile Groups of the UK (2013) Amphibian and Reptile Groups of the UK. *FrogLog* 105: 47–48.

Amphibian and Reptile Groups of the UK (2017) *Amphibian Disease Precautions: a Guide for UK Fieldworkers.* Advice Note 4 (revised). Amphibian and Reptile Groups of the United Kingdom.

Anon. (1985) First herpetofauna recorders' seminar. *Herpetofauna News* 1(1): 8.

Anon. (2016) Have a herpy Christmas…! *NARRS Newsletter* 13: 1.

Arak, A. (1988) Callers and satellites in the natterjack toad: evolutionarily stable decision rules. *Animal Behaviour* 36: 416–432. https://doi.org/10.1016/S0003-3472(88)80012-5

Arnold, H.R. (1973) *Provisional Atlas of Amphibians and Reptiles of the British Isles.* Abbots Ripton: Biological Records Centre.

Arnold, H.R. (1998) *Atlas of Amphibians and Reptiles in Britain.* London: HMSO.

Arnold, H.R. and Jefferies, D.J. (1997) Mammal Report for 1996. *Annual Report of Huntingdonshire Fauna and Flora Society* 49: 40–49.

Arnold, H.R. and Jefferies, D.J. (2000) Mammal Report for 1999. *Annual Report of Huntingdonshire Fauna and Flora Society* 52: 42–43.
Arntzen, J.W. and Teunis, S.F.M. (1993) A six year study on the population dynamics of the crested newt (*Triturus cristatus*) following the colonization of a newly created pond. *Herpetological Journal* 3: 99–110.
Arntzen, J.W., Wilkinson, J.W., Butôt, R. and Martinez-Solano, I. (2014) A new vertebrate species native to the British Isles: *Bufo spinosus* Daudin, 1803 in Jersey. *Herpetological Journal* 24: 209–216.
Ashby, K.R. (1969) The population ecology of a self-maintaining colony of the common frog (*Rana temporaria*). *Journal of Zoology* 158: 453–474. https://doi.org/10.1111/j.1469-7998.1969.tb02162.x
Atkins, W. (1998) 'Catch 22' for the great crested newt: observations on the breeding ecology of the great crested newt *Triturus cristatus* and its implications for the conservation of the species. *British Herpetological Society Bulletin* 63: 17–26.
Bacon, L. (2005) *Cambridge and Peterborough Provisional Mammals Atlas.* Cambourne: Cambridgeshire and Peterborough Biological Records Centre.
Baker, J.M.R. (1999) Abundance and survival rates of great crested newts (*Triturus cristatus*) at a pond in central England: monitoring individuals. *Herpetological Journal* 9: 1–8.
Baker, J.M.R. and Halliday, T.R. (1999). Amphibian colonization of new ponds in an agricultural landscape. *Herpetological Journal* 9: 55–63.
Baker, J., Beebee, T., Buckley, J., Gent, T. and Orchard, D. (2011) *Amphibian Habitat Management Handbook.* Bournemouth: Amphibian and Reptile Conservation.
Banks, B. and Beebee, T.J.C. (1986) A comparison of the fecundities of two species of toad (*Bufo bufo* and *B. calamita*) from different habitat types in Britain. *Journal of Zoology* 208: 325–337. https://doi.org/10.1111/j.1469-7998.1986.tb01898.x
Banks, B. and Beebee, T.J.C. (1988) Reproductive success of natterjack toads *Bufo calamita* in two contrasting habitats. *Journal of Animal Ecology* 57: 475–492. https://doi.org/10.2307/4919
Banks, B., Beebee, T.J.C. and Denton, J.S. (1993) Long-term management of a natterjack toad (*Bufo calamita*) population in southern Britain. *Amphibia-Reptilia* 14: 155–168. https://doi.org/10.1163/156853893X00327
Banks, B., Beebee, T.J.C. and Cooke, A.S. (1994) Conservation of the natterjack toad (*Bufo calamita*) in Britain over the period 1970–1990 in relation to site protection and other factors. *Biological Conservation* 67: 111–118. https://doi.org/10.1016/0006-3207(94)90355-7
Barker, F. (ed.) (2009) *Common Toads and Roads: Guidance for Planners and Highways Engineers (England).* Bournemouth: Amphibian and Reptile Conservation.
Barkham, P. (2021) Who doesn't love a turtle? The teenage boys on a mission – to rewild Britain with reptiles. https://theguardian.com/environment/2021/jan/10/who-doesnt-love-a-turtle-the-teenage-boys-on-a-mission-to-rewild-Britain-with-reptiles. Accessed 22 January 2022.
Beebee, T.J.C. (1973) Observations concerning the decline of the British amphibia. *Biological Conservation* 5: 20–24. https://doi.org/10.1016/0006-3207(73)90050-5
Beebee, T.J.C. (1975) Changes in status of the great crested newt *Triturus cristatus* in the British Isles. *British Journal of Herpetology* 5: 481–490.
Beebee, T.J.C. (1976) The natterjack toad (*Bufo calamita*) in the British Isles: a study of past and present status. *British Journal of Herpetology* 5: 515–521.
Beebee, T.J.C. (1979) Habitats of the British amphibians (2) suburban parks and gardens. *Biological Conservation* 15: 241–258. https://doi.org/10.1016/0006-3207(79)90046-6
Beebee, T.J.C. (1980a) Amphibian growth rates. *British Journal of Herpetology* 6: 107–108.
Beebee, T.J.C. (1980b) Conservation Committee Annual Report – 1979. *British Herpetological Society Bulletin* 1: 6–9.

Beebee, T.J.C. (1983) *The Natterjack Toad*. Oxford: Oxford University Press.
Beebee, T.J.C. (1986) Acid tolerance of natterjack toad (*Bufo calamita*) development. *Herpetological Journal* 1: 78–81.
Beebee, T.J.C. (1989) Natterjack toad (*Bufo calamita*) site register for the UK, 1970–1989 inclusive. Unpublished report. Brighton: University of Sussex.
Beebee, T.J.C. (1991) Purification of an agent causing growth inhibition in anuran larvae and its identification as a unicellular unpigmented alga. *Canadian Journal of Zoology* 69: 2146–2153. https://doi.org/10.1139/z91-300
Beebee, T.J.C. (1995) Amphibian breeding and climate. *Nature* 374: 219–220. https://doi.org/10.1038/374219a0
Beebee, T.J.C. (1996) *Ecology and Conservation of Amphibians*. London: Chapman and Hall.
Beebee, T.J.C. (1997) Changes in dewpond numbers and amphibian diversity over 20 years on chalk downland in Sussex, England. *Biological Conservation* 81: 215–219. https://doi.org/10.1016/S0006-3207(97)00002-5
Beebee, T.J.C. (2001) British wildlife and human numbers: the ultimate conservation issue? *British Wildlife* 1: 1–8.
Beebee, T.J.C. (2007) Thirty years of garden ponds. *Herpetological Bulletin* 99: 23–28.
Beebee, T.J.C. (2009) A comparison of single-sample effective population size estimators using empirical toad (*Bufo calamita*) population data: genetic compensation and size-genetic diversity correlations. *Molecular Ecology* 18: 4790–4797. https://doi.org/10.1111/j.1365-294X.2009.04398.x
Beebee, T.J.C. (2010) Ronald Maxwell Savage, 1900–1985: a tribute. *Herpetological Journal* 20: 115–116.
Beebee, T.J.C. (2012) Decline and flounder of a Sussex common toad (*Bufo bufo*) population. *Herpetological Bulletin* 121: 6–16.
Beebee, T.J.C. (2013a) Effects of road mortality and mitigation measures on amphibian populations. *Conservation Biology* 27: 657–668. https://doi.org/10.1111/cobi.12063
Beebee, T.J.C. (2013b) *Amphibians and Reptiles*. Naturalists' Handbooks 31. Exeter: Pelagic Publishing.
Beebee, T.J.C. (2014) Amphibian conservation in Britain: a 40-year history. *Journal of Herpetology* 48: 2–12. https://doi.org/10.1670/12-263
Beebee, T.J.C. (2015) The great crested newt: an ongoing conservation enigma. *British Wildlife* 26: 230–236.
Beebee, T. (2018a) *Climate Change and British Wildlife*. British Wildlife Collection 6. London: Bloomsbury.
Beebee, T.J.C. (2018b) Genetic contributions to herpetofauna conservation in the British Isles. *Herpetological Journal* 28: 51–62.
Beebee, T.J.C. and Buckley, J. (2001) *Natterjack Toad (*Bufo calamita*) Site Register for the UK 1970–1999 Inclusive*. Brighton: University of Sussex and Bournemouth: Herpetological Conservation Trust.
Beebee, T.J.C. and Buckley, J. (2014a) *Natterjack Toad (*Bufo calamita*) Site Register for the UK 1970–2009 Inclusive*. Bournemouth: Amphibian and Reptile Conservation Trust.
Beebee, T.J.C. and Buckley, J. (2014b) Relating spawn counts to the dynamics of British natterjack toad (*Bufo calamita*) populations. *Herpetological Journal* 24: 25–30.
Beebee, T.J.C. and Denton, J. (1996) *Natterjack Toad Conservation Handbook*. Peterborough: English Nature.
Beebee, T.J.C. and Griffin, J.R. (1977) A preliminary investigation into natterjack toad (*Bufo calamita*) breeding site characteristics. *Journal of Zoology* 181: 341–350. https://doi.org/10.1111/j.1469-7998.1977.tb03248.x
Beebee, T.J.C. and Griffiths, R.A. (2000) *Amphibians and Reptiles*. New Naturalist 87. London: HarperCollins.

Beebee, T.J.C. and Wong, A. L.-C. (1992) Prototheca-mediated interference competition between anuran larvae operates by resource diversion. *Physiological Zoology* 65: 815–831. https://doi.org/10.1086/physzool.65.4.30158541

Beebee, T.J.C., Flower, R.J., Stevenson, A.C. and seven others (1990) Decline of the natterjack toad (*Bufo calamita*) in Britain: palaeoecological, documentary and experimental evidence for breeding site acidification. *Biological Conservation* 53: 1–20. https://doi.org/10.1016/0006-3207(90)90059-X

Beebee, T.J.C., Buckley, J., Evans, I. and eight others (2005) Neglected native or undesirable alien? Resolution of a conservation dilemma concerning the pool frog *Rana lessonae*. *Biodiversity and Conservation* 14: 1607–1626. https://doi.org/10.1007/s10531-004-0532-3

Beebee, T.J.C., Wilkinson, J.W. and Buckley, J. (2009) Amphibian declines are not uniquely high amongst vertebrates: trend determination and the British perspective. *Diversity* 1: 67–88. https://doi.org/10.3390/d1010067

Bell, G.A.C. (1970) The distribution of amphibians in Leicestershire. *Leicester Literary ad Philosophical Society* 64: 122–142.

Bell, T. (1839) *A History of British Reptiles*. London: John van Voorst. https://doi.org/10.5962/bhl.title.10835

Bellairs, A. (1959) Malcolm Smith's contribution to herpetology. *British Journal of Herpetology* 2: 130–141.

Berger, G., Graef, F., Pallut, B. and three others (2018) How does changing pesticide use over time affect migrating amphibians: a case study on the use of glyphosate-based herbicides in German agriculture over 20 years. *Frontiers in Environmental Science* 6:6. https://doi.org/10.3389/fenvs.2018.00006

Berman, H.J. and Thomas, M. (1966) Amphibians and reptiles. *Annual Report of Huntingdonshire Fauna and Flora Society* 18: 19.

Berman, H.J., Johnson, D. and Barnes, J. (1967) Amphibians and reptiles. *Annual Report of Huntingdonshire Fauna and Flora Society* 19: 15–16.

Bickmore, D.P. and Shaw, M.A. (1963) *The Atlas of Britain and Northern Ireland*. Oxford: Clarendon Press.

Biggs, J., Ewald, N., Valentini, I. and six others (2014) Analytical and methodological development for improved surveillance of the great crested newt. Defra Project WC1067. Oxford: Freshwater Habitats Trust.

Borchert, J.R. (1948) The agriculture of England and Wales 1939–1946. *Agricultural History* 22: 56–62.

Boulenger, G.A. (1893) Specimens of *Rana temporaria* from Scotland. *Annals of Scottish Natural History* 1893: 202–204. https://doi.org/10.1080/00222939308677603

Boulenger, G.A. (1897) *The Tailless Batrachians of Europe*. Part I. London: Ray Society. https://doi.org/10.5962/bhl.title.5606

Boulenger, G.A. (1898) *The Tailless Batrachians of Europe*. Part II. London: Ray Society. https://doi.org/10.5962/bhl.title.5606

Bowley, A. (2013) The Great Fen – the challenges of creating a wild landscape in lowland England. *British Wildlife* 25: 95–102.

Bowley, A. (2020) *The Great Fen: a Journey Through Time*. Newbury: Pisces.

Boyd, C.E., Vinson, S.B. and Ferguson, D.E. (1963) Possible DDT resistance in two species of frogs. *Copeia* 1963: 426–429. https://doi.org/10.2307/1441363.

Brassley, P. (2000) Output and technical change in twentieth-century British agriculture. *Agricultural History Review* 48: 60–84.

Brotherton, P. (2020) Making space for nature – 10 years on. *Natural England Blog*. https://naturalengland.blog.gov.uk/2020/09/16/making-space-for-nature-10-years-on. Accessed 10 October 2021.

Buckley, J. and Beebee, T.J.C. (2004) Monitoring the conservation status of an endangered amphibian: the natterjack toad (*Bufo calamita*) in Britain. *Animal Conservation* 7: 221–228. https://doi.org/10.1017/S1367943004001428

Buckley, J. and Inns, H. (1998) Field identification, sexing and ageing. In A.H. Gent and S.D. Gibson (eds), *Herpetofauna Workers' Manual*. Peterborough: Joint Nature Conservation Committee, pp. 15–31.

Buckley, J., Beebee, T.J.C. and Schmidt, B.R. (2014) Monitoring amphibian declines: population trends of an endangered species over 20 years in Britain. *Animal Conservation* 17: 27–34. https://doi.org/10.1111/acv.12052

Buxton, A.S., Tracey, H. and Downs, N.C. (2021) How reliable is the habitat suitability index as a predictor of great crested newt presence or absence? *Herpetological Journal* 31: 111–117. https://doi.org/10.33256/31.2.111117

Buxton, A., Diana, A., Matechou, E., Griffin, J. and Griffiths, R. (2022) Reliability of environmental DNA surveys to detect pond occupancy by newts at a national scale. *Scientific Reports* 12: 1295–1314. https://doi.org/10.1038/s41598-022-05442-1

Campbell, L.H. and Cooke, A.S. (eds) (1997) *The Indirect Effects of Pesticides on Birds*. Peterborough: Joint Nature Conservation Committee.

Carrier, J.-A. and Beebee, T.J.C. (2003) Recent, substantial and unexplained declines of the common toad *Bufo bufo* in lowland England. *Biological Conservation* 111: 395–399. https://doi.org/10.1016/S0006-3207(02)00308-7

Carroll, E.A., Sparks, T.H., Collinson, N. and Beebee, T.J.C. (2009) Influence of temperature on the spatial distribution of first spawning dates of the common frog (*Rana temporaria*). *Global Change Biology* 15: 467–473. https://doi.org/10.1111/j.1365-2486.2008.01726.x

Carson, R. (1962) *Silent Spring*. Boston, MA: Houghton Mifflin.

Cayuela, H., Griffiths, R.A., Zakaria, N. and five others (2020) Drivers of amphibian population dynamics and asynchrony at local and regional scales. *Journal of Animal Ecology* 89: 1350–1364. https://doi.org/10.1111/1365-2656.13208

Central Science Laboratory (1998) *Assessing Pesticide Risks to Amphibians and Reptiles*. Research Project Final Report PN0921. London: Defra.

Chadwick, E.A., Slater, F.M. and Ormerod, S.J. (2006) Inter- and intraspecific differences in climatically mediated phenological change in coexisting *Triturus* species. *Global Change Biology* 12: 1069–1078. https://doi.org/10.1111/j.1365-2486.2006.01156.x

Chanin, P.R.F. and Jefferies, D.J. (1978) The decline of the otter *Lutra lutra* L. in Britain: an analysis of hunting records and discussion of causes. *Biological Journal of the Linnean Society* 10: 305–328. https://doi.org/10.1111/j.1095-8312.1978.tb00018.x

Clarke, E. (2022) The Habitats Regulations – under threat. Briefing document. London: Wildlife and Countryside Link.

Cooke, A.S. (1970) The effects of pp'-DDT on tadpoles of the common frog (*Rana temporaria*). *Environmental Pollution* 1: 57–71. https://doi.org/10.1016/0013-9327(70)90006-6

Cooke, A.S. (1971) Selective predation by newts on frog tadpoles treated with DDT. *Nature* 229: 275–276. https://doi.org/10.1038/229275a0

Cooke, A.S. (1972a) Indications of recent changes in status in the British Isles of the frog (*Rana temporaria*) and the toad (*Bufo bufo*). *Journal of Zoology* 167: 161–178. https://doi.org/10.1111/j.1469-7998.1972.tb01727.x

Cooke, A.S. (1972b) The effects of DDT, dieldrin and 2,4-D on amphibian spawn and tadpoles. *Environmental pollution* 3: 51–68. https://doi.org/10.1016/0013-9327(72)90017-1

Cooke, A.S. (1973a) The effects of DDT when used as a mosquito larvicide on tadpoles of the common frog *Rana temporaria*. *Environmental Pollution* 5: 259–273. https://doi.org/10.1016/0013-9327(73)90003-7

Cooke, A.S. (1973b) Response of *Rana temporaria* tadpoles to chronic doses of pp'-DDT. *Copeia* 1973: 647–652. https://doi.org/10.2307/1443062

Cooke, A.S. (1974a) The effects of pp'-DDT on adult frogs (*Rana temporaria*). *British Journal of Herpetology* 5: 390–396.
Cooke, A.S. (1974b) Differential predation by newts on anuran tadpoles. *British Journal of Herpetology* 5: 386–390.
Cooke, A.S. (1975a) Spawn site selection and colony size of the frog (*Rana temporaria*) and the toad (*Bufo bufo*). *Journal of Zoology* 175: 29–38. https://doi.org/10.1111/j.1469-7998.1975.tb01388.x
Cooke, A.S. (1975b) Pesticides and eggshell formation. *Symposia of the Zoological Society of London* 35: 339–361.
Cooke, A.S. (1976) Spawning dates of the frog (*Rana temporaria*) and the toad (*Bufo bufo*) in Britain. *British Journal of Herpetology* 5: 585–589.
Cooke, A.S. (1977) Effects of field applications of the herbicides diquat and dichlobenil on amphibians. *Environmental Pollution* 12: 43–50. https://doi.org/10.1016/0013-9327(77)90006-4
Cooke, A.S. (1978) Marking tadpoles with neutral red. *British Journal of Herpetology* 5: 701–705.
Cooke, A.S. (1979) The influence of rearing density on the subsequent response to DDT for tadpoles of the frog *Rana temporaria*. *Bulletin of Environmental Contamination and Toxicology* 21: 837–841. https://doi.org/10.1007/BF01685514
Cooke, A.S. (1980) Observations on how close certain passerine species will tolerate an approaching human in rural and suburban areas. *Biological Conservation* 18: 85–88. https://doi.org/10.1016/0006-3207(80)90072-5
Cooke, A.S. (1981a) Tadpoles as indicators of harmful levels of pollution in the field. *Environmental Pollution* 25: 123–133. https://doi.org/10.1016/0143-1471(81)90012-X
Cooke, A.S. (1981b) Amphibian growth rates. *British Journal of Herpetology* 6: 179–180.
Cooke, A.S. (1982) Amphibian and reptile report for 1981. *Annual Report of Huntingdonshire Fauna and Flora Society* 34: 35–38.
Cooke, A.S. (1983) Estimating numbers of natterjack tadpoles (*Bufo calamita*). *British Journal of Herpetology* 6: 299–300.
Cooke, A.S. (1984) The warty newt in Huntingdonshire. *Annual Report of Huntingdonshire Fauna and Flora Society* 36: 41–48.
Cooke, A.S. (1985a) The deposition and fate of spawn clumps of the common frog *Rana temporaria* at a site in Cambridgeshire 1971–1983. *Biological Conservation* 32: 165–187. https://doi.org/10.1016/0006-3207(85)90084-9
Cooke, A.S. (1985b) Frogs and collection in Cornwall. *British Herpetological Society Bulletin* 12: 39–40.
Cooke, A.S. (1985c) The warty newt at Shillow Hill: numbers and density. *Annual Report of Huntingdonshire Fauna and Flora Society* 37: 22–25.
Cooke, A.S. (1986a) Studies of the crested newt at Shillow Hill, 1984–1986. *Herpetofauna News* 6: 4–5.
Cooke, A.S. (1986b) The warty newt (*Triturus cristatus*) at Shillow Hill: ranging on arable land. *Annual Report of Huntingdonshire Fauna and Flora Society* 38: 40–44.
Cooke, A.S. (1988) Mortality of toads (*Bufo bufo*) on roads near a Cambridgeshire breeding site. *British Herpetological Society Bulletin* 26: 29–30.
Cooke, A.S. (1990) The impact of the Wildlife and Countryside Act, 1981, on the conservation of the crested newt (*Triturus cristatus*). *Annual Report of Huntingdonshire Fauna and Flora Society* 42: 28–37.
Cooke, A.S. (1991) The habitat of sand lizards (*Lacerta agilis*) at Merseyside. Research and Survey in Nature Conservation 41. Peterborough: Nature Conservancy Council.
Cooke, A.S. (1993a) The habitat of sand lizards (*Lacerta agilis*) on the Sefton Coast. In D. Atkinson and J. Houston (eds), *The Sand Dunes of the Sefton Coast*. Liverpool: National Museums and Galleries on Merseyside, pp. 123–126.

Cooke, A.S. (ed.) (1993b) *The Environmental Effects of Pesticide Drift*. Peterborough: English Nature.

Cooke, A.S. (1994) Fluctuations in night counts of crested newts at eight breeding sites in Huntingdonshire, 1986–1993. In A. Gent and R. Bray (eds), *Conservation and Management of Great Crested Newts*. Peterborough: English Nature, pp. 68–70.

Cooke, A.S. (1995a) A comparison of survey methods for crested newts (*Triturus cristatus*) and night counts at a secure site, 1983–1993. *Herpetological Journal* 5: 221–228.

Cooke, A.S. (1995b) Road mortality of common toads (*Bufo bufo*) near a breeding site, 1974–1994. *Amphibia-Reptilia* 16: 87–90. https://doi.org/10.1163/156853895X00226

Cooke, A.S. (1997a) Safeguarding herpetofauna from agrochemical use in the designed landscape. In R. Bray and A. Gent (eds), *Opportunities for Amphibians and Reptiles in the Designed Landscape*. English Nature Science Series 30. Peterborough: English Nature, pp. 66–70.

Cooke, A.S. (1997b) Monitoring a breeding population of crested newts (*Triturus cristatus*) in a housing development. *Herpetological Journal* 7: 37–41.

Cooke, A.S. (1999) Changes in status of the common frog and common toad. *Huntingdonshire Fauna and Flora Society 50th Anniversary Review*: 30–33.

Cooke, A.S. (2000) Monitoring a breeding population of common toads in a housing development. *Herpetological Bulletin* 74: 12–15.

Cooke, A.S. (2001a) Translocation of small numbers of crested newts (*Triturus cristatus*) to a relatively large site. *Herpetological Bulletin* 75: 25–29.

Cooke, A.S. (2001b) Invertebrate predation on larvae of the crested newt (*Triturus cristatus*). *Herpetological Bulletin* 77: 15–19.

Cooke, A.S. (2001c) A case study in the evolution of crested newt conservation. *Herpetological Bulletin* 78: 16–20.

Cooke, A.S. (2002) The influence of breeding success on adult length in a population of crested newts. *Herpetological Bulletin* 81: 8–11.

Cooke, A.S. (2003) Common frogs in a Cambridgeshire garden over a twenty year period *Herpetological Bulletin* 83: 16–21.

Cooke, A.S. (2011) The role of road traffic in the near extinction of common toads in Ramsey and Bury. *Nature in Cambridgeshire* 53: 45–50.

Cooke, A.S. (2019) *Muntjac and Water Deer: Natural History, Environmental Impact and Management*. Exeter: Pelagic Publishing. https://doi.org/10.53061/URGZ9475

Cooke, A.S. and Arnold, H.R. (1982) National changes in status of the commoner British amphibians and reptiles before 1974. *British Journal of Herpetology* 6: 206–207.

Cooke, A.S. and Arnold, H.R. (2001) Deepening a crested newt breeding site. *Annual Report of Huntingdonshire Fauna and Flora Society* 53: 35–38.

Cooke, A.S. and Arnold, H.R. (2003) Night counting, netting and population dynamics of crested newts (*Triturus cristatus*). *Herpetological Bulletin* 84: 5–14.

Cooke, A.S. and Cooke, R.A. (2008) The distribution of common toads in the south-west corner of the Fens. *Annual Report of Huntingdonshire Fauna and Flora Society* 60: 41–44.

Cooke, A.S. and Cooke, S.D. (1993) A technique for monitoring the final phase of metamorphosis in newts. *British Herpetological Society Bulletin* 42: 10–13.

Cooke, A.S. and Ferguson, P.F. (1974) The past and present status of the frog (*Rana temporaria*) and the toad (*Bufo bufo*) in Huntingdonshire. *Annual Report of Huntingdonshire Fauna and Flora Society* 26: 53–63.

Cooke, A.S. and Ferguson, P.F. (1975) Is the palmate newt a montane species? *British Journal of Herpetology* 5: 460–463.

Cooke, A.S. and Ferguson, P.F. (1976) Changes in status of the frog (*Rana temporaria*) and the toad (*Bufo bufo*) on part of the East Anglian Fenland in Britain. *Biological Conservation* 9: 191–198. https://doi.org/10.1016/0006-3207(76)90009-4

Cooke, A.S. and Frazer, J.F.D. (1976) Characteristics of newt breeding sites. *Journal of Zoology* 178: 223–236. https://doi.org/10.1111/j.1469-7998.1976.tb06009.x
Cooke, A.S. and Fulford, W.G. (1971) Observations on the feeding behaviour of a blind warty newt (*Triturus cristatus*). *British Journal of Herpetology* 4: 216.
Cooke, A.S. and Oldham, R.S. (1995) Establishment of populations of the common frog *Rana temporaria* and common toad *Bufo bufo* in a newly created reserve following translocation. *Herpetological Journal* 5: 173–180.
Cooke, A.S. and Scorgie, H.R.A. (1983) *The Status of the Commoner Amphibians and Reptiles in Britain.* Focus on Nature Conservation 3. Shrewsbury: Nature Conservancy Council.
Cooke, A.S. and Sparks, T.H. (2004) Population declines of common toads (*Bufo bufo*): the contribution of road traffic and monitoring value of casualty counts. *Herpetological Bulletin* 88: 13–26.
Cooke, A.S. and Zoro, J.A. (1975) The effects of p,p'-DDCN on tadpoles of the frog *Rana temporaria. Bulletin of Environmental Contamination and Toxicology* 13: 233–237. https://doi.org/10.1007/BF01721744
Cooke, A.S., Scorgie, H.R.A. and Brown, M.C. (1980) An assessment of changes in populations of the warty newt (*Triturus cristatus*) and smooth newt (*T. vulgaris*) in twenty ponds in Woodwalton Fen National Nature Reserve. *British Journal of Herpetology* 6: 45–47.
Cooke, A.S., Bell, A.A. and Haas, M. (1982) *Predatory Birds, Pesticides and Pollution.* Cambridge: Institute of Terrestrial Ecology.
Cooke, A.S., Banks, B. and Dennison, S. (1983) Stripeless natterjacks in England. *British Herpetological Society Bulletin* 7: 64–65.
Cooke, A.S., Morgan, D.H.W. and Swan, M.J.S. (1990) Frog collection with special reference to Cornwall. *British Herpetological Society Bulletin* 33: 9–11.
Cooke, A.S., Greig-Smith, P.W. and Jones, S.A. (1992) Consequences for vertebrate wildlife of toxic residues in earthworm prey. In P.W. Greig-Smith, H. Becker, P.J. Edwards and F. Heimback (eds), *Ecotoxicology of Earthworms.* Andover: Intercept, pp. 139–155.
Cooke, M.C. (1865) *Our Reptiles.* London: Robert Hardwicke.
Cooke, S.D. (1993) Metamorphosis of smooth newts (*Triturus vulgaris*) and great crested newts (*Triturus cristatus*) and the effect of scrub cover on breeding ponds. Unpublished undergraduate thesis, Nottingham Trent University.
Cooke, S.D., Cooke, A.S. and Sparks T.H. (1994) Effects of scrub cover of ponds on great crested newts' breeding performance. In A. Gent and R. Bray (eds), *Conservation and Management of Great Crested Newts.* English Nature Science 24. Peterborough: English Nature, pp. 71–74.
Corbett, K. (1989) *Conservation of European Reptiles and Amphibians.* London: Christopher Helm.
Corbett, K.F. and Beebee, T.J.C. (1975) The disappearing natterjack. *Oryx* 13: 47–49. https://doi.org/10.1017/S0030605300013004
Corn, P.S. (2000) Amphibian declines: review of some current hypotheses. In D.W. Sparling, G. Linder and C.A. Bishop (eds), *Ecotoxicology of Amphibians and Reptiles.* Pensacola: Society of Environmental Toxicology and Contamination, pp. 663–696.
Cowman, D.F. and Mazanti, S. (2000) Ecotoxicology of 'new generation' pesticides to amphibians. In D.W. Sparling, G. Linder and C.A. Bishop (eds), *Ecotoxicology of Amphibians and Reptiles.* Pensacola: Society of Environmental Toxicology and Contamination, pp. 233–268.
Creed, K. (1964) A study of newts in the New Forest. *British Journal of Herpetology* 3: 170–181.
Crick, M., Kirby, P. and Langton, T. (2005) Knottholes: the wildlife of Peterborough's claypits. *British Wildlife* 16: 413–421.

Cummins, C.P. (1986) Effects of aluminium and low pH on growth and development in *Rana temporaria* tadpoles. *Oecologia* 69: 248–252. https://doi.org/10.1007/BF00377630

Cummins, C.P. (1989) Interaction between the effects of pH and density on growth and development in *Rana temporaria* L. tadpoles. *Functional Biology* 3: 45–52. https://doi.org/10.2307/2389674

Cummins, C.P. (1990) Effects of acid waters on the survival of frogs. *Report of the Institute of Terrestrial Ecology 1989–1990*. Swindon: Natural Environment Research Council pp. 38–40.

Cummins, C.P., Greenslade, P.D. and Mcleod, A.R. (1999) A test of the effect of supplemental UV-B radiation on the common frog *Rana temporaria* L. during embryonic development. *Global Change Biology* 5: 471–479. https://doi.org/10.1046/j.1365-2486.1999.00242.x

Cunningham, A.A., Langton, T.E.S., Bennett, P.M. and four others (1996) Pathological and microbiological findings from incidents of unusual mortality of the common frog (*Rana temporaria*). *Philosophical Transactions of the Royal Society of London B* 351: 1539–1557. https://doi.org/10.1098/rstb.1996.0140

Cunningham, A.A., Garner, T.W.J., Aguilar-Sanchez, V. and seven others (2005) Emergence of amphibian chytridiomycosis in Britain. *Veterinary Record* 157: 386–387. https://doi.org/10.1136/vr.157.13.386

Davidson, C. (2004) Declining downwind: amphibian population declines in California and historical pesticide use. *Ecological Applications* 14: 1892–1902. https://doi.org/10.1890/03-5224

Davidson, C. and Knapp, R.A. (2007) Multiple stressors and amphibian declines: dual impacts of pesticides and fish on yellow-legged frogs. *Ecological Applications* 17: 587–597. https://doi.org/10.1890/06-0181

Davies, N.B. and Halliday, T.R. (1977) Optimal mate selection in the toad *Bufo bufo*. *Nature* 269: 56–58. https://doi.org/10.1038/269056a0

Davies, N.B. and Halliday, T.R. (1978) Deep croaks and fighting assessment in toads *Bufo bufo*. *Nature* 274: 683–685. https://doi.org/10.1038/274683a0

Davis, B.N.K. (1968) The soil macrofauna and organochlorine insecticide residues at twelve agricultural sites near Huntingdon. *Annals of Applied Biology* 61: 29–45. https://doi.org/10.1111/j.1744-7348.1968.tb04507.x

Davis, B.N.K. and French, M.C. (1969) The accumulation and loss of organochlorine insecticide residues by beetles, worms and slugs in sprayed fields. *Soil Biology and Biochemistry* 1: 45–55. https://doi.org/10.1016/0038-0717(69)90033-9

Defra (2011) *Biodiversity 2020: a Strategy for England's Wildlife and Ecosystem Services*. London: Defra.

Defra (2018) *A Green Future: Our 25 Year Plan to Improve the Environment*. London: Defra.

Defra (2022) *Nature Recovery Green Paper: Protected Sites and Species*. London: Defra.

Denniford, T. (2019) *Hampton: The Early Years*. Peterborough: Upfront.

Denton, J.S. and Beebee, T.J.C. (1993) Density-related features of natterjack toad (*Bufo calamita*) populations in Britain. *Journal of Zoology* 229: 105–119. https://doi.org/10.1111/j.1469-7998.1993.tb02624.x

Denton, J.S., Hitchings, S.P., Beebee, T.J.C. and Gent, A, (1997) A recovery program for the natterjack toad (*Bufo calamita*) in Britain. *Conservation Biology* 11: 1329–1338. https://doi.org/10.1046/j.1523-1739.1997.96318.x

de Silva, A. (2011) Some observations of malformation, eye disease, parasitic and viral infection and the effects of agrochemicals on amphibians in Sri Lanka. *FrogLog* 98: 24–25.

de Wijer, W., Watt, P.J. and Oldham, R.S. (2003) Amphibian decline and aquatic pollution: effects of nitrogenous fertiliser on survival and development of larvae of the frog *Rana temporaria*. *Applied Herpetology* 1: 3–12. https://doi.org/10.1163/157075403766451180

English Nature (2001) *Great Crested Newt Mitigation Guidelines.* Peterborough: English Nature.

Fahrig, L., Pedlar, J.H., Pope, S.E., Taylor, P.D. and Wegner, J.F. (1995) Effect of road traffic on amphibian density. *Biological Conservation* 73: 177–182. https://doi.org/10.1016/0006-3207(94)00102-V

Fair, J. (2007) In the eye of a storm. *BBC Wildlife* 25 (11): 29.

Fletcher, M.R., Hunter, K. and Barnett, E.A. (1995) *Pesticide Poisoning of Animals 1994: Investigations of Suspected Incidents in the United Kingdom.* London: MAFF.

Fletcher, M.R., Hunter, K., Barnett, E.A. and Sharp, E.A. (1997) *Pesticide Poisoning of Animals 1996: Investigations of Suspected Incidents in the United Kingdom.* York: MAFF.

Ford, R.L.E. (1954) *British Reptiles and Amphibians.* Young Naturalist Series. London: A. & C. Black.

Fortey, R. (2021) *A Curious Boy: the Making of a Scientist.* Glasgow: William Collins.

Frazer, J.F.D. (1956) Frog and toad breeding records for 1955. *British Journal of Herpetology* 2: 24–29.

Frazer, J.F.D. (1966) A breeding colony of toads (*Bufo bufo* (L.)) in Kent. *British Journal of Herpetology* 3: 236–252.

Frazer, J.F.D. (1978) Newts in the New Forest. *British Journal of Herpetology* 5: 695–699.

Frazer, J.F.D. (1983) *Reptiles and Amphibians in Britain.* New Naturalist 69. London: Collins.

Freshwater Habitats Trust (2013) *What We Want to Achieve. Strategic Framework, 2013–2020.* http://www.freshwaterhabitats.org.uk/wp-content/uploads/2013/09/FHT-Strategy-booklet_Sep13_web-version.pdf. Accessed 19 December 2021.

Freshwater Habitats Trust (2021a) Newt Conservation Partnership. https://freshwaterhabitats.org.uk/projects/newt-partnership. Accessed 19 December 2021.

Freshwater Habitats Trust (2021b) Saving Oxford's wetland wildlife eDNA surveys. https://freshwaterhabitats.org.uk/saving-oxfords-wetland-wildlife-edna-surveys. Accessed 19 December 2021.

Froglife (2013) *Amphibian Health and Disease.* Peterborough: Froglife.

Froglife (2021) Toads on roads: advice for planners and engineers. https://www.froglife.org/what-we-do/toads-on-roads/advice-for-planners-engineers. Accessed 20 December 2021.

Fry, G.L.A. and Cooke, A.S. (1984) *Acid Deposition and its Implications for Nature Conservation in Britain.* Focus on Nature Conservation 7. Shrewsbury: Nature Conservancy Council.

Fuller, R. and Gilroy, J. (2021) Rewilding and intervention: complementary philosophies for nature conservation in Britain. *British Wildlife* 32: 258–267.

Gent, A. and Bray, R. (eds) (1994) *Conservation and Management of Great Crested Newts.* English Nature Science 20. Peterborough: English Nature.

Gent, A. and Gibson, S. (1998) *Herpetofauna Workers' Manual.* Peterborough: Joint Nature Conservation Committee.

Gittins, S.P. (1983a) Population dynamics of the common toad (*Bufo bufo*) at a lake in mid-Wales. *Journal of Animal Ecology* 52: 981–988. https://doi.org/10.2307/4468

Gittins, S.P. (1983b) The breeding migration of the common toad (*Bufo bufo*) to a pond in mid-Wales. *Journal of Zoology* 199: 555–562. https://doi.org/10.1111/j.1469-7998.1983.tb05106.x

Gittins, S.P. (1983c) Road casualties solve toad mysteries. *New Scientist* 97: 530–531.

Gittins, S.P., Parker, A.G. and Slater, F.M. (1980) Population characteristics of the common toad (*Bufo bufo*) visiting a breeding site in mid-Wales. *Journal of Animal Ecology* 49: 161–173. https://doi.org/10.2307/4281

Gittins, S.P., Steeds, J.E. and Williams, R. (1982) Population age-structure of the common toad (*Bufo bufo*) at a lake in mid-Wales determined from annual growth rings in the phalanges. *British Journal of Herpetology* 6: 249–252.

Gittins, S.P., Kennedy, R.I. and Williams, R. (1985) Aspects of the population age-structure of the common toad (*Bufo bufo*) at Llandrindod Wells Lake, mid-Wales. *British Journal of Herpetology* 6: 447–449.

Gleed-Owen, C.P. (2000) Subfossil records of *Rana* cf. *lessonae, Rana arvalis* and *Rana* cf. *dalmatina* from Middle Saxon (c. 600–950 AD) deposits in eastern England: evidence for native status. *Amphibia-Reptilia* 21: 57–65. https://doi.org/10.1163/156853800507273

Goulson, D. (2014) *A Buzz in the Meadow*. London: Jonathan Cape.

Greig-Smith, P.W., Thompson, H.M., Hardy, A.R. and three others (1994) Incidents of poisoning of honeybees (*Apis mellifera*) by agricultural pesticides in Great Britain 1981–1991. *Crop Protection* 13: 567–581. https://doi.org/10.1016/0261-2194(94)90002-7

Griffiths, R.A. (1985) A simple funnel trap for studying newt populations and an evaluation of trap behaviour in smooth and palmate newts, *Triturus vulgaris* and *T. helveticus*. *Herpetological Journal* 1: 5–10.

Griffiths, R.A. (1986) Feeding niche overlap and food selection in smooth and palmate newts, *Triturus vulgaris* and *T. helveticus*, at a pond in mid-Wales. *Journal of Animal Ecology* 55: 201–214. https://doi.org/10.2307/4702

Griffiths, R. (1987) The amphibian communities project. *Herpetofauna News* 1(7): 5.

Griffiths, R.A. (1996) *Newts and Salamanders of Europe*. London: Poyser.

Griffiths, R.A. and Halliday, T.R. (2004) Global amphibian declines: is current research meeting conservation needs? *Herpetological Journal* 14: 165–166.

Griffiths, R.A. and Inns, H. (1998) Surveying. In A.H. Gent and S.D. Gibson (eds), *Herpetofauna Workers' Manual*. Peterborough: Joint Nature Conservation Committee, pp. 1–13.

Griffiths, R.A. and Williams, C. (2000) Modelling population dynamics of great crested newts (*Triturus cristatus*): a viability analysis. *Herpetological Journal* 10: 157–163.

Griffiths, R.A., Getliff, J.M. and Mylotte, V.J. (1988) Diel patterns of activity and vertical migration in tadpoles of the common toad *Bufo bufo*. *Herpetological Journal* 1: 223–226.

Griffiths, R.A., de Wijer, P. and May, R.T. (1994) Predation and competition within an assemblage of larval newts (*Triturus*). *Ecography* 17: 176–181. https://doi.org/10.1111/j.1600-0587.1994.tb00091.x

Griffiths, R.A., Sewell, D. and McCrea, R.S. (2010) Dynamics of a declining amphibian metapopulation: survival, dispersal and impact of climate. *Biological Conservation* 143: 485–491. https://doi.org/10.1016/j.biocon.2009.11.017

Grossenbacher, K. (1988) *Verbreitungsatlas der Amphibien der Schweiz*. Bern Natural History Museum.

Grossenbacher, K. (2002) First results of a 20-year study on common toad *Bufo bufo* in the Swiss Alps. *Biota* 3: 43–48.

Hagström, T. (1979) Population ecology of *Triturus cristatus* and *T. vulgaris* (Urodela) in SW Sweden. *Holarctic Ecology* 2: 108–114. https://doi.org/10.1111/j.1600-0587.1979.tb00688.x

Hall, R.J. and Henry, P.F.P. (1992) Assessing the effects of pesticides on amphibians and reptiles: status and needs. *Herpetological Journal* 2: 65–71.

Halley, J.M., Oldham, R.S. and Arntzen, J.W. (1996) Predicting the presence of amphibian populations with the help of a spatial model. *Journal of Applied Ecology* 33: 455–470. https://doi.org/10.2307/2404977

Halliday, T.R. (1974) Sexual behavior of the smooth newt, *Triturus vulgaris* (Urodela: Salamandridae). *Journal of Herpetology* 8: 277–292. https://doi.org/10.2307/1562896

Haubrock, P.J. and Altrichter, J. (2016) Northern crested newt (*Triturus cristatus*) migration in a nature reserve: multiple incidents of breeding season displacements exceeding 1km. *Herpetological Bulletin* 138: 31–33.

Hayes, T.B., Haston, K., Tsui, M., and three others (2003) Atrazine-induced hermaphroditism at 0.1 ppb in American leopard frogs (*Rana pipiens*): laboratory and field evidence. *Environmental Health Perspectives* 111: 568–575. https://doi.org/10.1289/ehp.5932

Hayes, T.B., Khoury, V., Narayan, M. and eight others (2010) Atrazine induces complete feminization and chemical castration in male clawed toads (*Xenopus laevis*). *Proceedings of the National Academy of Sciences* 107: 4612–4622. https://doi.org/10.1073/pnas.0909519107

Hazelwood, E. (1970) Frog pond contaminated. *British Journal of Herpetology* 4: 177–185.

Hellmich, W. (1962) *Reptiles and Amphibians of Europe*. London: Blandford Press.

Hels, T. and Buchwald, E. (2001) The effect of road kills on amphibian populations. *Biological Conservation* 99: 331–340. https://doi.org/10.1016/S0006-3207(00)00215-9

Hemelaar, A.S.M. and van Gelder, J.J (1980) Annual growth rings in phalanges of *Bufo bufo* (Anura, Amphibia) from the Netherlands and their use for age determination. *Netherlands Journal of Zoology* 30: 129–135. https://doi.org/10.1163/002829680X00069

Herpetofauna Consultants International (2000) *Hampton Nature Reserve: Monitoring of Landscape Change with General Observations of Site Characteristics*. Halesworth: HCI.

Heusser, H. (1970) Laich-Fressen durch Kaulquappenals mögliche Ursache spezifischer biotroppräferenzen und kurzer Laichzeiten bei europäischen Froschlurchen (Amphibia, anura). *Oecologia* 4: 83–88. https://doi.org/10.1007/BF00390615

Heusser, H. (1971) Differenzierendes Kaulquappen-Fressen durch Molche. *Experientia* 27: 475. https://doi.org/10.1007/BF02137323

Heusser, H. (1972) Intra- und interspezifische Crowding-Effecte bei Kaulquappen einheimischer Anuren-Arten. *Vierteljahrsschrift der Naturforschenden Gesellschaft in Zürich* 117: 121–128

Hill, I.D.C., Rossi, C.A., Petrovan, S.O. and three others (2019) Mitigating the effects of a road on amphibian migrations: a Scottish case study of road tunnels. *Glasgow Naturalist* 27 Supplement: 25–36. https://doi.org/10.37208/tgn27s06

Hilton-Brown, D. and Oldham, R.S. (1991) *The Status of the Commoner Amphibians and Reptiles in Britain 1990 and Changes During the 1980s*. Contract Survey Report 131. Peterborough: Nature Conservancy Council.

Hitchings, S.P. and Beebee, T.J.C. (1997) Genetic substructuring as a result of barriers to gene flow in urban common frog (*Rana temporaria*) populations: implications for biodiversity conservation. *Heredity* 79: 117–127. https://doi.org/10.1038/hdy.1997.134

HM Treasury (2021) *The Economics of Biodiversity: the Dasgupta Review*. Headline messages. London: HM Treasury.

Holman, A. (1998) *Pleistocene Amphibians and Reptiles in Britain and Europe*. Oxford: Oxford University Press. https://doi.org/10.1093/oso/9780195112320.001.0001

Horner, H.A. and Macgregor, H.C. (1985) Normal development in newts (*Triturus*) and its arrest as a consequence of an unusual chromosome situation. *Journal of Herpetology* 19: 261–270. https://doi.org/10.2307/1564180

House of Commons Environmental Audit Committee (2021) *Biodiversity in the UK: Bloom or Bust?* HC 136. London: House of Commons.

Hows, M., Pilbeam, P., Conlan, H. and Featherstone, R. (2016) *Cambridgeshire Mammal Atlas*. Cambridgeshire Mammal Group.

Humphreys, E., Toms, M., Newson, S., Baker, J. and Wormald, K. (2011) An examination of reptile and amphibian populations in gardens, the factors influencing garden use and the role of a 'Citizen Science' approach for monitoring their populations within this habitat. BTO Research Report 572. Thetford: British Trust for Ornithology.

Inns, H. (2009) *Britain's Reptiles and Amphibians*. Old Basing: WILDGuides.

Jackson, B.G. (1965) Agriculture. In J.A. Steers (ed.), *The Cambridge Region*. London, British Association for the Advancement of Science, pp. 85–111.

Jarvis, L.E. (2015) Factors affecting body condition in a great crested newt *Triturus cristatus* population. *Herpetological Bulletin* 134: 1–5.

Jefferies, D.J. and Woodroffe, G.L. (2008) Otter *Lutra lutra*. In S. Harris and D.W. Yalden (eds), *Mammals of the British Isles*, 4th edition. Southampton: Mammal Society, pp. 437–447.

Jehle, R., Thiesmeier, B. and Foster, J. (2011) *The Crested Newt*. Bielefeld: Laurenti-Verlag.

Jenyns, I. (1835) *A Manual of British Vertebrate Animals*. Cambridge: Pitt.

Joint Nature Conservation Committee (2019) Conservation status assessment for the species: S1166 Great crested newt (*Triturus cristatus*). Fourth Report by the UK under Article 17 of the Habitats Directive. Peterborough: JNCC.

Kelly, G. (2004) Literature/archive research for information relating to pool frogs *Rana lessonae* in East Anglia. English Nature Research Reports 480. Peterborough: English Nature.

Kelly, P.J. and Wray, J.D. (1971) The educational uses of living organisms. *Journal of Biological Education* 5: 213–218. https://doi.org/10.1080/00219266.1971.9653710

Kennedy, V.H., Horill, A.D. and Livens, F.R. (1990) *Radioactivity and wildlife*. Focus on Nature Conservation 24. Peterborough: Nature Conservancy Council.

Kiesecker, J.M. (2002) Synergism between trematode infection and pesticide exposure: a link to amphibian limb deformities in nature? *Proceedings of the National Academy of Sciences* 99: 9900–9904. https://doi.org/10.1073/pnas.152098899

Kupfer, A. and Kneitz, S. (2000) Population ecology of the great crested newt (*Triturus cristatus*) in an agricultural landscape: dynamics, pond fidelity and dispersal. *Herpetological Journal* 10: 165–171.

Langton, T.E.S. (ed.) (1989) *Amphibians and Roads*. Shefford: ACO Polymer Products.

Langton, T. (1990) Herpetofauna Groups of Great Britain. *Herpetofauna News* 2(2): 2–5.

Langton, T. (1995) Introducing Froglife. *Herpetofauna News* 2(8): 1.

Langton, T. (2009) Great crested newt *Triturus cristatus*: 30 years of implementation of international wildlife conventions, European and UK law in the United Kingdom 1979–2009. Unpublished report to the European Commission.

Langton, T., Beckett, C. and Foster, J. (2001) *Great Crested Newt Conservation Handbook*. Halesworth: Froglife.

Lawton, J.H., Brotherton, P.N.M., Brown, V.K. and 12 others (2010) *Making Space for Nature: a Review of England's Wildlife Sites and Ecological Network*. Report to Defra. London: Defra.

Lewis, B., Griffiths, R.A. and Barrios, Y. (2007) Field assessment of great crested newt *Triturus cristatus* mitigation projects. Natural England Research Report 001. Peterborough: Natural England.

Lewis, B., Griffiths, R.A. and Wilkinson, J.W. (2017) Population status of great crested newts (*Triturus cristatus*) at sites subjected to development mitigation. *Herpetological Journal* 27: 133–142.

Macgregor, H.C. (1995) Crested newts: ancient survivors. *British Wildlife* 7: 1–8.

Manigold, D.B. and Schulze, J.A. (1969) Pesticides in selected western streams – a progress report. *Pesticides Monitoring Journal* 3: 125–135.

Mann, R.M., Hyne, R.V., Choung, C.B. and Wilson, S.P. (2009) Amphibians and agricultural chemicals: review of the risks in a complex environment. *Environmental Pollution* 157: 2903–2927. https://doi.org/10.1016/j.envpol.2009.05.015

Martin, J. (2000) *The Development of Modern Agriculture*. London: Springer Nature. https://doi.org/10.1057/9780230599963.

Matos, C., Petrovan, S., Ward, A.I. and Wheeler, P. (2017) Facilitating permeability of landscapes impacted by roads for protected amphibians: patterns of movement for the great crested newt. *Peerj S*: e2922. https://doi.org/10.7717/peerj.2922

McGrath, A.L. and Lorenzen, K. (2010). Management history and climate as key factors driving natterjack toad population trends in Britain. *Animal Conservation* 13: 483–494. https://doi.org/10.1111/j.1469-1795.2010.00367.x

Medlock, J.M., Handsford, K.M., Anderson, M., Mayho, R. and Snow, K.R. (2012) Mosquito nuisance and control in the UK – a questionnaire-based survey of local authorities. *European Mosquito Bulletin* 30: 15–29.

Meek, R. (2022) Long-term changes in four populations of the spiny toad, *Bufo spinosus*, in Western France; data from road mortalities. *Conservation* 2(2): 248–261. https://doi.org/10.3390/conservation2020017

Mellanby, K. (1967) *Pesticides and Pollution*. New Naturalist 50. London: Collins.

Meyer, A.H., Schmidt, B.R. and Grossenbacher, K. (1998) Analysis of three amphibian populations with quarter-century long time-series. *Proceedings of the Royal Society of London B* 265: 523–528. https://doi.org/10.1098/rspb.1998.0326

Moore, H.J. (1954) Some observations on the migration of the toad (*Bufo bufo bufo*). *British Journal of Herpetology* 1: 194–224.

Moore, N.W. (1966). An assessment of the discussions. *Journal of Applied Ecology* 3 (supplement): 291–295. https://doi.org/10.2307/2401471

Moore, N.W. (1991) Development of dragonfly communities and the consequences of territorial behaviour: a 27 year study of small ponds at Woodwalton Fen, Cambridgeshire, United Kingdom. *Odonatologica* 20: 203–231.

Moore, N.W. (2001) Changes in the dragonfly communities at the twenty ponds at Woodwalton Fen, Cambridgeshire, United Kingdom, since the study of 1962–1988. *Odonatologica* 30: 289–298.

Natural England (2016) New licensing policies: great for wildlife – great for business. https://www.gov.uk/government/news/new-licensing-policies-great-for-wildlife-great-for-business. Accessed 11 April 2021.

Natural England, JNCC, Natural Resources Wales, NatureScot and Northern Ireland Environment Agency (2021) *Nature Positive 2030 – Summary Report*. Peterborough: JNCC.

Nature Conservancy Council (1983) *The Ecology and Conservation of Amphibian and Reptile Species Endangered in Britain*. Shrewsbury: Nature Conservancy Council.

Nature Conservancy Council (1989) *Guidelines for the Selection of Biological SSSIs*. Peterborough: Nature Conservancy Council.

Newbold, C. (1975) Some ecological effects of two herbicides, dichlobenil and diquat, on pond ecosystems. Unpublished PhD thesis, University of Leicester.

Newt Conservation Partnership (2021) *Monitoring Report Year 2*. https://freshwaterhabitats.org.uk/wp-content/uploads/2021/02/NCP-Monitoring-Report_Feb21.pdf. Accessed 16 January 2022.

Newton, I. (1979) *Population Ecology of Raptors*. Berkhamsted: Poyser.

NHBS (2021) Amphibian and Reptile Conservation Trust: q&a with Dr Tony Gent. https://www.nhbs.com/blog/arc-trust-dr-tony-gent. Accessed 9 April 2021.

O'Brien, D., Hall, J.E., Miró, A. and three others (2021) Reversing a downward trend in threatened peripheral amphibian (*Triturus cristatus*) populations through interventions combining species, habitat and genetic information. *Journal for Nature Conservation* 64: 126077. https://doi.org/10.1016/j.jnc.2021.126077

Oldham, R.S. (1963a) Notes on the breeding behaviour of *Rana temporaria*. *British Journal of Herpetology* 3: 79–80.

Oldham, R.S. (1963b) Homing behaviour in *Rana temporaria* Linn. *British Journal of Herpetology* 3: 116–127.

Oldham, R.S. (1985) Toad dispersal in agricultural habitats. *Bulletin of the British Ecological Society* 16: 211–215.

Oldham, R.S. (1999a) Amphibians and agriculture: double jeopardy. In M. Whitfield, J. Matthews and C. Reynolds (eds), *Aquatic Life Cycle Strategies: Survival in a Variable Environment*. Plymouth: Marine Biological Association, pp. 105–124.

Oldham, R.S. (1999b) The impact of a road development on a toad population: a test of density dependence in the terrestrial habitat. *Bulletin of the British Ecolcgical Society* 30: 29–37.

Oldham, R.S. and Swan, M.J.S. (1991) Conservation of amphibian populations in Britain. In A. Seitz and V. Loescheke (eds), *Species Conservation: A Population-Biological Approach*. Basel: Birkhauser Verlag, pp. 141–158. https://doi.org/10.1007/978-3-0348-6426-8_10

Oldham, R.S., Latham, D.M., Hilton-Brown, D. and three others (1997) The effect of ammonium nitrate fertiliser on frog (*Rana temporaria*) survival. *Agriculture Ecosystems and Environment* 61: 69–74. https://doi.org/10.1016/S0167-8809(96)01095-X

Oldham, R.S., Keeble, J., Swan, M.J.S. and Jeffcote, M. (2000) Evaluating the suitability of habitat for the great crested newt (*Triturus cristatus*). *Herpetological Journal* 10: 143–155.

O'Rourke, D.P. (2007) Amphibians used in research and teaching. *Institute of Laboratory Animal Research* 48: 183–187. https://doi.org/10.1093/ilar.48.3.183

Orbell, R. and Orbell, S. (2015) Amphibian and reptile report 2014. *Annual Report of Huntingdonshire Fauna and Flora Society* 67: 54–57.

Orbell, R. and Orbell, S. (2016) Amphibian and reptile report 2015. *Annual Report of Huntingdonshire Fauna and Flora Society* 68: 59–63.

Orchard, D., Tessa, G. and Jehle, R. (2019) Age and growth in a European flagship amphibian: equal performance at agricultural ponds and favourably managed aquatic sites. *Aquatic Ecology* 53: 37–48. https://doi.org/10.1007/s10452-018-09671-3

Osborn, D., Cooke, A.S. and Freestone, S. (1981) Histology of a teratogenic effect of DDT on *Rana temporaria* tadpoles. *Environmental Pollution* 25: 305–314. https://doi.org/10.1016/0143-1471(81)90091-X

Parker, P., Kirby, P., Darling, T. and FitzJohn, P. (2011) King's Dyke Nature Reserve: making space for new ponds in the Peterborough clay pits. https://freshwaterhabitats.org.uk/wp-content/uploads/2013/09/KD-case-study-20131.pdf. Accessed 6 March 2022.

Pechmann, J.H.K. and Wilbur, H.M. (1994) Putting declining amphibian populations in perspective: natural fluctuations and human impacts. *Herpetologica* 50: 65–84.

Peltzer, P.M., Lajmanovich, R.C., Attademo, A.M. and four others (2007) Population and health of common toads across agricultural lands: implications in worldwide declines. *FrogLog* 84: 4–6.

Perring, F. (1966) Where have all the frogs gone? *Wild Life Observer* 19: 10–11.

Petrovan, S.O. and Schmidt, B.R. (2016) Volunteer conservation action data reveals large-scale and long-term negative population trends of a widespread amphibian, the common toad (*Bufo bufo*). *PLoS ONE* 11(10): e0161943. https://doi.org/10.1371/journal.pone.0161943

Petrovan, S.O. and Schmidt, B.R. (2019) Neglected juveniles; a call for integrating all amphibian life stages in assessments of mitigation success (and how to do it). *Biological Conservation* 236: 252–260. https://doi.org/10.1016/j.biocon.2019.05.023.

Pickford, D.B., Jones, A., Velez-Pelez, A. and four others (2015) Screening breeding sites of the common toad (*Bufo bufo*) in England and Wales for evidence of endocrine disrupting activity. *Ecotoxicology and Environmental Safety* 117: 7–19. https://doi.org/10.1016/j.ecoenv.2015.03.006

Pierce, B.A. (1993) The effects of acid precipitation on amphibians. *Ecotoxicology* 2: 65–77. https://doi.org/10.1007/BF00058215

Pinder, L.C.V., House, W.A. and Farr, I.S. (1993) Effects of insecticides on freshwater invertebrates. In A.S. Cooke (ed.), *The Environmental Effects of Pesticide Drift*. Peterborough: English Nature, pp. 64–75.

Pochini, K.M. and Hoverman, J.T. (2017) Reciprocal effects of pesticides and pathogens on amphibian hosts: the importance of exposure order and timing. *Environmental Pollution* 221: 359–366. https://doi.org/10.1016/j.envpol.2016.11.086

Preece, R.C. and Sparks, T.H. (2012) *Fauna Cantabrigiensis*. London: Ray Society.

Prestt, I. (1970) Organochlorine pollution of rivers and the heron (*Ardea cinerea* L.). IUCN Technical Meeting 11, Papers and Proceedings 1: 95–102.

Prestt, I., Cooke, A.S. and Corbett, K. (1974) British amphibians and reptiles. In D.L. Hawksworth (ed.), *The Changing Fauna and Flora of Britain.* London: Academic Press, pp. 229–254.

Price, S.J., Leung, W.T.M., Owen, C.J. and six others (2019) Effects of historic and projected climate change on the range and impacts of an emerging wildlife disease. *Global Change Biology* 25: 2648–2660. https://doi.org/10.1111/gcb.14651

Ratcliffe, D.A. (1967) Decrease in eggshell weight in certain birds of prey. *Nature* 215: 208–210. https://doi.org/10.1038/215208a0.

Ratcliffe, D.A. (1980) *The Peregrine Falcon.* Calton: Poyser.

Raye, L. (2017) Frogs in pre-industrial Britain. *Herpetological Journal* 27: 368–378.

Reading, C.J. (1984) Interspecific spawning between common frogs (*Rana temporaria*) and common toads (*Bufo bufo*). *Journal of Zoology* 203: 95–101. https://doi.org/10.1111/j.1469-7998.1984.tb06046.x

Reading, C.J. (1988) Growth and age at sexual maturity in common toads (*Bufo bufo*) from two sites in southern England. *Amphibia-Reptilia* 9: 277–288. https://doi.org/10.1163/156853888X00369

Reading, C.J. (1998) The effect of winter temperatures on the timing of breeding activity in the common toad *Bufo bufo. Oecologia* 117: 469–475. https://doi.org/10.1007/s004420050682

Reading, C.J. (2007) Linking global warming to amphibian declines through its effect on female body condition and survivorship. *Oecologia* 151: 125–131. https://doi.org/10.1007/s00442-006-0558-1

Reading, C.J. and Clarke, R.T. (1983) Male breeding behaviour and mate acquisition in the common toad. *Journal of Zoology* 201: 237–246. https://doi.org/10.1111/j.1469-7998.1983.tb04273.x

Reading, C.J. and Jofré, G.M. (2021) Declining common toad body size correlated with climate warming. *Biological Journal of the Linnean Society* 134: 577–586. https://doi.org/10.1093/biolinnean/blab101

Relton, J. (1972) Disappearance of farm ponds. *Monks Wood Experimental Station Report* 1969–1971: 32.

Robinson, S.A., Richardson. S.D., Dalton, R.L. and four others (2017) Sublethal effects on wood frogs chronically exposed to environmentally relevant concentrations of two neonicotinoid insecticides. *Environmental Toxicology and Chemistry* 36: 1101–1109. https://doi.org/10.1002/etc.3739

Robinson, S.A., Richardson. S.D., Dalton, R.L. and five others (2019) Assessment of sublethal effects of neonicotinoid insecticides on the life-history traits of 2 frog species. *Environmental Toxicology and Chemistry* 38: 1967–1977. https://doi.org/10.1002/etc.4511

Rowe, G. and Beebee, T.J.C. (2003) Population on the verge of a mutational meltdown? Fitness costs of genetic load for an amphibian in the wild. *Evolution* 57: 177–181. https://doi.org/10.1111/j.0014-3820.2003.tb00228.x

Russell, R.W. and Hecnar, S.J. (1996) The ghost of pesticides past? *FrogLog* 19: 1.

Salazar, R.D., Montgomery, R.A., Thresher, S.E. and Macdonald, D.W. (2016) Mapping the relative probability of common toad occurrence in terrestrial lowland farm habitat in the united Kingdom. *PLoS ONE* 11(2): e0148269. https://doi.org/10.1371/journal.pone.0148269

Savage, R.M. (1935) The influence of external factors on the spawning date and migration of the common frog, *Rana temporaria temporaria. Proceedings of the Zoological Society of London* 1935: 49–98. https://doi.org/10.1111/j.1469-7998.1935.tb06232.x

Savage, R.M. (1961) *The Ecology and Life History of the Common Frog.* London: Pitman. https://doi.org/10.5962/bhl.title.6538

Sayer, C., Hawkins, J. and Greaves, H. (2022) Restoring the ghostly and the ghastly: a new golden age for British lowland farm ponds? *British Wildlife* 33: 477–487.

Schmidt, B.R. (2004) Declining amphibian populations: the pitfalls of count data in the study of diversity, distributions, dynamics and demography. *Herpetological Journal* 14: 167–174.

Schwartz, A.L.W., Shilling, F.M. and Perkins, S.E. (2020) The value of monitoring wildlife roadkill. *European Journal of Wildlife Research* 66: 18. https://doi.org/10.1007/s10344-019-1357-4

Scorgie, H.R.A. and Cooke, A.S. (1979) Effects of the triazine herbicide cyanatryn on aquatic animals. *Bulletin of Environmental Contamination and Toxicology* 22: 135–142. https://doi.org/10.1007/BF02026920

Semlitsch, R.D. (1990) Effects of body size, sibship, and tail injury on the susceptibility of tadpoles to dragonfly predation. *Canadian Journal of Zoology* 68: 1027–1030. https://doi.org/10.1139/z90-149

Sheail, J. (1998) *Nature Conservation in Britain. The Formative Years.* London: Stationery Office.

Simms, C. (1969) Indications of the decline in breeding amphibians at an isolated pond in marginal land. *British Journal of Herpetology* 4: 93–96.

Simpson, S. (2021a) ARC/ARG UK Herpetofauna Workers Meeting 2021. *Natterjack* 224: 1–2.

Simpson, S. (2021b) Rewilding and introductions: working together. *Natterjack* 224: 7–8.

Slater, F. (2002) Progressive skinning of toads (*Bufo bufo*) by the Eurasian otter (*Lutra lutra*). *IUCN Otter Specialist Group Bulletin* 19: 25–29.

Smalling, K.L., Fellers, G.M., Kleeman, P.M. and Kuivila, K.M. (2013) Pesticide accumulation in California Sierra Nevada frogs. *FrogLog* 108: 18.

Smith, M. (1951 and 1969) *The British Amphibians and Reptiles.* 1st and 4th editions. New Naturalist 20. London: Collins..

Smith, M. (1953) The shortage of toads and frogs. *Country Life* 114: 770–771.

Smith, P.H. and Skelcher, G. (2019) Effects of environmental factors and conservation measures on a sand-dune population of the natterjack toad (*Epidalea calamita*) in north-west England: a 31-year study. *Herpetological Journal* 29: 146–154. https://doi.org/10.33256/hj29.3.146154

Snell, C. (1994) The pool frog – a neglected native. *British Wildlife* 6: 1–4.

Snell, C. (2006) Status of the common tree frog in Britain. *British Wildlife* 17: 153–160.

Snell, C. (2016) The northern pool frog. *British Wildlife* 28: 2–11.

Sparling, D.W., Linder, G. and Bishop, C.A. (2000) *Ecotoxicology of Amphibians and Reptiles.* Pensacola: Society of Environmental Chemistry and Toxicology.

Sparks, T.H. and Carey, P.D. (1995) The responses of species to climate over two centuries: an analysis of the Marsham phenological record, 1746–1947. *Journal of Ecology* 83: 321–329. https://doi.org/10.2307/2261570

Sparks, T., Tryjanowski, P., Cooke, A., Crick, H. and Kuźniak, S. (2007) Vertebrate phenology at similar latitudes: temperature responses differ between Poland and the United Kingdom. *Climate Research* 34: 93–98. https://doi.org/10.3354/cr034093

Steward, J.W. (1969) *The Tailed Amphibians of Europe.* Newton Abbot: David & Charles.

Swan, M.J.S. (1987) National amphibian survey – much work still to be done. *Herpetofauna News* 1(10): 7.

Swan, M. and Oldham, R.S. (1993) *Herptile Sites: National Amphibian Survey.* English Nature Research Report 38. Peterborough: English Nature.

Swan, M.J.S. and Oldham, R.S. (1994) Amphibians and landscape composition. In J.E. Dover (ed.), *Fragmentation in Agricultural Landscapes.* Proceedings of the 3rd UK

Conference of the International Association of Landscape Ecology, Preston, pp. 176–183.

Taylor, H.R.H. (1948) The distribution of reptiles and amphibia in the British Isles with notes on species recently introduced. *British Journal of Herpetology* 1: 1–38.

Taylor, H.R.H. (1963) The distribution of amphibians and reptiles in England and Wales, Scotland and Ireland and the Channel Islands: a revised survey. *British Journal of Herpetology* 3: 95–115.

Teacher, A.G.F., Cunningham, A.A. and Garner, T.W.J. (2010) Assessing the long-term impact of *Ranavirus* infection in wild common frog populations. *Animal Conservation* 13: 514–522. https://doi.org/10.1111/j.1469-1795.2010.00373.x

Tew, T., Biggs, J. and Gent, A. (2018) 'District licensing' for great crested newts – delivering a big idea. *In Practice* 100: 35–39.

van Dam, H. (1988) Acidification of three moorland pools in the Netherlands by acid precipitation and extreme drought periods over seven decades. *Freshwater Biology* 20: 157–176. https://doi.org/10.1111/j.1365-2427.1988.tb00439.x

van Gelder, J.J. (1973) A quantitative approach to the mortality resulting from traffic in a population of *Bufo bufo* L. *Oecologia* 13: 93–95. https://doi.org/10.1007/BF00379622

von Bülow, B. and Kupfer, A. (2019) Monitoring population dynamics and survival of northern crested newts (*Triturus cristatus*) for 19 years at a pond in central Europe. *Salamandra* 55: 97–102.

Vasudev, V., Krishnamurthy, S.V. and Gurushankara, H.P. (2007) Organophosphate pesticides – a major threat to anuran populations in an agroecosystem of Western Ghats, India. *FrogLog* 83: 8–9.

Wagner, N., Reichenbecher, W., Tiechmann, H., Tappeser, B. and Lötters, S. (2013). Questions concerning the potential impact of glyphosate-based herbicides on amphibians. *Environmental Toxicology and Chemistry* 32: 1688–1700. https://doi.org/10.1002/etc.2268

Walker, P. (2008) St Neots Islands (Illand) Common and Lammas Meadow. *Huntingdonshire Fauna and Flora Society 60th Anniversary Review*: 40–44.

Ward, R., Liddiard, T., Goetz, M. and Griffiths, R. (2016) Head-starting, re-introduction and conservation management of the agile frog on Jersey, British Channel Isles. In P.S. Soorae (ed.), *Global Re-Introduction Perspectives 2016. Case-Studies from Around the Globe*. Gland: Switzerland, pp 40–44.

Weber, J.-M. (1990) Seasonal exploitation of amphibians by otters (*Lutra lutra*) in north-east Scotland. *Journal of Zoology* 220: 641–651. https://doi.org/10.1111/j.1469-7998.1990.tb04740.x

Wells, T. (2008) Introduction. *Huntingdonshire Fauna and Flora Society 60th Anniversary Review*: 1–8.

Wheeler, P.M., Ward, A.I., Smith, G.C., Croft, S. and Petrovan, S.O. (2019) Careful considerations are required when analysing mammal citizen science data – a response to Massimino *et al. Biological Conservation* 232: 274–275. https://doi.org/10.1016/j.biocon.2019.01.021

White, G. (1978) *The Natural History of Selborne*. London: Book Club Associates.

Wildlife Trust for Bedfordshire, Cambridgeshire and Northamptonshire (2021) *The Cambridge Nature Network*. https://www.wildlifebcn.org/sites/default/files/2021-05/CambridgeNatureNetworkSummaryDigitalVersion.pdf. Accessed 26 May 2022.

Wilkinson, J.W. (2015) *Amphibian Survey and Monitoring Handbook*. Exeter: Pelagic Publishing.

Wilkinson, J.W. and Arnell, A.P. (2013) NARRS Report 2007–2012: establishing the baseline. Amphibian and Reptile Conservation Research Report 13/01. Bournemouth: ARC.

Wilkinson, J.W. and Buckley, J. (2012) Amphibian conservation in Britain. *FrogLog* 101: 12–13.

Wilkinson, J.W., Wright, D. and Arnell, A. (2011) Assessing population status of the great crested newt in Great Britain. Contract Report 080. Peterborough: Natural England.

Williams, P.J., Biggs, J., Whitfield, M. and six others (2010) *The Pond Book: a Guide to the Management and Creation of Ponds.* Oxford: Freshwater Habitats Trust.

Williamson, T. (2022) Rewilding: a landscape-history approach. *British Wildlife* 33: 423–429.

Wood, T.J. and Goulson, D. (2017) The environmental risks of neonicotinoid pesticides: a review of evidence post 2013. *Environmental Science and Pollution Research* 24: 17285–17325. https://doi.org/10.1007/s11356-017-9240-x

Wycherley, J. and Anstis, R. (2001) *Amphibians and Reptiles of Surrey.* Pirbright: Surrey Wildlife Trust.

Yalden, D.W. (1967) Reptiles and amphibians in the London area. *London Naturalist* 46: 68–75.

Young, S.L. and Beebee, T.J.C. (2004) An investigation of recent declines in the common toad. English Nature Research Report 584. Peterborough: English Nature.

Zakaria, N.B. (2017) Long-term population ecology of the great crested newt in Kent. PhD thesis, University of Kent.

Index

Figure numbers are in **bold** and table numbers are in *italics*.